Calculating a Natural World

Inside Technology

edited by Wiebe E. Bijker, W. Bernard Carlson, and Trevor Pinch

A list of the series appears at the back of the book.

Calculating a Natural World

Scientists, Engineers, and Computers during the Rise of U.S. Cold War Research

Atsushi Akera

The MIT Press
Cambridge, Massachusetts
London, England

First MIT Press paperback edition, 2008

© 2007 Massachusetts Institute of Technology

For information on quantity discounts, please email special_sales@mitpress.mit.edu.

Set in stone serif and stone sans by SNP Best-set Typesetter Ltd., Hong Kong. Printed
and bound in the United States of America.

Library of Congress Cataloging-in-Publication Data

Akera, Atsushi.
    Calculating a natural world : scientists, engineers, and computers during the
rise of US Cold War research / Atsushi Akera.
        p. cm.
    Includes bibliographical references and index.
    ISBN 978-0-262-01231-7 (hc.: alk. paper)—978-0-262-51203-9 (pb.: alk. paper)
    1. Electronic data processing—United States—History. 2. Computers—United
States—History. I. Title.

    QA76.17.A42 2006
    004′.0973—dc22

                                                                        2006046247

10 9 8 7 6 5 4 3 2

Contents

Preface

I would like to dedicate this book to the various people who helped me with it, beginning with my family. Although I am still not sure that they fully understand my interest in the humanities, my parents have always patiently supported my interests. Hopefully, they will see the imprint of the knowledge they helped to convey in this book. My deepest gratitude goes to my wife, Rachel, and my children, Seiji and Kai, who have also patiently supported my efforts. Growing up in an academic household may not always be easy, but I hope it will give Seiji and Kai a respect for liberal values.

Tom Hughes has been the source of my intellectual inspiration. He taught me, and he encouraged me to go after the "grand" projects. My fondest memories of our many wonderful conversations in Chestnut Hill are preserved in the preface to my dissertation, and remain as alive and true today. Given my focus on practice, perhaps he will recognize why I chose not to work so directly in his footsteps. Nevertheless, many of his ideas, and the spirit behind them, can be found throughout this work.

Other faculty members who have significantly supported my work and interests include Henrika Kuklick, who, in addition to conveying many important insights about social theory, took up the work of fostering the graduate student community at Penn. Walter Licht introduced me to the labor and social historical perspectives that undergird this work. Sharon Traweek fostered my interest in ethnography, and Charlie Weiner my interest in history while I was an undergraduate in MIT's STS Program. These are all very wonderful people, and our field is blessed to have so many caring individuals.

I also remember quite fondly life on Camac Street, where all the messy ideas for this book first amassed themselves. My thanks to Elisa Becker, who shared in my life there (including coffee from Fantes and muffin tops from Jamieson's), and to the rest of the graduate student community at Penn. Equally important have been Manfred Fischbeck, Brigitta Hermann, and the rest of Group Motion, as well as Darla Johnson, Dee McCandless, and my friends in Austin, who helped bring balance to an academic way of life. Meanwhile, this book remains dedicated to Val. My only regret is that it is not quite the book he would have liked to see.

I have been fortunate in that those of us who study the history of computing have become quite an extensive, and cohesive, community. My special thanks to all my senior colleagues—Bob Seidel, Bill Aspray, Larry Owens, David Alan Grier, Mike Mahoney, Martin Campbell-Kelly, Paul Ceruzzi, Tim Bergin, Mike Williams, Rik Nebeker, J. A. N. Lee, Arthur Norberg, and Steve Lubar among them—who have welcomed young scholars with open arms. Among this "young" crowd, I am especially indebted to Pap Ndiaye, Tom Haigh, Nathan Ensmenger, Chigusa Kita, Patricia Hemmis, Janet Abbate, David Mindell, and Paul Edwards for their comments and conversations. This book would not be as rich as it is without the support and contribution of these individuals. I also thank all my colleagues at Rensselaer. They have upheld the interdisciplinary character of STS, and this work has matured through their ideas and encouragement.

I would also like to also express my sincere gratitude to Larry Cohen, who first thought to take on my manuscript, and to Sara Meirowitz, Paul Bethge, and others at The MIT Press for seeing it through to production. The series editors, Trevor Pinch, Wiebe Bijker, and especially Bernie Carlson, have been tremendously supportive, and permissive of my unrefined interest in social theory. They have helped me to discipline my use of theory so that it integrates and enhances historical narrative, rather than serving as a digression.

There are of course many others who helped this project through its various stages. Especially important have been Beth Kaplan, Carrie Seib, and their predecessors at the Charles Babbage Institute; Helen Samuels, Elizabeth Andrews, and Nora Murphy at the MIT Institute Archives; Dawn Sanford at the IBM Archives; Erica Mosner at the IAS Archives; Nancy Miller and Mark Lloyd at the University of Pennsylvania Archives; Michael Ryan of the Annenberg Rare Book and Manuscript Library; and Debbie Douglas and Frank Conahan at the MIT Museum. I am also indebted to the archival staffs of the Library of Congress, the National Archives, the Smithsonian Institution's Archives Center, and the Bentley Historical Library.

Chapters 6 and 7 are modestly revised versions of previously published articles, notably "Voluntarism and the Fruits of Collaboration: The IBM User's Group, Share" (*Technology and Culture* 42, 2001: 710–736, © Society for the History of Technology) and "IBM's Adaptation to Cold War Markets: Cuthbert Hurd and His Applied Science Field Men" (*Business History Review* 76, 2002: 767–802, © 2002 President and Fellows of Harvard College).

Calculating a Natural World

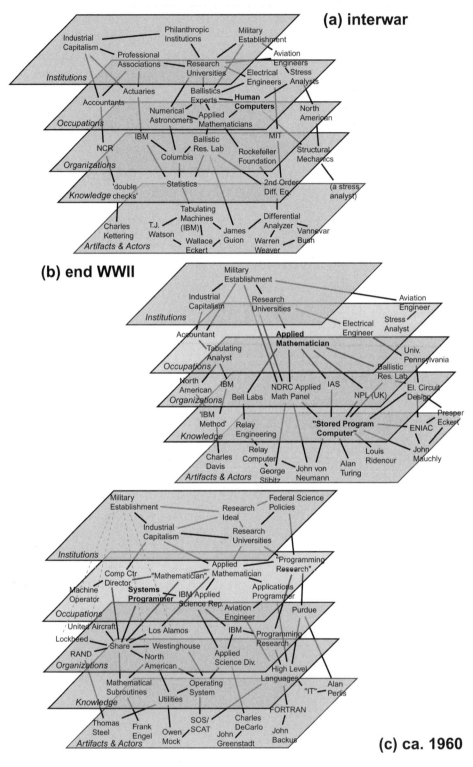

Figure I.1

Introduction

At one level, this is a rather conventional history of computing that focuses on the era of the big machines—the machines that made the reputations of IBM and many academic laboratories. At another level, it is a history of the emerging infrastructure of Cold War research in the United States. Its purpose is to demonstrate how these two histories are intertwined.

Contrary to common caricatures of Cold War science and engineering, research during the Cold War years was not only about research of military significance; it was also about a wide range of technical advances that were used to sustain the dualistic economy of the United States during the Cold War. This book's underlying thesis is that the intensity of technological innovation during the Cold War years resulted neither from military foresight nor from academic influence, but rather from a fundamental pluralism in the demands that were placed upon research. The different aims and objectives of military, academic, and commercial institutions not only drew researchers in these three directions but also created a language of needs, interests, opportunities, and priorities that opened up many avenues for research. We have become accustomed to speaking about a "military-industrial complex." In this book I pay considerable attention to productive tensions and differences, as well as to the alliances that enabled all the historical actors to proceed with their work. I argue that it was the tensions as much as the convergences that led to one of the most intense periods of scientific discovery and technological innovation that the U.S. has seen.[1]

A topic as broad as Cold War research requires a specific point of entry. This is where the history of computing serves as a kind of metonymic device—a part that stands for a whole—and thus as a historical lens through which to observe larger changes in U.S. research institutions and their practices. The electronic computer is an ideal artifact for such a study. An expensive technology that was brought to fruition precisely during the early Cold War years, it touched upon the broadest aspects of Cold War research policies. Moreover, computers were what we have come to call the quintessential "dual-use" technology. From the outset, computers were weighed for both military and commercial significance, and hence they straddled the very institutional boundaries that are central to this study. It is also significant that computers became valuable research instruments in nearly all disciplines. They were, quite literally, part of the infrastructure of Cold War research. In this book, the history of computing will be used as a means of understanding something larger about technical and institutional innovation during the Cold War.

Figure I.1 provides a sort of cognitive map for the book. The three diagrams in it are based on a particular "ecological" view of knowledge. They are designed to highlight the metonymic relations between esoteric knowledge and its supporting institutions.[2] Though they are highly selective (and hence reductive), they are meant to depict, in turn, the socio-technical organization of computing during the interwar years (diagram a), computing at the end of World War II (diagram b), and computing immediately before the development of computer time sharing (diagram c). These diagrams depict rather clearly how the metonymic relations between knowledge and social institutions are not based on any simple relation of one-to-one correspondence. The layered representation suggests, first of all, that esoteric knowledge is upheld through many intermediate forms of social organization, including occupations, disciplines, and corporate organizations. Moreover, each body of knowledge is upheld by an assortment of artifacts and actors (and skilled practices, which are not directly represented in these diagrams). The "spatial" dimension within each layer also makes it possible to think about the "regionalization" of experience—about how military institutions differ, for instance, from commercial and academic ones, or how different disciplines went about pursuing research in different ways. Meanwhile, the implicit networks that tie together the different layers and their various elements help us to identify some of the social arrangements and technical convergence that emerge during the course of research. (Diagram c depicts, for instance, the collaboration by several U.S. corporations to create a computer users' group called Share, and diagram b depicts the general consensus that emerged around the concept of the stored-program computer after World War II.) In the end, the subtle differences in the overall topology of these networks are the principal subject of this book. An important challenge for the book is to explain how the historical arrangements for computing went from the relatively hierarchal organization of knowledge depicted in diagram b to the more distributed organization of computing and computer-development research depicted in diagram c.

It has become quite common in the post-Cold War era to speak about historical changes in the overall "ecology" of research.[3] Figure I.1, however, is based on a particular notion of an "ecology of knowledge" advanced by the historian of science Charles Rosenberg (1997a/1979) and, somewhat independently, by a group of sociologists working in the symbolic-interactionist tradition.[4] This concept is quite foundational to my book's academic aims, and later in this introduction I will provide a more precise, more theoretical articulation of this concept and of its relationship to the representational scheme employed in figure I.1. However, I will begin with a more conventional focus on history and historiography.

As a history of the early era of digital computers, this book draws on numerous existing studies. Especially important, and mentioned here in chronological order by subject, are James Cortada's *IBM, NCR, Burroughs and Remington Rand and the Industry They Created, 1965–1956* (1993), Nancy Stern's *From ENIAC to UNIVAC* (1981), William Aspray's *John von Neumann and the Origins of Modern Computing* (1990), Kent Redmond and Thomas Smith's *Project Whirlwind* (1980), and Paul Edwards's *The Closed World* (1996). There are also many dedicated studies of IBM, including Emerson Pugh's *Building IBM* (1995). I also draw on Arthur Norberg and Judy O'Neill's *Transforming Computing Technology* (1996). I rely upon and revisit these studies in an effort to produce a more synthetic account of the early history of digital computers.

Those familiar with the historiography of computing will recognize that such a synthesis has long been wanting. However, in view of my emphasis on the metonymic relations between knowledge and its supporting institutions, this historical synthesis is also necessary. All the studies cited above are meticulously researched and valuable. Nevertheless, as long as the history of computing has been built around the study of specific computer systems and institutions, each work has tended to highlight certain aspects of Cold War research while deemphasizing others. Were these simply different facets of the research conducted during this era—research conducted in separate, autonomous regions of an ecology of knowledge—these studies would collectively document all that occurred. However, if the thesis is as I have stated it, namely that the immense productivity of research during the Cold War era resulted from the productive tensions between institutions, it is necessary to take a more integrated approach that pays close attention to the complex interplay between institutions and across the layers of an ecology of knowledge. The basic question I raise in this study can be answered only through a multi-site study that brings together earlier accounts, and in some less-than-obvious ways.

The overall work this book does in tying the history of computing to an institutional history of Cold War research places it squarely within the constructivist tradition in the history of science and technology. By now there have been several decades of solid scholarship on the so-called social construction of science and technology, summarized by Jan Golinski in *Making Natural Knowledge: Constructivism and the History of Science* (1998).

Constructivism remains the dominant mode of analysis in historical and sociological studies of science and technology. Nevertheless, there continues to be a lack of unanimity, and even explicit discontent regarding its central principles. Many of the disagreements have revolved around recurrent bouts with the relativistic implications of constructivist scholarship. Critiques have ranged from the casual dismissal of "merely social" explanations that fail to account for the political intent embedded in technological artifacts to more formal efforts to understand the role of nature and material artifacts in determining historical outcomes.[5] The most noted work in the latter genre is that of Bruno Latour (1993). However, Latour's writings have themselves generated considerable controversy, specifically with regard to the relativistic implications of what others have taken to be his effort to ascribe human agency to inanimate objects and other non-human entities.[6] Meanwhile, the infusion of new postmodernist theories and methodologies into the social study of science and technology has broadened the scope of criticism by posing an explicitly political and culturally relativistic agenda. It was amidst this intellectual confusion that the Sokal hoax—an incident in which a physicist fabricated an article that was published by a leading journal of postmodernist analysis—blew open the debate over the relativistic implications of constructivist scholarship.[7]

To those not familiar with the debate, the tension between realism and relativism might seem an idle academic conversation. However, consider the the so-called von Neumann architecture. It has dominated computer design for six decades. But computer scientists would be the first to admit that other designs could have offered equal or better performance for a broad range of applications. Yet it was the consensus that emerged around a single design, based on a fast central processor and a fast memory system, that facilitated immense improvements in circuit speed and density. While there have been many attempts to build fast computers using alternative designs, the consensus on a specific approach brought forth industrial investments and other commitments that made rapid improvement of the technology possible. Whatever the technical advantages of having a central processor, the overall phenomenon of the von Neumann architecture cannot be understood except as something integrating natural and social processes. Understanding the structure of this integration, and how it occurs, remains the principal challenge for constructivist scholarship.

I will continue my historiographic overview by describing my book in relation to three threads in the debates over constructivist scholarship.

The first thread has to do with the "mutual shaping" thesis: the idea that we ought to explore the mutual relationship between knowledge and its supporting institutions, rather than a relationship in which "society" simply shapes "science" and "technology." Although there have been many calls for full examination of this mutual dependency, one of Latour's major critiques has been his suggestion that historians and sociologists have tended to embrace the notion of social construction without a full commitment to this symmetry.[8] This has been echoed more recently by Gabrielle Hecht (1998) with respect to the history of technology.

The relative dearth of constructivist studies that support the "mutual shaping" thesis is somewhat ironic. Peter Berger and Thomas Luckmann's book *The Social Construction of Reality* (1966), which launched much of the interest in constructivism across the humanities and the social sciences, gave the term "social construction" a broader meaning. Drawing directly on phenomenological traditions in interpretive sociology, Berger and Luckmann were at least as interested in how social institutions become articulated through local action as in how prevailing interests and institutions affect local outcomes—the trope to which most science and technology studies scholars turn when speaking about the social construction of a fact or an artifact. I am interested in fully supporting the "mutual shaping" thesis, and in documenting not only the emergence of the digital computer but also the emergence of a new research infrastructure.

This book builds on the work of Gabrielle Hecht (1998), Donald MacKenzie (1990), Ken Alder (1997), and others who have studied the mutual relationship between new technology and emergent social institutions.[9] In this respect, its most obvious counterpart is Paul Edwards's 1996 book *The Closed World: Computers and the Politics of Discourse in Cold War America*. Edwards's work has been instrumental in opening up the question of how military and other interests influenced computer designs and the question of how technical developments in computing influenced Cold War military institutions. Going well beyond the realm of discursive analysis, Edwards offers a compelling account of how the computer both shaped and was shaped by a "closed system" metaphor that had grave implications at the highest levels of military strategy. Edwards also lays out a specific concept of "mutual orientation," a concept on which I draw in this book.[10] I agree with Edwards's methods and with his historical interpretation; the phenomenon he describes was clearly an important element in Cold War research. My interest, however, lies in using the computer to identify other, pluralistic dimensions of Cold War research that were not themselves governed by a "closed system" metaphor.

The second historiographic thread has to do with a kind of intellectual retrenchment that has followed in the wake of the so-called Science Wars that erupted over the relativistic implications of constructivist scholarship.[11] I consider it important not to simply return to old "internalist" agendas that focus on scientific and engineering knowledge and processes without giving serious consideration to their surrounding context. Nevertheless, those who have turned to more detailed study of scientific knowledge, instruments, and practices—and historians of technology who have likewise responded to the call to pay greater attention to artifacts and the engineering workplace—have done so in a genuine attempt to understand the material and ideational basis on which the various fields of science and engineering are constructed. This corresponds to a study of the lower layers of an ecology of knowledge (see figure I.1), and it offers a valuable extension to constructivist methodologies.[12] In this respect, my book draws considerable inspiration from David Mindell's *Between Human and Machine* (2002).

Nominally, Mindell's book is about the history of control-systems engineering. However, as its title implies, the book sets out to extend constructivist analyses by paying careful attention to the most esoteric, technical dimensions of an emerging field. Mindell documents the persistent resistance to closure that occurred within the very field that produced the notion of a closed system. Thus, in contrast with the relatively smooth process by which mutual orientation occurs in Edwards's account, Mindell traces, quite exquisitely, the complex and often friction-ridden interplay of institutions, ideas, artifacts, and practices that eventually gave birth to modern control-systems engineering. This is a perspective that is absolutely essential to my work. Moreover, Mindell's study does not simply document a disciplinary convergence. Very different disciplinary traditions in information theory, control systems, cybernetics, and digital computers design emerged from World War II, held together only temporarily through the wartime work of building control systems for anti-aircraft guns. In form as well as in substance, Mindell's book is also a multi-site study conducive to an ecological view of knowledge. It is significant, too, that Mindell's book ends with an account of the synthesis of artifacts and ideas, and of individuals and organizations that established computing as a coherent body of knowledge. This is exactly where my book begins.

In a sense, my book is situated halfway between Edwards's and Mindell's. Edwards uses the notion of mutual orientation to tell a story about a Cold War order. Mindell employs a similar approach to tell a story about technical disciplines and disciplinary organization. I pay equal attention to the technical forms and the institutional forms that emerged from efforts to align digital computers with various institutional values and priorities.

Third, this book addresses a historiographic issue having to do with the roles of human actors and intentional action in shaping knowledge and social institutions. As I have already noted, I pay close attention to how "natural" events, material artifacts, and imagined physical relations—an atomic explosion, a recalcitrant memory tube, the mathematics of hydrodynamic equations—contributed to the course of history. I also pay close attention to historical contingencies, and to the unintended consequences of even the best-laid plans. Yet I remain committed to exploring how human actors come to attribute meaning to the sociotechnical world through their everyday encounters with it. Indeed, it has been observed that my work tends to balance biography with technical and institutional history. The stories I tell in this book are often about how individuals, sometimes acting singly and sometimes collectively, shape social institutions through the very act of extending an institution into a local site of practice. I also describe how they do this because of their technical interests, which helped to define these new institutional forms and which are in turn upheld by them.

The work that best exemplifies the kind of analysis pursued in this book is outside the realm of the history of science and technology. It is Olivier Zunz's *Making America Corporate* (1990). Zunz also describes a major institutional transformation in U.S. history, namely the rise of major corporations at the beginning of the twentieth century. In asking to what extent the managers, executives, and agents of the modern corporation had the flexibility to interpret the demands of industrial capitalism and to shape the corporation in their own image, Zunz situates his interpretation of white-collar workers precisely between that of Richard Hofstadter and that of C. Wright Mills. I take a similar social-historical approach in asking how all those involved with Cold War research—researchers, university administrators, military program managers, and others—gave definition to the institutions in which they worked and to the technical subjects that interested them.

However, there is an important difference between the history of the corporation and the history of Cold War research. Those working within modern corporations were bound by a hegemonic ideology of industry and profit during an era marked by a significant surge in the global system of capitalism. Managers and executives may have played a vital part in defining the practices—"technical" practices, including those of cost accounting and the management of sales districts—that were constitutive of this system. Nevertheless, these individuals' innovations were always placed against a common measure of profit. Because of a widely diffused set of norms that were already integral to capitalism as a social institution, eclectic practices almost always gave way to more profitable ones. Cold War research presented no such unified vision. The institutional pluralism of Cold War research ensured that there was no single standard by which to measure the productivity of a research program. As a consequence, industrial laboratories remained confused about the aims of basic research. Research universities in the United States remained unclear about their obligation to serve the practical demands of the civilian economy and the

nation's military arsenal. Military institutions were often swayed by rhetoric favoring fundamental research. Meanwhile, those who chose to align
themselves with new areas of specialization often found no simple path
by which to advance a professional or disciplinary identity. Operating with
such ambiguity, researchers and research administrators were able to
pursue far more eclectic strategies in defining not only their research
agendas but also the identities of their institutions and their occupations.
In the process, they defined what it meant to do Cold War research.

Consider, for instance, two of the arrangements described later in this
book. During the late 1940s, the National Bureau of Standards set up a
computer-procurement service on behalf of other federal agencies. This
allowed the NBS, a civilian government laboratory, to divert military funds
to private manufacturers interested in developing a commercial market for
electronic computers. And during the 1960s, computer programmers
employed by the academic computing center at the University of
Michigan created a space for themselves to pursue research in computer
science. They did so by promising to deliver a central computing facility
that would enhance engineering education, support scientific research, and
actualize a university administrator's desire to centralize resource planning
within the university. In each of these cases, the course of events would
come to determine the specific roles that a federal research laboratory and
the central computing facility of a second-tier research university would
play in research having to do with digital computers. Nevertheless, the
significant indeterminacy in the final outcome suggests that Cold War
research was not written according to some fixed and immutable template.
Social history clearly takes on new meaning when carried out in a pluralistic vein.

"Calculation," as employed in this book's title, is a double entendre that
plays on the notion of centers of calculation that Latour introduced in
Science in Action (1987). Although I will qualify this somewhat in discussing
an ecological view of knowledge, I share with Latour a general interest in
the intentional actions that bring about both social and technical change.
In this respect, the "natural" world remade by digital computers was as
much about the new social arrangements for carrying out research as the
knowledge of the natural world that it generated.

In any event, it was the diverse institutional arrangements for Cold War
research, as realized through the eclectic strategies pursued by researchers

and all the others involved in the research enterprise, that gave rise to the immense technical productivity and institutional fluidity of the Cold War years. However, this should not be taken on a simply celebratory note. The concept of pluralism as employed in political theory carries with it the connotation of a hegemonic order established in spite of differences. Although I do not explicitly use the term in this manner (I use it primarily to connote difference), it is quite possible that Cold War research, despite all its differences, represented a consensus on a particular vision that precluded other alternatives. I leave this for others to study. At a more mundane level, Cold War research was simply expensive. Tensions and differences often produced redundant, over-ambitious, and incoherent research programs. Yet the demands of the Cold War allowed novel solutions that failed in one context to be utilized in another. As is often noted in the literature on R&D management, the efficiency of innovation is not easily measured. I leave the question of how we can properly measure the social utility of innovation during the Cold War to others.

Summaries of Chapters

Chapter 1 supplies some necessary background by depicting the overall ecology of computing expertise between the interwar years and the end of World War II. This chapter also provides an opportunity to document what the "circulation of knowledge" looked like at a time when computing, not yet a unified field, was an agglomeration of locally integrated artifacts and practices sustained within separate institutional niches for commercial accounting, scientific computing, and engineering analysis.

Chapter 2 follows the peripatetic career of John Mauchly, one of several individuals recognized for having invented the electronic computer. Displaced from his studies of molecular physics by circumstances related to the Great Depression, Mauchly sought new opportunities in a desperate attempt to maintain a productive research career. In the process, he was exposed to a rather wide array of knowledge and artifacts, including the scintillation counters used by physicists, the manual methods of computation used in studies of radio propagation, and the statistical techniques employed by social scientists and numerical meteorologists. At the University of Pennsylvania, he came across the knowledge of "best practices" in radio electronics, which, when combined with his prior knowledge, enabled him to design a large-scale digital electronic computer. In Mauchly's case, the circulation of knowledge resulted from how he came to embody relevant knowledge as he traveled through the institutional ecologies described in chapter 1.

Each of the subsequent chapters focuses on a specific institutional ecology that surrounded a particular computing or computer-development project. Chapter 3 focuses on John von Neumann and on the institutional ecology that was created by the prominence of mathematical physicists and applied mathematicians after World War II. The Institute for Advanced Study became, quite literally, a center of calculation for postwar computing work. The rapid convergence of interest around von Neumann's ideas provides an opportunity to demonstrate how a strong sociotechnical network can come to characterize, and therefore remain important to, an ecological view of knowledge. On the other hand, a close study of von Neumann's contemporaries reveals the limits of von Neumann's ideas and how different institutions and different alliances continued to pull research on digital computers in different directions. Chapters 4–6 are built on this thesis. Especially in chapters 4 and 5, I look at projects that began substantially within von Neumann's institutional fold only to diverge significantly from the vision for postwar research held by von Neumann and his mathematical colleagues.

In chapter 4, I consider the computing-related research programs at the National Bureau of Standards. Both of the two major initiatives at the NBS—a federal computer-procurement service and a program dedicated to the engineering of computer components—were originally tied quite closely to von Neumann and his work. However, in responding to the many different opportunities afforded by Cold War research, both groups came to diverge considerably from their original missions. The NBS was the first U.S. research laboratory to successfully operate a stored-program computer. By comparing the work done at the NBS against better-endowed projects, this chapter provides a way to consider more precisely how institutional location, organizational structure, and individual initiative shape the process of innovation.

In chapter 5, I revisit Jay Forrester's work on the Whirlwind computer at the Massachusetts Institute of Technology. I take a close look at the complex interplay of the project's technical, fiscal, and institutional dimensions, and at how this interplay led Forrester to articulate a research program that was very different from von Neumann's. I also consider how the very difficulties that were associated with Project Whirlwind gave definition to the institutional relationship between MIT and the Office of Naval Research.

In chapters 6 and 7, I turn to the commercial sector, and specifically to two case studies involving IBM's entry into electronic computing. Many studies written primarily from the perspective of business history have documented how IBM came to dominate the computer industry for 30 years. What is not so often noted in the literature is that IBM was, at the end of World War II, simply a mid-size firm specializing in manufacturing, leasing, and servicing office machinery. As others have noted, IBM's installed base of office equipment provides the most compelling explanation for its postwar prominence.[13] Nevertheless, new institutional alliances between business and government, and between science and industry, required IBM and other firms to extend their corporate cultures in numerous ways to take advantage of the new opportunities. In chapter 6, I look at IBM's efforts to absorb certain elements of a scientific culture in its attempt to support the sale of computers to the postwar scientific community. That chapter is based on a study of IBM's Applied Science Department, established in 1949. In chapter 7, I consider how IBM's customers contributed to this relationship through participation in a users' group known as Share. The Applied Science Department and Share were suspended within and helped constitute the curious amalgam of corporate, academic, and occupational priorities that was becoming an increasingly familiar feature of Cold War research.

In chapters 8 and 9, I return to the university as a site of Cold War transformations, comparing the academic computing center at MIT and that at the University of Michigan. Chapter 8 focuses on the initial efforts at MIT and Michigan to establish a central academic computing facility; chapter 9 gives an account of the two universities' entries into research on computer time sharing. In both chapters, I trace the considerable tension concerning whether computers were to function as a service facility, a commercial technology, or a subject of research. These chapters reveal how MIT and Michigan, with their different institutional locations, different resources, and different talents, resolved this tension in different ways. Historically, MIT has received much of the credit for the work described in these two chapters. However, it is quite evident that Michigan contributed directly to technical innovation and to what I refer to throughout the book as the "diversity" of Cold War research.

The case studies in this book constitute a broad survey of early developments in computing; they also describe a wide cross-section of the different kinds of institutions and organizations that contributed to Cold War research. But the principal value of the book lies in how these elements are brought together in a single study.

On the other hand, it is important to recognize the limits of any metonymic strategy. Computers were not important to every aspect of Cold War research, nor are the case studies in this book sufficient to demonstrate the computer's reach in scientific and military affairs. Moreover, one would expect that a very different technology, such as nuclear weapons or nuclear energy, would reveal a very different and presumably more authoritarian set of arrangements. But stated in the converse this simply suggests that the computer offers an especially valuable window on the most pluralistic dimensions of Cold War research. Even if digital computers turn out to be an extreme example, a historical study that unveils this diversity ought to provide a valuable foundation for a more robust study of Cold War science and engineering. My hope, in any event, is to demonstrate how an ecological view of knowledge, with an emphasis on institutional pluralism, can be a valuable approach to the study of innovation.

The Ecology of Knowledge

This book is written to be accessible to an audience larger than that of academic historians. Nevertheless, insofar as the notion of an ecology of knowledge serves as a central organizing principle for the book, I will spend some time providing a more formal definition of this concept. Those not specifically interested in this academic discussion may proceed directly to chapter 1.[14]

As I noted earlier, the notion of an ecology of knowledge can be traced back to Charles Rosenberg's essay "Toward an Ecology of Knowledge: Discipline, Context and History" (1997a/1979) and to a series of studies produced by a group of sociologists of science working in the symbolic-interactionist tradition. Important among the latter are Susan Leigh Star and James Griesemer's "Institutional Ecology, 'Translations' and

Boundary Objects" (1989), which in some respects is unfortunately known more for its idea of "boundary objects" than for its notion of institutional ecology; Adele Clarke and Joan Fujimura's introduction to *The Right Tools for the Job* (1992), which explicitly calls for bringing a more ecological perspective into science studies; and the 1995 volume *Ecologies of Knowledge*, edited by Susan Leigh Star. These latter works all draw upon the ecological outlook that served as a foundation for much of the work of the Chicago School of Sociology, and especially the ecological framework for studying social institutions advanced by C. Everett Hughes and Anselm Strauss.[15] The works by Star and her colleagues and by Rosenberg helped to establish and sustain the productive synthesis that made constructivism the dominant mode of analysis in the history and sociology of science.[16]

Of these works on the ecology of knowledge, Rosenberg's early essays do the most to set up the metonymic perspective that is central to the present book. In his somewhat later essay "Woods or Trees? Ideas and Actors in the History of Science" (1997b/1988), in a passage in which he points out the importance of studying the "structure of integration" between internal and external accounts of science,[17] Rosenberg writes:

The woods-and-trees metaphor is particularly apt in this connection. The contemporary ecologist's conception of a particular woods assumes and necessitates an understanding of the trees—their species, their climatic and nutritional needs; woods and trees are in this sense indistinguishable in some sense as interactive system and as a linked research agenda. We too must disentangle and specify in our particular sphere the relationships between the actor's perceptions and strivings, his or her institutional climate, the soil that nurtures—or fails to nurture—particular career and intellectual options.[18]

This passage is quite important for the present book, for it establishes not only a clear metaphor for visualizing the metonymic relationship between knowledge and its supporting institutions but also a delicate balance between biography and institutional context, even while tying both to a distinct research agenda.

The more recent work in the sociology of science nevertheless advances Rosenberg's insights in several important respects. First, insofar as Star and her colleagues have pursued their work in the laboratory-studies tradition, their work embodies many of the recent advances pertaining to the material culture of science, which again corresponds to the lower layers of the representation provided in figure I.1 above. Though present, this interest in material culture receives less attention in Rosenberg's descriptions of an ecology of knowledge.

In working from a symbolic-interactionist perspective, Star and her colleagues are also more aware of the multiplicity of perspectives that together constitute any particular domain of knowledge—an "arena," in their terminology. Thus, whereas Rosenberg provides some sense of the distance that separates a profession from a discipline, Star and Griesemer's 1989 article and many of the works that appear in Star's 1995 compilation point more directly to the co-production of knowledge involving a wide assortment of actors. While this emphasis on co-production builds more or less directly on Howard Becker's *Art Worlds* (1982), what is perhaps noteworthy is the explicit sense of spatiality and geographic extension it adds to the metaphor of an ecology of knowledge. Thus, when I speak about John Mauchly's peripatetic career, or about how each chapter describes a specific institutional ecology within the general arena of computing expertise, I draw explicitly on both the literal and the cognitive sense of space and distance as established by the symbolic interactionists.

If I extend Star and Rosenberg's insights, it is by developing the metaphor of an ecology of knowledge even further, both through the use of representation and through a more formal articulation of the notion of metonymy as applied to the relationship between esoteric knowledge and social institutions.[19] I have written elsewhere about the representational scheme employed in figure I.1, discussing it specifically in relation to earlier work in actor-network theory.[20] Here I continue to refer to the representation because it serves as a useful aid to the further theoretical articulation of an ecology of knowledge.

There is room for debate as to whether diagrams such as those in figure I.1 necessarily benefit academic historians. All representations have reductive aspects and, if taken literally, may simplify what can better be told through historical narrative. Nevertheless, an ecological view of knowledge, especially one grounded in the material culture of science and engineering, can produce a level of complexity that presses against the limits of narrative forms, and, perhaps as important, against the limits of an audience's capacity to synthesize the arguments presented in the text. Precisely to the extent to which ecologists have turned to representation to make complex, interdependent relationships in nature tractable to analysis, the layered representation in figure I.1 is designed to do so for an ecology of knowledge.

Moreover, I specifically refer to the representation at this point because I have found it to be useful in developing a more precise understanding of metonymic relationships. From the standpoint of a formal study of language, the notion of metonymy is not simply a part-whole relationship; it is sustained through the "syntagmatic" dimension of language and the relationship of "contiguity." As described by Roman Jakobson in a series of essays in *On Language* (1990a), the syntagmatic dimension of language pertains to a logical succession of word choices that lead to a meaningful sentence or utterance.[21] Words that exhibit such coherence are said to exhibit a relationship of contiguity. But significantly, although Jakobson's focus in "Parts and Wholes in Language" is on language, he draws on earlier work by Ernest Nagel (1963), who suggests that metonymic relations also can be built through semiotic associations with objects and processes in the world. This makes it possible to view syntagm and contiguity relations as existing through spatial arrangements, and not just those that appear in the more linear constructions of language. This is, in fact, the way semiotics has been applied in studies of paintings, advertisements, and other visual, representational objects.[22]

It is this last set of observations that I draw on in developing a more fundamental understanding of the metonymic relations that constitute an ecology of knowledge. Certainly it could be said that as a leaf is to a tree, and a tree to a forest, computers were constitutive of the discipline of control-systems engineering, which in turn was constitutive of a particular closed-world discourse. Yet for each of the "institutions" described in figure I.1 (such as the "military establishment" depicted in diagram a) it was also necessary for a wide variety of other entities—actors, artifacts, esoteric knowledge, organizations—to be arrayed beneath the institutional entity in suitable relations of contiguity so as to establish a coherent body of practice. Much of this book is about how local actors build the contiguity relations—stable relationships of ideas, actors, objects, and practice—that sustained new disciplinary identities and the broader institutions of Cold War research.

On the other hand, one important thing about figure I.1, as I noted earlier, is that none of its diagrams depict simple, hierarchal relationships. They each exhibit differing degrees of pluralism, even diagram b. Here symbolic-interactionist perspectives are again valuable in that they press against the implicit notion of hermeneutic closure that accompanies a linguistic definition of metonymy. Clearly there are always different ways of interpreting the world. There are local differences in meaning and interpretation, even as these regionalized differences are mobilized to constitute a broader social institution such as Cold War research. Indeed, as I argued more casually above, it is these differences of interpretation that created the room for negotiation in constructing new contiguity relationships, namely the novel institutional arrangement and research programs that constituted the many different approaches to Cold War research.

The idea that interpretive differences foster negotiation points to the performative dimensions of language. As first established by Victor Turner (1967) in cultural anthropology, the performative idiom emphasizes how meanings are always being constituted and reconstituted through action. Even in the most sacred, scripted forms of ritual, the specific meaning of each performance varies, albeit to different degrees, because of the intersubjective encounters with an audience inhabiting different circumstances. The performative idiom has been embraced by some science and technology studies scholars. Andrew Pickering espouses it in his philosophical study of scientific practice (1995), and a similar emphasis on "process-oriented" accounts can be found in work written in the symbolic-interactionist tradition. This would first suggest that each of the interconnecting lines in the diagrams in figure I.1 is not some abstract, metonymic association sustained through a formal definition in language but rather an association sustained through a well-defined body of practice. The significance of tabulating machines to the actuarial profession, or that of digital computers to the image of the postwar research university, must be sustained through their regular and routine use via an established body of practice. This is equally true for any of the lateral associations that exist within a layer; for instance, IBM's relations with its scientific customers relied on its having, or having built, an elaborate body of practice for sustaining this relationship.

To the extent that this book is based on a search for such practices, it is informed by an ethnomethodological approach to science and technology studies, a position generally associated with the work of Michael Lynch (1993). On the other hand, insofar as I am interested in technical and institutional innovation, I depart from much of ethnomethodological tradition by focusing not only on the *stable* body of practice that defines a community of practitioners but also on the *mutability* of these practices in the course of (in fact, in ways that define) technical and institutional change. And although ethnomethodologists are correct to suggest that all "accountable" actions—actions that can be identified and described by a social analyst—must be social in origin, this does not bar the possibility that there may be a set of accountable second-order (or even higher-order) practices that communities have for transforming their own practices.

Those interested in a dynamic view of an ecology of knowledge must seek to identify practices that explicitly aim to bring about technical and institutional change. An analytical framework for doing so can be found in any of a number of sources; Giddens's structuration theory (1976, 1979, 1984) and the closely related arguments of Pierre Bourdieu (1977) are perhaps the most notable. Giddens's formal articulation of the duality of structure, in which he regards social structure as both the cause and the product of human actions, is certainly foundational to this discussion. Moreover, Giddens's work supports the thesis that the actions that produce social transformations must already be encoded as social structures in the exact same manner as the cultivated practices that ensure social reproduction. However, Bourdieu's more grounded approach provides a better sense of how socially defined actions can produce institutional transformation. This is somewhat ironic, insofar as Bourdieu's emphasis, at least in his early writings, was on social reproduction. Yet embedded in Bourdieu's descriptions of (for instance) gift exchange among the Kabyle is a more vivid sense of the opportunistic and interpretive strategies that might result in social transformation as well as reproduction.[23]

Giddens's critics would be right to point out that his ideas can be a gloss on the subject of social structure and change. In spite of raising the specter of practice—"action," in his terminology—Giddens fails to give much thought to the actual practices that contribute to social or technical transformation. Figure I.1 makes it clear that what remains weak in Giddens's structuration theory is its failure to describe the myriad forms of practice that metonymically constitute the relations between knowledge and its supporting institutions. But in this respect the recent advances in science studies may make it possible to add flesh to what Giddens regards as an ontological enterprise.[24]

Moreover, Giddens's theory is valuable because it appears to offers a way to advance constructivist scholarship by developing the structural metaphor between knowledge and its social institutions—something advocated not only by Rosenberg, but also by Ian Hacking in his general critique of constructivist scholarship.[25] The representational strategy employed in figure I.1 can in fact be taken to be an explicit attempt to develop the metaphor in this direction, although I follow Rosenberg in overlaying an ecological metaphor on the structural metaphor that Giddens and Hacking used.

Put more abstractly, the representational scheme employed in figure I.1 is an attempt to integrate a theory of semiotics with a theory of practice via Giddens's structuration theory. It serves as a vehicle for mapping out a complex semiotic field as both sustained by and transformed through established bodies of practice. It is important in this respect that the reference to semiotic theory here would be to those approaches associated with Charles Saunders Peirce as opposed to Ferdinand de Saussure. From the standpoint of an ecological view of knowledge, Peirce's notion of indexicality points crucially to both the pragmatic and the material foundations of language.[26]

Such an approach should also help to build a bridge to Wittgenstein's phenomenological position, as best described by Michael Lynch (1993). At the same time, the dynamic view of an ecology of knowledge presented here should offer at least one answer to Lynch's challenge to constructivists, namely the need to fully document how socially defined practices that exist outside (or even within) a realm of disciplinary practice can affect the internal conduct of that discipline. It should also demonstrate the reverse effect, namely how social institutions gain their definition through the appropriation and extension of locally generated practices (which would be consistent with the "mutual shaping" thesis).

Still, the practical challenge remains, namely that of identifying the myriad bodies of practice that constitutes and reconstitutes esoteric knowledge and its supporting institutions. It might seem on the surface that the practice of scientific discovery and technological innovation would be quite different from the rituals used to uphold the familiar social structures studied by traditional anthropologists, and hence difficult to observe. Yet, although retrospective accounts of discovery and invention—e.g., the typical history of a heroic scientist or inventor—routinely recast individual actions in a functionalist vein, ethnographically oriented accounts of what scientists and engineers do have found the same kind of accountable forms of improvisation found within other domains of social practice. This is clearly consistent with the current understanding of, for instance, Thomas Edison's inventive practices.[27] Meanwhile, Robert Thomas (1994) and other organizational sociologists who have adapted ethnographic techniques to the study of R&D organizations have likewise been able to describe how individual engineers and managers come to define purposive action within an institutional setting rife with ambiguous and conflicting rules for organizational conduct.

It is important to recognize that there are practical limits to historical methodology. The work of extracting information about practice is difficult enough to do using ethnomethodological techniques; historical sources are often not up to the task. Driven as well by the need to tell a sufficiently broad story, each of the chapters in this book moves back and forth between different levels of narrative, from an explicit effort to describe the higher-order practices that produce technical and institutional change, to descriptive accounts of historical changes in practice, to more conventional narratives that pay no particular attention to practice. This having been said, my book is substantially based on a study of the situated practices that produce technical and institutional innovation.

In order to support this focus on transformative practice, I introduce here, on a highly contingent basis, a system of classification for categorizing these practices as defined in relation to their effects on an ecology of knowledge. These categories should be viewed not as mutually exclusive, but as different forms of practice that may be invoked concurrently.

The most straightforward category of transformative practice is that of *syntagmatic extension*. As is already integral to the notion of metonymy, institutions extend their reach when stable assemblages of knowledge, artifacts, and instrumental practices are taken from one context and reproduced within another. The work performed by IBM's sales representatives, for instance, would be a body of practice specifically designed to accomplish such an outcome. *Interpretive extension* pertains to practices that explicitly set out to modify or extend an existing arrangement. The historical studies of Edison's inventive practices and Donald MacKenzie's

(1990) account of the strategies that Charles Stark Draper developed for extending the reach of inertial guidance systems both point to how technical and institutional innovation may result from cultivated practices for achieving this kind of incremental change.

What also becomes quite evident in this study is how innovation occurs when established bodies of practice are brought together at the juncture of multiple institutions and disciplines. That such strategies for the *recombination* of knowledge and institutions were an important part of Cold War research is demonstrated by the popularity of "interdisciplinary" strategies since the end of World War II. Interdisciplinary strategies were certainly important to postwar developments in computing.

Finally, it is essential to consider how certain metonymic associations may be lost or forgotten during the process of extending an institution or a body of knowledge. Especially during the interwar years, when computing was not yet a unified field, a good deal of innovation occurred through the highly selective appropriation of the knowledge and associated practices advanced within a different sphere. Though computing as a whole became more of a unified field after World War II, such acts of appropriation continued to characterize the field. It is important to consider how these efforts to appropriate knowledge required a degree of *dissociation* from prior contexts and how this contributed to the diversity of the resulting technical and institutional forms.

There are clearly established theoretical analogues for these different categories of practice. Latour's notion of translation, as developed in *Science in Action* (1987), maps rather readily onto the notion of syntagmatic extension. His focus, like mine, is on sociotechnical networks, and on the tactics researchers use to interpret the interests of other actors so as to support the seemingly effortless transfer of knowledge and artifacts from site to site. Likewise, MacKenzie's (1990) notion of trajectory encompasses the idea that researchers employ carefully cultivated strategies for extending technological artifacts and their institutional relevance. Dissociation, meanwhile, would be integral to Galison's (1997) notion of intercalation, which accounts for the breaks that routinely occur between scientific theories, instruments, and experimental traditions. I make these connections not to overstate the theoretical importance of my work, but to demonstrate that an ecological view of knowledge can encompass important ideas in science and technology studies.

To reiterate, I do not regard this system of classification as a formal means of describing all aspects of innovation. This is not the appropriate place to work out such a system. I simply introduce this contingent classification based on the recurring forms of practice I encountered while working through the historical material. But I should perhaps also note that these categories map a bit too readily onto familiar evolutionary processes—syntagmatic extension to reproduction, interpretive extension to adaptation, recombination to mutation, and dissociation to selection. This suggests that they may be products of the undue influence of an evolutionary paradigm that remains difficult to separate from an ecological metaphor. The ecology of knowledge, when viewed in terms of a formal network, must have a network topology quite different from the ecology of nature, owing to the very different way in which ideas, as opposed to genetic information, are held and reproduced. Still, I provide this system of classification for readers to keep in the back of their minds while working through the historical episodes I present.

Figure 1.1
Vannevar Bush and his differential analyzer. Courtesy of MIT Museum.

1   An Ecology of Knowledge: Technical and Institutional Developments in Computing, 1900–1945

Before World War II, computing was not yet a unified field; it was a loose agglomeration of local practices sustained through various institutional niches for commercial accounting, scientific computing, and engineering analysis. I paint this first portrait on a rather broad canvas, using the metaphor of an "ecology of knowledge" to map out the different and yet related approaches to computing onto a single conceptual field. In describing how this landscape changed over the course of several decades, I also provide an initial sense of how a multitude of technical exchanges involving esoteric knowledge and its sustaining contexts gave rise to both technological and institutional innovation. This chapter also supplies an essential background for the rest of the book, while serving also as a historiographic introduction.

The historical questions that drive this chapter concern how esoteric knowledge circulates within a loosely constituted technical field and how new disciplinary identities emerge from this circulation. The early technical exchanges in computing were facilitated by many local networks of common association assembled around specific disciplines, institutions, and computing machines. Although these early exchanges gave rise to some groups with broader interests in computing, this centripetal tendency was offset by the centrifugal influence of broader historical and institutional developments. It was the centralizing effects of the mobilization of science during World War II that remade computing into a more coherent subject. Despite the wide variety of innovations that emerged from wartime research, certain individuals, especially those affiliated with the National Defense Research Committee, came to occupy a vantage point from which they could view computing as a unified discipline.

This chapter's breadth also makes it possible for me to draw on recent conversations about the "circulation of knowledge." As noted by Warwick Anderson (2002), an interest in the circulation of knowledge emerged from the literature on science and colonialism, where it continues to be a valuable means of challenging naive accounts about the diffusion of Western knowledge. The focus has been on the "complexities of contact" that characterize all exchange relationships.[1] Anderson observes how these studies extend Bruno Latour's notion of circulation, as described in *Science in Action* (1987). In addition, the concept is broadly consistent with other studies of technical exchange networks in the science-studies literature, from symbolic-interactionist studies to Peter Galison's metaphor of a "trading zone." As indicated by these other studies, the notion of circulation can be productively applied to many topics outside of colonial and postcolonial studies. The concept has in fact already been applied to the history of technology through John Law's notion of heterogeneous engineering, albeit also in a colonial context.[2]

Consistent with my interest in the metonymic relationship between knowledge and its sustaining institutions, the notion of circulation pays explicit attention to social relations as well as to the knowledge that gets circulated. In fact, I take the notion of circulation to be integral to the metaphor of an ecology of knowledge. The concept supports the sense of geographic extension found within any ecological metaphor; it provides a way to visualize how new patterns of institutional and technological interdependence emerged through such extension. At the end of this chapter, I also consider why the notion of circulation is better suited to the technical exchanges characteristic to the early history of computing than the notion of a "trading zone" advanced by Galison.

Early Landscape

Commerce

At the start of the twentieth century, many different approaches to computing were scattered across the U.S. institutional landscape. The country was undergoing rapid industrialization and urbanization, as upheld through high levels of immigration. New systems of commerce, along with the ascent of older immigrant populations out of the throes of poverty, laid an important foundation for a new white-collar workforce. The skyscrapers in New York and Chicago and the brilliant displays offered by new urban department stores were the most overt symbols of the nation's newfound wealth. Yet hidden within these structures was an army of white-collar workers who administered the ever-increasing volume of financial transactions (Zunz 1990). The changing nature of accounting and commerce provided one important site for new developments in computing.

Even in the realm of commerce, there was no single approach to computing. Banks remained conservative. Though somewhat less conservative than their British counterparts, U.S. banks continued to hire white, male, non-immigrant workers into their senior clerical positions. Banks had to expand their capacity for managing transactions, but they also adhered to hiring practices that upheld the outward sign of moral restraint. A bank's principal commodity, after all, was money. Banks based in the United States did begin to hire women by the early 1900s, but they were sent to the back offices and given highly segmented and routinized tasks. As suggested by Sharon Strom (1992), feminization, routinization, and mechanization went hand in hand; it was this part of the bank's operations that initially gave way to mechanization through such devices as Comptometers and bookkeeping machines.[3]

Restaurant owners and retailers held to no such illusion about the moral character of their employees. Still, even here there were competing systems of cultural meanings. On the one hand, an expanding culture of genteel leisure and business respectability demanded an expanding rank of waiters with an outward appearance of politeness and respectability. Likewise, the saleswomen who worked at department stores had to have sufficient composure to be regarded as reliable arbiters of middle-class taste. However, cash transactions left room for abuse. Waiters could roam the floor with a mixture of tips and diverted receipts; the social dynamics of the sales floor also offered room for abuse—or at least fears of abuse. Entertainment and retail were already cast as social domains of lesser moral restraint, and the cultural yoke of servitude only served to augment suspicions. Business historians have carefully documented how the National Cash Register Company and other firms succeeded by introducing new transaction-accounting methods into the U.S. retail industry. Nevertheless, these firms also capitalized on a retailer's fears of embezzlement. One of NCR's famous inventors, Charles Kettering, spent three weeks cavorting with waiters to learn of such things as erasures and double checks. By the early 1900s, NCR had multiple departments specializing in tailoring its machines for the specific intersection of workers, customers, and transactions that characterized each retail trade.[4]

For utility companies, moral restraint among the regular employees was somewhat less of a concern. Closely supervised workrooms provided an effective and traditional means of curbing embezzlement. But a major concern for U.S. utilities was that of establishing a different kind of accountability in the face of Progressive Era politics. The famous Philadelphia Rates Case of 1911 was primarily about investors and their watered-down stocks. Nevertheless, accusations of bureaucratic inefficiency followed on the heels of corruption. Frederick Taylor and his method of scientific management were called upon to probe the operations of the Philadelphia Electric Company. Utility companies continued to experience renewed pressure to justify their rates and to demonstrate operating efficiency. Not surprisingly, these companies were among the first to adopt a feminized workforce and specialized billing machines.[5]

Science and Engineering

The aforementioned changes were facilitated by broadly shared perceptions about gender, wage work, efficiency, and accountability. Yet, although it is possible to speak of a time "when women were computers," both the social and technical foundations for computing were broader than that. Progressive reforms played themselves out not only in the political and economic domain but also in the halls of science. Backed by a renewed wave of imperialism and nationalism, academic institutions in the United States were quick to adopt the German model of the research university. Beginning with Johns Hopkins University (established in 1876) and the University of Chicago (established in 1892), a new focus on research and objective knowledge emerged across the U.S. intellectual landscape, bankrolled by new philanthropic institutions whose wealth originated in commerce.[6]

Although new philanthropic organizations such as the Carnegie Corporation and the Rockefeller Foundation built their agendas around the notion of excellence, it was not uncommon for a U.S. research institution to remain committed to the idea of applications. For instance, the Carnegie Institution of Washington's work on terrestrial magnetism was seen both as a contribution to physical knowledge and as an asset to radio propagation. The research directors of U.S. observatories and the director of the federal Bureau of Standards (established in 1901) continued to situate their institutions within a rhetoric of public interest. In general, the system of science patronage during the interwar years did not so much remove U.S. scientists from new European directions in research, as it defined the approaches they were most likely to take. As a consequence, leading European research centers could label U.S. research programs as more "experimental," as often reinforced by the greater willingness of philanthropic institutions to invest in highly visible research apparatus and machinery.[7]

U.S. contributions to science were in fact often built around new research instruments that brought greater precision to the sciences. But even here, European developments often led those in the United States. One of the more elaborate scientific instruments built during the late nineteenth century was a pair of analog computing devices assembled by the British scientist William Thomson (later Lord Kelvin). Between 1876 and 1879, Thomson built a tidal analyzer and tidal predictor that combined the physical properties of a mechanical integrator with the abstract mathematics of Jean-Baptiste-Joseph Fourier. Yet, if European scientists were the first to develop a valuable instrument, U.S. scientists were quick to improve these devices using techniques derived from U.S. industrial processes. Albert Michelson, although known more for his work in interferometry (which itself relied on precision industrial processes), used precise metal springs produced through new manufacturing processes to improve upon Thomson's work. In Michelson's harmonic analyzer, the springs could be used to extract the harmonic components of a periodic function out to the eightieth harmonic.[8]

Numerical astronomy was another field with a demonstrated need for precision. As was the case with the discovery of Neptune, astronomers could make important discoveries by analyzing minute differences between calculated and observed phenomena. More significant from the standpoint of the total volume of computation was the nautical ephemeris, an aid to navigation consisting of a daily table of the moon's position used to compute a ship's longitude. The classic three-body problem, represented by the mutual gravitational attraction between the sun, earth and moon, presented quite a formidable mathematical problem. Ernest Brown, a British trained mathematician at Yale University, produced a definitive set of equations for the lunar orbit, which were published in 1919 as *Tables of the Moon.* The British Nautical Almanac Office (NAO) computed daily lunar positions for its annual publication using Brown's equations. However, this work remained mostly un-mechanized; widespread cultural attitudes about mathematical labor, along with the limits of existing machinery, meant that most of the work was carried out by current and retired NAO staff members with the aid of logarithmic tables.[9]

Nor was precision always a requirement in the sciences. When a phenomenon under study did not have the regularity of a known physical law, or when a scientific instrument itself introduced certain limits to precision, the need for precise computation could fall outside the disciplinary practice of a scientific field. Still, an emphasis on computation emerged within some new specialties. Experimental apparatus in new fields, such as molecular spectroscopy and particle physics, generated graphical data that required analysis. Other fields, such as meteorology and terrestrial magnetism, began generating copious volumes of data. Although analog methods of computation prevailed in some of these areas, numerical techniques prevailed in those topics where the data came to be tabulated in numerical form. Scientists first turned to their graduate students, staff members, and wives to perform these calculations. It was not long before mechanical calculators, built mostly for commercial accounting work, became a standard fixture in U.S. laboratories and observatories. The Millionaire and Marchant calculators, which had special features such as semi-automatic operation, gained followers in the scientific community.[10]

Despite these developments in the sciences, it was the engineers who made the most extensive, if less elaborate, use of computation. The intense concentration of industry at the start of the twentieth century drove the scale of industrial machinery up, even as national markets drove firms to greater competition. This created new pressures to improve the efficiency of capital equipment. Proponents of modern engineering belittled earlier "rule-of-thumb" engineering practices and began promulgating new analytical methods for the design of new artifacts, from electrical power systems to military armaments. The largest industries forged close ties with research universities in an attempt to formally analyze their machinery and their technological systems. While some firms refused to extend special recognition to college trained engineers even as late as World War I, by the interwar years many firms chose to place their faith in formally trained engineers (some with advanced degrees).[11]

Precision was less likely to be a concern for most engineers. The limits of an engineered artifact limited the need for precision. Transformers lost power, materials varied in strength, and parts could not be manufactured beyond a tolerance of 0.0001 inch. There was rarely an economic need for extending precision beyond two or three figures, nor did engineers feel the pressure to establish objectivity through the kind of precise figures work by which a bank could tally a nine-digit balance down to the exact dollars and cents. Moreover, most engineering design calculations were non-repetitive. This combination of physical needs and computational requirements made the slide rule the most common computing instrument for engineers. Because of the specific kinds of engineering problems that arose in each industry, instrument makers began designing specialized slide rules to serve each need.[12]

However, there were a number of industries whose increasingly complex, interdependent systems required new tools for analysis. When Bell Telephone Laboratories was established in 1925, it created a mathematics group to help analyze signal transmission along its long distance telephone lines. New aeronautical engineering facilities, such as the Guggenheim Aeronautical Laboratory at the California Institute of Technology, employed numerical computation to deal with the voluminous data generated by its new wind tunnel. Meanwhile, academic laboratories at Stanford University, at the Massachusetts Institute of Technology, and elsewhere began developing new analytical techniques for dealing with the nation's expanding utilities. But despite the economic rationale for these new associations between universities and industry, the relationship was rarely designed to be simply instrumental for the latter. Even academic programs that were quite dependent on industrial sponsorship drew simultaneously from the broader culture of academic reform to create new knowledge that was distinct from that of an industrial enterprise.[13]

Take, for instance, the changes in the electrical engineering program at the Massachusetts Institute of Technology. When regional power grids spread across the Eastern Seaboard during the 1920s, a serious crisis developed as electrical transients and other phenomena began to cripple and shut down portions of the grid. Regional utilities turned to MIT in the hopes of understanding the crisis.[14] Under the leadership of Vannevar Bush, MIT's electrical engineering department set out to tackle the associated problems through a combination of simulation and formal analysis.

But Bush also tied graduate education to research. This was already a familiar practice within the sciences, and it offered the quickest way to draw together a significant research program in analytical engineering. Bush's students carried out much of the work of devising new analytical methods and instruments, though often inspired by one of Bush's insights.

Of Bush's two research strategies, in the short run, simulation had the greater impact. Harold Hazen's network analyzer, which modeled an electrical network in miniature, was widely adopted by the electrical utilities as a way of dealing with recalcitrant power grids. In contrast, mathematical modeling remained a limited tool for studying complex systems. The complexity of real-world systems all too easily outstripped the capacity of formal analysis. Yet formal analysis constituted the bulk of the work performed by Bush's students. Mathematical modeling lent itself to discrete engineering problems that were well suited to thesis research, and it was always possible to justify the work in terms of its usefulness in dealing with a particular class of problems.[15]

It was, moreover, the work of formal analysis that produced a substantial program for building new mathematical instruments. Electrical transients, as with other dynamic physical systems, can be represented as a system of second-order differential equations. Insofar as electrical components had linear properties that enabled them to be subject to formal analysis, these same components had stable electrical properties that made them suitable instruments for *doing* analysis. Bush received his training in mechanical—not electrical—engineering. But mechanical devices had similar, stable mathematical properties; Bush's thesis work was built around a mechanical integrator similar to the one used by William Thomson. The mathematical properties of artifacts were something Bush was acutely aware of in promulgating a program for engineering analysis.[16]

It was also Hazen who first realized that it was possible to build an apparatus to solve a second-order differential equation. In addition to using a mechanical integrator, Hazen employed a standard watt-hour meter to handle the second integral equation necessary for solving a second-order differential equation. This active device provided the necessary signal amplification for the system to converge upon a solution. However, Hazen's apparatus had drawbacks. The use of electrical current to model a mathematical variable placed constraints on the acceptable range of problems that could be posed to the machine, since high currents would

damage the apparatus and were dangerous. It was Bush himself who intervened by drawing on his former expertise. In 1927, Bethlehem Steel engineer C. W. Nieman had published an article describing a mechanical torque
amplifier used to manipulate heavy machinery in a steel mill. Bush recognized that a torque amplifier could be used for a far more reliable analyzer
that relied exclusively on mechanical parts. Through further extensions,
Bush and his students built a device that could solve entire systems of
second-order and higher-order differential equations. Publicly disclosed in
1930, it was in this form that the differential analyzer gained its general
acceptance.[17]

Immediately after the publication of his article, Bush began to receive
inquiries not only from representatives of the electrical industry but also
from scientists and applied mathematicians working in a variety of fields,
from atomic physics to quantum mechanics to military ballistics. The
study of dynamic physical systems was hardly the exclusive domain of
electrical engineers. Nor did the progressive reform of U.S. research institutions bring new forms of mathematics only to electrical engineering. As
requests to use or reproduce the differential analyzer began arriving from
dozens of laboratories, Bush began to shift from seeing the differential analyzer as a research instrument to seeing it as an object of research.[18]

The attention that the differential analyzer drew among physical scientist gave Bush the opportunity to place his work at the intersection of
industrial and foundation patronage. The head of the Rockefeller Foundation's new Natural Sciences Division, Warren Weaver, was an applied mathematician known for his contributions to mathematical physics. Although
Weaver funded few U.S. projects in the physical sciences during the 1930s,
a research instrument of demonstrated value to many scientific fields
proved to be an exception. In 1935, Bush received a substantial grant from
the Rockefeller Foundation to develop a more powerful and flexible differential analyzer that could be made available to the entire scientific community. With a supplemental grant from the Carnegie Corporation in
1939, Bush and his successor, Samuel Caldwell, were able to create a Center
of Analysis at MIT.[19]

Mathematics

Insofar as mathematics serves as an intermediary between computing
and its applications, it is worth considering separately how the history of

mathematics contributed to early developments in computing. In some respects, many of the most important advances in modern mathematics occurred under the wings of physics. James Clerk Maxwell's unified theory of electromagnetism, the crowning achievement of the Cavendish Laboratory in Cambridge, was built on mathematical foundations. Likewise, early speculations about quantum theory were built upon the work of the Göttingen mathematician Felix Kline, and subsequent work by David Hilbert and Richard Courant provided a foundation for the further articulation of quantum mechanics during the 1920s. These developments helped to tie mathematics to the progressive reform of science and engineering.[20]

In spite of—or perhaps because of—these connections, the progressive reform of mathematical education also became tied to a rhetoric of pure and abstract knowledge that specifically disavowed any relation to the physical sciences. Even the work by Hilbert and his colleagues at Göttingen hinged on a distinctly abstract turn in mathematics that was rooted in Germany's history of humanistic reforms. Universities in the United States followed suit. By the 1920s, abstract studies in topics such as topology, group theory, and differential geometry had spread not only across research universities on the East Coast but also to leading state universities in the Midwest and beyond.[21] This meant that the U.S. mathematics community was not, in itself, a very conducive site for new developments in computing. "Pure" mathematics, which relied on formal proofs, had little need for computation. Although work in mathematical physics did contribute directly to new advances in computing, these developments occurred in physics departments, not in mathematics departments. Warren Weaver, the Rockefeller Foundation officer, made his contributions while working in the physics department at the University of Wisconsin. Numerical astronomy was another field that made use of applied mathematics. However, in the United States numerical astronomers specifically split off from the American Mathematical Society because of the latter's turn toward abstract math. Most scientific and engineering fields received little support from research mathematicians. It was for this reason that major innovations in computing apparatus emerged from analytically inclined engineering departments, such as the one at MIT.[22]

An important exception was the field of military ballistics. During World War I, many U.S. research mathematicians temporarily abandoned their abstract investigations to do war work. Forest Moulton, a prominent numerical astronomer at the University of Chicago, was asked to assemble a team of mathematicians within the Army's Office of the Chief of Ordnance because of the close relationship between celestial mechanics and the uninterrupted flight of an artillery shell. However, unlike the movement of the planets, a shell's trajectory is affected by air resistance. This presented a somewhat more difficult mathematical problem, in that it required the solution of a non-linear differential equation. Moulton and his staff succeeded in devising an overall procedure for producing artillery range tables. The non-linear component of the ballistics equation was analyzed experimentally using a wind tunnel operated under the general supervision of physicists at the Bureau of Standards. The actual range tables were then produced by a separate group of mathematicians assembled at the Aberdeen Proving Ground in Maryland. Mathematical interest in military ballistics continued in peace. Gilbert Bliss returned from his stint at Aberdeen to his position at the University of Chicago, where he went on to study the calculus of variations. This work was of significant value to the mathematicians who went on to pursue ballistics research during World War II.[23]

Another important exception, though more relevant to the work in Europe than in the United States, was the work on mathematical tables. Mathematical tables, like ballistic tables, are tables of numbers that provide pre-prepared solutions to mathematical equations—most notably transcendental functions like $f(x) = \sin(x)$ and $f(x) = \ln(x)$—that do not lend themselves to easy computational solution. While the tables themselves are quite tedious to produce—they involve computing numerical solutions for every value of an equation at an interval precise enough for the intended application (e.g., $\sin(x)$ for $x = 0.014, 0.015, \ldots$)—their advantage lies in the fact that once a table is computed and published, there is no need to repeat the same calculation.

During this period, the underlying field of *numerical analysis* presented little in the way of interesting research questions for mathematicians. Nevertheless, as scientists and engineers began to make increased use of mathematical tables, various initiatives emerged for producing mathematical tables of greater accuracy and precision. One of the most notable and early efforts of this sort consisted of Gaspard Prony's work during the early years of post-revolutionary France. In appealing to republican ideology and the values of the Enlightenment, Prony set out to demonstrate that women could perform the rigorous mathematical labor necessary to produce the logarithmic tables needed for a new geographical (military ordnance) survey of the French Republic. Equally notable were Charles Babbage's efforts during the 1820s. Driven more by the nautical and industrial demands of the British Empire, Babbage drew on the mechanical technologies of the British industrial revolution to design a machine that could mechanize mathematical tables production. Babbage never got his difference engine to work. Nevertheless, efforts to produce more precise and accurate mathematical tables persisted into the twentieth century, most notably under the auspices of the British Association for the Advancement of Science.[24]

Although industrial expansion in the United States during the early part of the twentieth century created similar pressures to generate new mathematical tables, mathematical tables work was quite limited there. The United States simply lacked the necessary institutional infrastructure for mathematical training and supervision, especially because its research mathematicians had retreated into abstract math. U.S. scientists and engineers were avid users of mathematical tables, but few such tables were produced in the United States.

Statistics was the one other branch of mathematics with a special relation to computing. Early in the twentieth century, statistics gained a strong foothold in the physical and the social sciences. Statistics became an essential tool for studying urban populations, education, epidemiology, and other subjects dear to the hearts of progressive reformers. Statistics also was integral to the actuarial calculations of the burgeoning insurance industry, and statistical work associated with cost accounting was becoming integral to the rising management profession's sense of occupational identity. Moreover, insurance companies, railroads, and public utilities turned to statistics in an effort to define their businesses with respect to an expanding regulatory state.[25]

Tabulating machines emerged in response to the demand for social and economic statistics. Beginning with the 1890 U.S. Census, Herman Hollerith created a business in the assembly of statistical machinery. His firm merged with several others to become the Calculating-Tabulating-Recording company in 1911. It became the International Business Machines Corporation in 1924.[26]

On the eve of the Great Depression, IBM helped establish the first mechanized academic statistical bureau at Columbia University. Columbia was well known for its work on education, and Benjamin Wood, the head of the university's Bureau of Collegiate Educational Research, developed an interest in educational testing. Seeking a machine that could automatically tabulate examination papers, Wood approached IBM's president, Thomas J. Watson, with the hope that the firm might donate some of its equipment and engineering talent. Always attuned to new markets, Watson recognized the value of such an association. Watson gave Wood a significant suite of IBM equipment. He also retained Wood as a consultant and authorized IBM engineers to assist him in the design of new machinery. This arrangement again brought together commercial and academic interests. It also integrated computing machines with the ideals of welfare capitalism and the progressive hope of self improvement through education.[27]

In all these different ways, it should be clear that computing by the end of the 1920s was performed in a variety of different ways, for a variety of different reasons, at a variety of different sites. Although what I have described here is far from an exhaustive account, it should have made apparent how alternative approaches to computing emerged in different, if inter-related, institutional ecologies. Economic needs, cultural values, and social ideologies all contributed to the design of new computing machinery and computing practices. So too did physical laws, disciplinary configurations, prevailing labor processes, and the physical properties of esoteric artifacts.

Within this loose configuration of knowledge, artifacts, and practice, technical exchange and innovation occurred through a number of different modes. At times things happened as a part of broad-based institutional strategies, as was the case with the relationship between feminization, routinization and mechanization. At other times, innovation occurred more as a result of individual initiative. Vannevar Bush and Benjamin Wood assembled new knowledge and associations through opportunistic strategies forged at the confluences of institutions. Part and parcel to some of these opportunistic developments were the product-development strategies of IBM and NCR and the intellectual agendas defined by academic disciplines such as mathematical physics, electrical engineering, and numerical astronomy. Despite all the seemingly fortuitous developments in computing, each development was defined through a coherent and tractable logic.

It should also be apparent that new institutional arrangements accompanied technological innovation, even during the early work. Forest Moulton helped reorient military and academic institutions toward an area of mutual interest. Bush subtly reoriented his work toward foundation patronage, even as he refashioned his mathematical instruments to serve both industrial and scientific interests. Wood built a bridge between IBM and the academic community. Each of these arrangements would play a significant role in subsequent historical developments.

Innovations During the 1930s

Although many different approaches to computing emerged during the first decades of the twentieth century, few if any of these approaches remained fixed and stable. Many sites continued to pursue new innovations as a result of technical or socioeconomic needs and opportunities. At the broadest level, the Great Depression, followed by the early stages of U.S. war preparations, created an array of new reasons for further advances in computing. The innovations described in this section should make it quite evident how new knowledge unfolds through the appropriation, extension, and integration of prior knowledge. Nor was the circulation limited to esoteric knowledge, narrowly defined. As will be evident in each of the examples below, the ecology of knowledge for computing was transformed as much by the appropriation of earlier institutional arrangements as by esoteric knowledge.

Here, in order to achieve the necessary focus, I will narrow the scope of inquiry to two slices of computing history. The first is defined by a specific output (mathematical tables), the second by a specific artifact (vacuum tubes and their close cousin, gas-filled triodes). These two topics cut across a reasonably wide range of developments during the 1930s. They also provide an important foundation for the wartime developments described in the next section. Of more immediate interest, these two slices through history provide a robust look at just how extensive the circulation of knowledge can be in a loosely defined technical field.

Mathematical Tables

As the progressive reform of scientific institutions continued into the 1930s, there emerged a specific concern about the lack of mathematical tables work in the United States. Albert Bennett, a Brown University mathematician who served as an advisor to the Army's ballistics wind tunnel experiments back in World War I, convinced the National Research Council to establish a new Committee on Bibliography of Mathematical Tables and Other Aids to Computation. This continued to be a conservative solution that relied on publishing a bibliography of existing tables.

However, the Great Depression brought with it an unexpected resource. All of a sudden there were large numbers of highly skilled mathematicians desperate for work. In 1935, the Works Projects Administration announced that it would extend its relief program to unemployed professionals. Bennett proposed to organize a Mathematical Tables Project that would use the WPA relief rolls to generate new mathematical tables. In doing so, Bennett turned to his wartime association with the Bureau of Standards (now the National Bureau of Standards) to secure a federal sponsor for the project as required by the WPA.[28]

The actual project was set up at New York University, not in Washington or at Brown University, under the technical supervision of a mathematical physicist, Arnold Lowan. The work on mathematical tables still presented no major research interest. While the work required fairly sophisticated knowledge of numerical analysis in order to devise procedures suitable for human computers, the work tended to be derivative. It also carried a heavy administrative burden. However, Lowan was a European émigré who had completed his studies in the United States because of the turmoil in Europe. More generally, the general circumstances of the Great Depression ensured that someone of Lowan's background was available to take up this position.[29]

The Mathematical Tables Project drew extensively from earlier developments in computing, and it did so in rather unexpected ways. Lowan drew on his knowledge of applied mathematics to devise a simple, manual self-checking procedure that could help to ensure the accuracy of his mathematical tables. He also did so in part because the WPA stipulated that its funds be used to pay for workers, not equipment. The WPA was, after all, a relief program. However, the simplicity of Lowan's new procedure meant that he did not have to hire trained mathematicians to do the actual computation. Although Lowan hired several unemployed mathematicians into the supervisory positions he had to offer, the vast majority of those who joined the MTP were unemployed clerical workers who were displaced by the collapse of New York's commercial economy.[30]

It would be unfair to refer to the aforementioned workers as an unskilled workforce. The real challenge of mathematical tables work lay not in the difficulty of the mathematics but in organizing a workforce capable of performing tedious calculations with a minimum of error. Whereas Prony turned to republican ideology to cultivate the required work ethic, and Babbage to mechanization to circumvent the issue, Lowan relied on highly disciplined and seasoned clerical workers who could accept the tedium of doing manual arithmetic up to the WPA's limit of 39 hours a week. This meant that Lowan and Bennett were finally able to draw together the institutional infrastructure necessary for doing work on mathematical tables in the United States. By 1940, Lowan had produced a dozen new mathematical tables with a staff of over 300 human computers.[31]

Lowan's project simultaneously bore the stamp of the interwar ascent of science. The MTP, by charter, was supposed to focus on tables of the broadest possible use to U.S. scientists and engineers. However, the task of identifying tables of broad utility was itself a significant challenge. It was also a task that was open to interpretation. Lowan began by soliciting the scientific community for advice. Predictably, this produced a number of highly specialized requests. For instance, the physicist Hans Bethe asked Lowan to tabulate a specific fourth-order non-linear differential equation for evaluating a point-source model of stellar structure in studies of stellar cooling. (Bethe would eventually receive a Nobel Prize for his work on nuclear reactions.) This request piqued Lowan's interests because non-linear differential equations had to be solved using the relatively unfamiliar method of numerical integration; this made the work a valid research topic. Lowan did seek guidance from NBS director Lyman Briggs, who remained responsible for the project. But insofar as Briggs himself was a proponent of fundamental research, he offered no objection. The novel resources offered by the Great Depression transformed the boundary of what was considered computable in the realm of physics and applied mathematics.[32]

Related developments occurred in the field of numerical astronomy. The British Nautical Almanac Office had been busy producing annual tabulations of lunar and planetary orbits using human computers. These tables differed from mathematical tables only in that, like the table requested by Bethe, they offered a particular solution to a specific phenomenon. In 1925, Leslie Comrie, a recent PhD in astronomy from Cambridge University, arrived to become the deputy superintendent of the NAO. Comrie immediately recognized that the NAO might be able to make use of the latest accounting machines, whose functions had been augmented to support certain accounting operations. He set out to improve the NAO's computing operations through a program of mechanization. As Comrie became more familiar with accounting equipment, he discovered the even greater capabilities of the tabulating machine. Its reconfigurable units and plug boards had evolved even more directly in relation to the increasingly complex clerical operations of commercial enterprise. Close study revealed that these machines could support the sequence of mathematical operations found in numerical astronomy, including the very laborious process of harmonic synthesis used to produce daily lunar positions for the *Nautical Almanac*. In 1929, Comrie leased a set of machines from IBM's British affiliate, the British Tabulating Machines Company, and launched a major initiative to compute the lunar table out to the year 2000.[33]

It was during the early 1930s that Comrie's technique was picked up in the United States by the Columbia University astronomer Wallace Eckert. Eckert obtained his PhD under Ernst Brown at Yale, and therefore Brown's knowledge of Comrie's work transferred to Eckert quite directly along disciplinary lines. Somewhat more contingently, Eckert landed a position in Columbia University's astronomy department, where he had previously served as an astronomer's assistant. Eckert was immediately drawn to Wood's Statistical Bureau. For his part, Wood welcomed Eckert's interest. Eckert's training as a numerical astronomer gave him one of the most advanced training in applied mathematics that anyone could receive at the time in the United States. As his affiliation with the Statistical Bureau grew, Eckert took advantage of the special relationship Wood had with IBM. Eckert asked for and received a significantly modified machine in 1934. Watson continued to expand his interest in supporting new work in the sciences. Through further discussions, it was decided that this and other equipment would be made available to the entire U.S. astronomical community as part of the new Thomas J. Watson Astronomical Computing Bureau at Columbia University.[34]

IBM's growing affiliation with Columbia served as a stepping stone for an even larger collaboration with Harvard University. Howard Aiken began his career as an engineer with gas and electric utilities. Influenced by the increasingly analytic traditions of the profession, Aiken entered the PhD program in physics at the University of Chicago in 1932, then transferred to Harvard a year later. Although Aiken elected to study physics because of the allure of higher knowledge, he drifted to Harvard's Cruft Laboratory, a departmental laboratory devoted to analytical studies of communications engineering. This dual desire to pursue higher knowledge and knowledge of practical relevance remained a productive tension in his career.[35]

Like Vannevar Bush, Aiken straddled the boundary between science and engineering, though beginning somewhat from the other side. From the standpoint of physics, Aiken was convinced that the "future of [the] physical sciences rest in mathematical reasoning directed by experiment."[36] From the standpoint of analytical engineering, Aiken was familiar with the broad array of engineering problems that could be solved through formal analysis. Moreover, in working within the Cruft Laboratory, Aiken grew more aware that many of these problems transcended the boundaries of these two disciplines. Studies of the ionosphere, for example, involved complex equations whose solution was essential to a physical understanding of the upper atmosphere. And yet these equations provided a necessary foundation for the future of television and radio.

Aiken's thesis work on space charge conduction (an important phenomenon that occurs within a vacuum tube) also had both theoretical and practical applications. The work also pointed to the possibility of a computational solution. However, the complexity of the problem—again a non-linear differential equation—precluded a solution using manual methods of calculation. But the inability to solve such a problem ran up against the very principles of an analytical engineering laboratory. All around him, meanwhile, were the electrical devices he needed to build a powerful computing machine.

From 1937 on, Aiken spent much of his energy designing an "automatic calculating machine." He framed this work in terms of the broad needs of the sciences, rather than the narrow demands of his own thesis topic. Indeed, Aiken completed his thesis through formal analysis—a difficult, classical solution to a differential equation in closed form. At the same time, Aiken continued to become more enamored by the fact that a minor difference between an experimental and theoretically computed result could result in scientific discoveries. Driven more directly by the agenda of the sciences than Bush, Aiken made precision his chosen strategy. He turned to numerical rather than analog means of computation.

It was during his early investigation of automatic computing machines that Aiken came across Charles Babbage's work. Babbage had worked on a more ambitious machine known as the "analytical engine," which unlike the difference engine was designed to solve much more general problems in numerical computation. Working from a relatively limited description of this machine, Aiken integrated what else he had learned about calculating devices to draft rough specifications for a fully automatic computing machine. Then he began approaching prospective sponsors. Through the help of a member of the Harvard Business School who was trained in numerical astronomy, Aiken gained access to one of IBM's senior inventors, the engineer James Bryce. Although Aiken was then but a graduate student, the project had its appeal; so too did the thought of an affiliation with Harvard. Bryce encouraged Aiken to work closely with IBM's engineers in Endicott, New York.[37]

Simon Schaffer has considered how the state of mechanical technology during the early industrial revolution, and the specific distribution of this knowledge between Babbage and his skilled artisans, conspired to prevent Babbage from completing his difference engine. Similar issues would eventually create a rift between Harvard and IBM. However, in this instance, Aiken was able to draw on the mechanical skills of IBM engineers long enough to bring his machine to fruition. By March 1939, IBM and Harvard had signed an agreement whereby IBM promised to donate a machine built to Aiken's general specifications. Although war production and other factors contributed to a substantial delay, Aiken's machine was delivered to Harvard in late 1943.[38]

Babbage had written in his memoirs that the analytical engine would have to be built by someone in a future generation. Others would portray Aiken as having heard Babbage speaking to him across the generations. Still, a close study of the Harvard Mark I reveals the priorities of a twentieth-century engineer turned physicist rather than those of a nineteenth-century mathematician and political philosopher. Whereas Babbage spoke of a "mill" and a "store" in describing his analytical "engine," Aiken spoke of the Mark I as an "arithmetical computing plant." The Mark I also exceeded Babbage's analytical engine in precision and complexity. The machine was designed with 72 electromechanical counters, each with 23 digits of precision. By ganging together two of the counters, the Mark I could have 46 digits of precision. The Mark I, in other words, was a manifestation of Aiken's commitment to analytical precision. Apart from its practical uses, Aiken could also admire his sleek-looking creation as a highly intricate control instrument abstracted away from the mundane purposes of industrial capital.[39]

Finally, there was the Army's continued interest in artillery range tables. Although the research mathematicians assembled at Aberdeen mostly disbanded at the end of World War I, routine production of ballistic tables continued on a limited basis. Then the escalating tensions in Europe provided the necessary justification for revitalizing ballistics research within the Army. Sometime around 1932, James Guion, the Army officer in charge of the human computers at Aberdeen, began considering new ways to accelerate the production of ballistic tables. A member of the military establishment, Guion, unlike Lowan, was not kept from exploring mechanical alternatives. The mathematical relationship between differential equations and the exterior ballistics equation led Guion to Vannevar Bush and his work on the differential analyzer. While the non-linear differential equations of exterior ballistics could not be solved directly on a differential analyzer, there was an indirect method where the non-linear component of the equation could be supplied as an input into the machine—much in the way that this was already done with the numerical procedures employed by Guion's staff.[40]

After further investigation, Guion approached Bush with the offer of a contract. But the mathematics intervened. Although the work of building a new input mechanism and the new class of mathematical equations that such an extension would allow were of some interest, Bush had already turned his attention to the more interesting challenges of the machine supported by the Rockefeller Foundation. Still, because he worked within an academic context, Bush offered his knowledge freely.

It was at this point that the University of Pennsylvania's Moore School of Electrical Engineering entered the picture. Based on past collegial connections and a shared academic interest, a member of the Moore School approached Bush with a similar inquiry. Specifically, Irven Travis, an established instructor at the Moore School, decided to pursue a PhD and

approached Bush with a request to work on improving the differential analyzer for his thesis research. Travis hoped to use WPA funds to help build
this new analyzer. The WPA's requirement for federal supervision perhaps
helped to bring the three organizations together. Travis agreed to conduct
the design studies necessary to meet Aberdeen's needs. Bush provided
Travis with the schematic diagrams for the differential analyzer, and lent
Travis his chief draftsman. After making some design improvements, Travis
supervised the construction of two very similar analyzers, one which was
delivered to Aberdeen and the other retained by the Moore School.[41]

Electronics

Another significant development during the 1930s was the use of electronic devices for counting and calculation. During the early part of the
twentieth century, electronics matured at the juncture of electrical engineering, communications engineering, and physics. In fact, the word "electronics" originally referred to the physical study of the behavior of
electrons inside vacuum tubes. Lee DeForest's original work on the audion
resulted from his effort to apply his knowledge of gas discharge physics to
wireless telegraphy. Studies of vacuum tube physics remained an important part of the work at Bell Telephone Laboratories well into the 1930s.
However, in 1930, owing in part to the influence of the magazine *Electronics*, radio amateurs appropriated the term to mean the design of circuits employing vacuum tubes and similar devices.[42]

Citing the early work in electronics as a kind of precursor to electronic
computing does pose the risk of an "asymmetric" explanation that
accounts for the past in terms of subsequent developments; almost none
of the early work in electronics was geared toward supporting the numerical abstractions of the digital domain. But it is important that the vast
popularity of radio during the late 1920s and the 1930s guaranteed the dispersion of basic electronics skills across a very broad spectrum of scientists
and engineers. Still, nearly all the work in radio electronics was based on
exploiting the linear properties of a vacuum tube's capacity to amplify a
signal.

An important exception to this arose in physics, when a group of experimental physicists decided to regard the vacuum tube not as an object but
an instrument of research. Early in the twentieth century, many advances
in experimental physics were driven by new imaging devices derived from

the "cloud chambers" built by the British physicist C. T. R. Wilson. However, by the mid 1930s a number of experimentalists had broken with tradition in their effort to define a new "logic" tradition that relied on statistical inference. Through the clever arrangement of detectors and accompanying "coincidence circuits," these physicists gained the ability to infer the existence of subatomic particles through particular events that their circuits were designed to detect and isolate. The speed of a subatomic particle required the use of vacuum tubes in these coincidence circuits.[43]

The high incidence rates of certain kinds of events also made the audible signals generated by these detectors too frequent to count. This led to electronic "scaling circuits" that could reduce these audible clicks by a factor of two or more. These scaling circuits could, in effect, count. Nevertheless, as embedded within a research tradition that was built around relatively simple statistical inferences, the logic tradition in physics itself generated no extensive need for calculation—at least not at electronic speeds. On the other hand, coincidence circuits and scaling circuits were widely described in journals such as the *Review of Scientific Instruments*, and they became a familiar part of the material culture of experimental physics.

It was Vannevar Bush who first re-appropriated the physicist's circuits to perform something closer to computation. In 1932, Bush was made Vice President and Dean of Engineering at MIT on the basis of his technical and administrative accomplishments. As his focus shifted from research to research administration, Bush developed an interest in the ever-expanding volume of scientific publications. His own success, after all, was based on drawing together the knowledge of different disciplines. Bush, an engineer, envisioned an engineered solution, albeit one that drew on the material culture of physics. At the heart of his solution was the realization that the physicist's coincidence circuit could be used not only to identify a coincidence generated by natural phenomena, but coincidences generated by coded information. Bush proposed to build a machine that would recognize keywords encoded onto the edge of microfilm containing journal articles. This would eventually become the basis of the Memex, an imaginary machine that Bush described in a 1945 *Atlantic Monthly* article. Between 1938 and 1940, Bush set out to build his first prototype, the Selector, with the support of the National Cash Register Company and the Eastman Kodak Company. Meanwhile, U.S. war preparations allowed Bush to simultaneously apply his ideas to the military art of cryptography. Bush closed a deal with the Navy's Director of Naval Communications before he began working with NCR and Kodak, and delivered a machine called the Comparator in late 1939.[44]

As Bush and Caldwell's knowledge of electronics grew, they decided to employ such circuits in the Rockefeller Differential Analyzer (RDA). One of the main limitations of the original differential analyzer was that it took a lot of time to configure the machine to represent a specific mathematical equation. The RDA was designed to reconfigure itself automatically. The precise details of the RDA have yet to be disclosed in the historical literature. But what is clear from published accounts is that, although the RDA continued to use mechanical integrators to perform the actual computation, it used electronic circuits to interpret the governing mathematical equations as encoded onto IBM punched cards.[45]

This work also led Caldwell to explore the arithmetic capability of electronic devices, with the assistance of National Cash Register. The idea of building a Rapid Arithmetic Machine using electronic devices originated with MIT. However, National Cash Register also supported this work directly as part of a new corporate strategy. NCR's new president, Edward Deeds, recognized that the company was suffering from market saturation and that it had a rather weak patent portfolio with respect to future office machinery. Advanced engineering therefore emerged as an important tactic in NCR's efforts to capture new markets. Even as NCR began collaborating with MIT, Deeds set up a new electronics laboratory headed by an NCR engineer, Joseph Desch. Desch beat MIT in producing and patenting novel computing circuitry. He had a working calculator by 1940.[46]

Similar developments took place inside IBM. Taking his inspiration from an electronic counter described in *Electronics* magazine, Byron Phelps, a member of the IBM electrical laboratory at Endicott, set out to build a scale-of-ten ("decade") ring counter that could be substituted for the mechanical counters IBM was then using. Phelps, concerned with cost and reliability, used diodes and a binary-coded decimal representation to assemble a ring counter that required only five vacuum tubes. But further explorations by Phelps and his supervisor, Ralph Palmer, point to some of the constraints faced by engineers working in a laboratory whose mission remained closely bound to a firm's product line. In 1942, Phelps and Palmer set out to build an electronic multiplier. They also built an electronic "cross-footing" punch that could add or subtract two numbers located on a single card and punch the results back onto the same card, all within one card cycle. These were both familiar procedures, which IBM knew to be a part of a firm's typical accounting operations because of its extensive experience with "methods work." Accelerating these procedures would improve the throughput of IBM machinery, and thereby give the firm a competitive advantage.[47]

John Atanasoff, a physicist at Iowa State College, pursued a different path. Like Aiken, Atanasoff had been an engineer. He had earned a PhD in mathematical physics at the University of Wisconsin. Although mathematical physics had spread as far as Wisconsin, it had not extended to Iowa State. Atanasoff was therefore initially hired as a mathematics instructor teaching mathematics to science and engineering undergraduates. This too put Atanasoff in a position similar to Aiken's, insofar as he too could see the varied uses of mathematics in physics and in engineering. But unlike Aiken, who turned his attention to precision, Atanasoff turned his attention to large systems of linear equations. The solution of such equations was emerging as an important tool in electrical engineering, in quantum physics, and in molecular physics, his chosen specialty.[48]

Like others before him, Atanasoff set out to build a machine that would solve equations of a specific kind. Atanasoff imagined a device that would use the common method of Gaussian elimination to solve systems of linear equations. This required that he begin by storing all the coefficients from each linear equation in a form that would be accessible to arithmetic circuitry. This he accomplished by placing 50 rows of capacitors across the circumference of a rotating drum, where each row contained one of 50 binary digits from each of the 30 coefficients representing one linear equation. By drawing numbers from two rotating drums and depositing the results back onto the drums, it would be possible to perform the successive subtractions of the Gaussian elimination algorithm with each rotation of the drums. The apparatus was also designed to take advantage of the parallelism inherent to the method of Gaussian elimination.[49]

The arithmetic circuits posed a greater design challenge. Trained in physics, Atanasoff was familiar with scaling circuits and coincidence circuits. He first joined the others who had evaluated scaling circuits as an electronic alternative to a mechanical counter. However, it was during a moment of inspiration in 1937 or 1938 that Atanasoff decided to use coincidence circuits to do arithmetic. This required Atanasoff to realize, quite independent of Claude Shannon's similar and famous observation with regard to electrical relays, that electronic circuits could perform binary arithmetic using Boolean algebra—a subject with which Atanasoff was familiar because of his mathematical background. Working from this idea, Atanasoff built his electronic computer (with the assistance of Clifford Berry, a graduate student in electrical engineering) between 1940 and 1942.[50]

Atanasoff's computer has received considerable historical attention, not least because certain aspects of its design were replicated in subsequent digital electronic computers. However, in order to avoid an anachronistic interpretation it is important to understand Atanasoff's system for what it was. In more ways than one, Atanasoff's design was the product of necessity, not prescience. Atanasoff had considered using electrical relays and even punched-card machinery. However, the rate at which he hoped to draw numbers off of the rotating drums made it necessary to turn to electronic circuitry, much as IBM engineers turned to electronics to improve the cross-footing punch. More important, Atanasoff's computer was never designed to store all the equations from a system of linear equations, only the two equations being subject to one pass of the Gaussian elimination algorithm. This meant that successive results had to be sent to an intermediate storage medium and fed back into the computer in proper succession, manually. Unlike Aiken, Atanasoff had few resources other than a $5,000 grant from the Research Corporation, a non-profit organization that had been set up to encourage academic faculty to pursue patentable inventions. This precluded any attempt to produce a more extensive device. Atanasoff's machine was never intended to be a fully automatic computer. It was built in the tradition of other mechanical aids to calculation, much in the way that scientists made use of semi-automatic calculating machines such as the Millionaire.[51]

The work that was done on mathematical tables and on electronic computing devices in the 1930s points to the extensive circulation of knowledge that occurred in computing during this period. With mathematical tables, it was especially clear how different institutional locations presented different opportunities. Lowan drew on the disciplined labor afforded by the WPA, Eckert on the engineering abilities of IBM, Aiken on the technical challenges posed by the Cruft Laboratory at Harvard, and Guion and Travis on the knowledge and academic commitments of MIT.

Each of these men proceeded in a way that supported specific organizational priorities: the military interests of Aberdeen, the commercial and philanthropic interests of IBM, the disciplinary interests of the Moore School. Yet each solution was also affected by some element of historical contingency, whether at the level of the Great Depression or at the level of the individual choices used to define what an interesting scientific or engineering problem was. In each case, new approaches to the production of mathematical tables emerged from a creative combination of knowledge and artifacts, esoteric skills and practices, and institutional needs and resources.

With electronic devices, it was possible to trace the circulation of knowledge directly by following a specific device. An atomic physicist's coincidence circuits and scaling circuits were adopted by electrical engineers, by those who designed accounting and tabulating machines, and by a molecular physicist. This occurred through mostly familiar modes of disciplinary communication. The blurred boundaries between physics and electronics and between electrical and mechanical engineering also facilitated the broad dissemination of knowledge about electronics. So too did new institutional relationships, such as that between MIT and NCR. More generally, knowledge about these two esoteric circuits traveled along familiar networks of association among and beyond those who were interested in computing.

These technical exchanges also created new institutional relationships, both with respect to mathematical tables and with respect to electronic devices. The relationship between Lowan and the scientific community, the relationship between IBM and prestigious academic institutions, and the relationship between Aberdeen and the Moore School would be significant for subsequent developments. Yet, despite the technical exchanges and an expanding network of affiliations, computing did not yet emerge as a unified field. The unifying effects of the technical exchanges were offset by the diversity of initiatives designed to meet different institutional needs. The continued fragmentation of the field was nowhere more evident than in the machinery, all of which continued to be built for specific purposes.

Nevertheless, the ecology of knowledge for computing was changing. In addition to the most apparent new institutional affiliations, intersecting disciplinary networks in physics, applied mathematics, and analytical engineering could now be found within the general realm of computing. In other words, those interested in numerical analysis or electronic counters had begun to discover each other's work, regardless of their institutional location or disciplinary origin. If these networks were not yet woven into whole cloth, the situation was ripe for some kind of synthesis.

Mobilization for War

The National Defense Research Committee

Describing the synthesis that the science mobilization effort brought to U.S. scientific, military, and industrial research traditions, A. Hunter Dupree wrote of a "great insaturation." According to Dupree, this synthesis occurred as a result of efforts to mobilize civilian scientists by the National Defense Research Committee and by its successor, the Office of Scientific Research and Development.[52] There has been some discussion about whether the new system of federal patronage was as conservative as it was transformative.[53] But regardless of one's position in this debate, it is worth considering what this "insaturation" looked like from the standpoint of a specific technical field. Here I trace the ways in which technical and institutional synthesis both did and did not occur in computing as a result of wartime research.

The history I have told so far has some bearing on the broader history of science mobilization, for it was Vannevar Bush who created the NDRC and the OSRD. Bush left MIT in 1939 to become the president of the Carnegie Institution of Washington. With war looming in Europe, Bush became involved with wide ranging conversations about science mobilization. Bush and his cohort—MIT President Karl Compton, Harvard University President James Conant, and National Academy of Sciences President Frank Jewett among them—grew concerned about the limited scope of science mobilization during World War I, which had occurred under the National Research Council, an organization created for this purpose.[54]

Bush moved to sidestep the NRC. As elegantly summarized by Daniel Kevles, Bush put forward a research organization based not on the disciplinary organization of the NRC, but on specific military applications, including ordnance, explosives, and radar. This meant that the National Defense Research Committee was conducive to interdisciplinary projects that were based on drawing together relevant knowledge as they remained scattered across different institutions and disciplines. Still, it is evident from the historical record that Bush and his co-founders (including Compton, Conant, Jewett, and Richard Tolman) never intended the NDRC to be the only effort to mobilize science during World War II. Military agencies already had their own laboratories and network of private contractors. Bush and his colleagues saw the NDRC's mission as that of mobilizing those parts of the national research infrastructure—universities and certain science-based industrial laboratories—that, for reasons of academic and commercial autonomy, remained beyond the reach of the military.[55]

Although Bush is regarded as having led a well-coordinated science mobilization effort, he pursued a policy that explicitly decentralized decision making. Drawing on academic traditions of committee work, as well as on the structure of the National Advisory Committee on Aeronautics (which he had chaired since 1939), Bush organized the National Defense Research Committee as a series of divisions and sections, each with direct responsibility for a specific area of military application. Even at the sectional level, technical decisions were made by a committee of leading researchers from universities and industry, as occasionally supplemented by a liaison assigned by an interested military agency. This decentralized structure carried over into the OSRD, established in late 1942, when it absorbed the NDRC with minimal changes to its overall approach to research administration.[56] The resulting patchwork of technical and institutional jurisdictions generated both collaboration and conflict. The main aim of this section will be to explore how computing emerged as a more unified field amidst this continued tension between the centralization and the decentralization of knowledge.

Much of the NDRC's early work in computing occurred under the aegis of its work on anti-aircraft weapons, which has been described by David Mindell (2002). The interest in building better gun directors (the "fire-control" problem as referred to by the NDRC) emerged in no small part because of Bush's own concerns about air warfare. From the use of the differential analyzer in military ballistics research, Bush was aware that anticipating where to aim a gun when faced with a fast-moving target was a major strategic concern. When Warren Weaver (by then a close friend and colleague) wrote to offer help, Bush asked him to head up the fire-control section (NDRC Section D-2). Bush had faith in Weaver's administrative capabilities. The choice also favored a more formal approach to engineering analysis.[57]

After an initial survey of the prevailing technology, Weaver began to enlist people and organizations as committee members and as contractors. To deal with the engineering aspects of the problem, Weaver brought in various members of the so-called fire-control fraternity. This "fraternity" included Sperry Gyroscope, Ford Instruments, and Barber Coleman. Weaver also drew in members of MIT's electrical engineering faculty and engineers at Bell Telephone Laboratories. To deal with the mathematical aspects, Weaver enlisted Bell Labs' mathematics department, MIT's Center of Analysis, and the two computing groups at Columbia University.[58]

By combining people from different backgrounds, Weaver ensured that the existing tensions between different disciplines and organizations would be reproduced within the boundaries of his own project. These tensions might have been manageable had Weaver effectively delineated everyone's responsibilities. In hindsight, it was clear that the fire-control problem could have been divided into three parts, one having to do with the target computer, one with the servomechanisms for the gun, and one with how to integrate the components amid electrical noise and other uncertainties. However, this breakdown of the problem required new knowledge about control-systems engineering. That knowledge did not yet exist; in fact, it emerged from the NDRC's work on fire-control systems. Swayed by his own disciplinary inclinations, Weaver put the mathematicians in a position of considerable authority. He gave them the power to specify what the engineers did and to evaluate the results of their efforts. As the engineers began to amass more and more relevant knowledge through their grounded experiments, predictable tensions followed.[59]

It was the mathematicians who were demoted in the ensuing conflict. During a major reorganization of the NDRC in late 1942, Vannevar Bush replaced Warren Weaver with Harold Hazen, who by then was chairing MIT's Electrical Engineering Department. Bush asked Hazen to head the NDRC's new fire-control division, Division 7. Weaver remained the head of Section 7.5, which was responsible for fire-control analysis. Yet it was also clear that the mathematicians no longer held greater authority.[60]

But this historical twist was not so bad from the standpoint of mobilizing mathematics. Bush knew that the reorganization would place someone of Weaver's stature in an unacceptably minor administrative position. The

NDRC had already fielded several requests for mathematical assistance from the military agencies. Bush asked Weaver to create a new unit called the Applied Mathematics Panel. Set up in November 1942, the AMP was charged with providing mathematical advice and services directly to military agencies, and to all NDRC divisions. This meant placing mathematicians even more clearly in the service of other scientists and engineers. Nevertheless, once Weaver was relived of his obligation to focus on fire control, he found himself free to mobilize the U.S. mathematics community far more extensively than he had done so far.[61]

Warren Weaver's strategy for mobilizing mathematicians had four components. First, following the formal structure of the NDRC, Weaver assembled a panel consisting of Thornton Fry (Bell Labs), E. J. Moulton (Columbia University), Richard Courant (New York University), and Samuel Wilks (Princeton University). Mina Rees, a Hunter College mathematics faculty member with a PhD from the University of Chicago, served as the panel's official technical aide. Second, as was the case for Arnold Lowan, securing interesting mathematical problems was a challenge. Weaver therefore established a formal liaison with each NDRC division and with each interested military agency or office. Third, Weaver asked academic research mathematicians to serve as voluntary advisors in their respective areas of mathematical expertise. When a problem required more than a quick consultation, arrangements were made for a full OSRD contract with these individuals. Fourth, Weaver absorbed several existing academic computing groups, including Lowan's Mathematical Tables Project and the Statistical and Astronomical Computing Bureaus at Columbia University, into the AMP's administrative field. The Applied Mathematics Panel met every week in Weaver's office, and occasionally at Bell Labs, to coordinate resources with requests.[62]

As Larry Owens has written, it is important not to exaggerate the accomplishments of the Applied Mathematics Panel. The panel was established relatively late into the war, and the NDRC's policy of judging worth on the basis of apparent military priorities alienated many members of the U.S. mathematics community. Many felt neglected.[63] Conversely, insofar as Weaver had to work with research mathematicians who, by choice, had little real-world experience, their solutions tended to be trapped in mathematical abstractions that were not immediately useful in battle. For instance, Weaver ended up supporting a considerable body of work by

Richard Courant, a former director of Göttingen's Institute of Mathematics who was expelled by the Nazis. Courant and his group at New York University pursued various studies, including the relevance of partial differential equations to shock waves, to non-linear waves, and to the entry of an artillery shell. They also developed formal models of gas dynamics as it pertained to rockets and jet propulsion. Yet the level of abstraction by which Courant pursued this work meant that most of the work was not relevant to the designing of weapons systems until after World War II, when digital computers made such models practicable.[64]

All the same, the Applied Mathematics Panel presided over more than 200 studies, many of them initiated at the request of an NDRC division or one of the armed services. These included studies of the hydrodynamics of underwater explosions, of the statistical quality control of munitions, of calculations pertaining to the atomic pile at the Metallurgical Laboratory in Chicago, and of optimal flight patterns for the strategic bombing of Japan—hardly trivial matters. Of more direct relevance to computing, Weaver wound up replicating Columbia University's statistical and astronomical computing bureaus many times over. By war's end, there were fourteen such facilities churning out technical computations using IBM machinery. Regardless of the wartime efficacy of its studies, the Applied Mathematics Panel diffused new knowledge of applied mathematics and computing into many domains of military science and engineering.[65]

The NDRC and Electronics

The NDRC was also quick to consider the value of interwar electronics developments. One of its earliest projects was to build a high-speed scintillation counter for the atomic-pile experiments at Chicago's Metallurgical Laboratory. The NDRC had absorbed the Uranium Committee, which President Franklin D. Roosevelt had established to study the feasibility of the atomic bomb. George Harrison, MIT's dean of science and the head of the NDRC section responsible for instrumentation research, enlisted one of the engineers, Wilcox Overbeck, who was working on MIT's Rapid Arithmetic Machine. He also enlisted Joseph Desch of NCR. In the end it was Desch who, in working closely with University of Chicago physicists, assembled a scintillation counter that could operate at the unprecedented rate of 1 megahertz.[66]

Through his administrative relation with Harrison, Weaver became aware of Desch's work. He proceeded to evaluate the possible uses of electronic devices in fire-control systems. This was in late 1941. He assigned this task to George Stibitz, a technical aide to Section D-2 and a mathematical physicist working under Thornton Fry at Bell Labs. But as Stibitz's enthusiasm grew, Weaver's began to wane. While NCR seemed to possess considerable capabilities in electronics, the firm was reticent about disclosing its knowledge. After all, electronics was integral to NCR's corporate strategy. The Navy Bureau of Ordnance was supporting an interesting project at RCA headed by Vladimir Zworykin. Zworykin had sketched the general outlines of an electronic numerical computer, but had then turned his attention to such a computer's components (including a new single-tube decade ring counter). Though this choice was driven primarily by concerns about the reliability of components, the approach was well suited to a tube manufacturer. But this choice led Weaver to conclude that Zworykin's work was not far enough along to be useful for the war effort. Wary of allowing the NDRC to become an orphanage for abandoned military projects, Weaver wrote: "Impulse electronic computing devices is not sufficiently advanced to warrant the attempt, at this time, to incorporate such devices in an overall design for a [gun] predictor."[67]

Pressed by Zworykin and by Harrison, Weaver allowed one more round of evaluation. But he permitted his skepticism to prevail. In asking all interested parties to submit complete block diagrams for an electronic gun director in two weeks' time, Weaver ensured that the results of this evaluation were a foregone conclusion. On the other hand, in following standard procedure, Weaver required all prospective contractors to submit multiple copies of their proposal for distribution to the other applicants. This ensured that the latest knowledge about electronic computing was circulated broadly among Section D-2's contractors. MIT, RCA, Bell Labs, and Eastman Kodak all chose to participate. So did NCR, apparently on the premise that it had more to lose from being excluded than from the technical exchange.[68]

Relay Computers and Cryptography

There were two other wartime efforts to build new computing machinery. The first was a series of electromechanical computers built by George Stibitz, the technical aide to NDRC Section D-2. As indicated earlier, Fry's department at Bell Labs was established to help with the mathematical analysis of signal line transmission and other phenomena of interest to AT&T. This group often made use of complex number arithmetic (involving numbers of the form $a + bi$, where $i = \sqrt{-1}$) in routine analyses of transmission line characteristics, filter networks, resonance, and other electrical phenomena. Stibitz realized that electrical relays—standard components of telephone switching systems—could be used to perform arithmetic. In 1940, before he joined the NDRC, Stibitz (with help from a Bell Labs engineer) had built a Complex Number Calculator. He had demonstrated this machine at the annual meeting of the American Mathematical Society.[69]

According to Mindell (2002), NDRC Section D-2's abandonment of its interest in digital electronics ensured that electromechanical devices would dominate all of its work in digital computing. At first, Weaver and his successor (Hazen) continued to rely on MIT's differential analyzer and on the IBM facilities at Columbia University. However, when new computing capacity was needed to test the fire-control systems the NDRC had developed, Stibitz's talents were called into play. Stibitz first built a machine called the Punched Tape Dynamic Tester, which was used to generate an analog flight profile from a set of binary numbers punched on a strip of paper tape. Feeding this flight profile into a gun director made it possible to evaluate the director's performance in relation to the ideal results.[70] However, this test apparatus generated its own need for calculation. In order to generate the continuous numerical data that represented a flight profile, human computers had to perform extensive interpolation from a sampled data set. As many as 20,000 data points were required for a single flight profile. With 60 flight profiles per test, this meant producing more than a million data points to test a single gun director. In July 1943, Stibitz received a contract to build an electromechanical interpolator.[71]

At one level, Stibitz's Relay Interpolator extended the tradition of a machine built for a specific purpose. However, Stibitz proceeded to separate the control circuits from the arithmetic elements in a way that was quite different from that used in the differential analyzer or that used in Atanasoff's electronic computer. This followed from the way human computers performed interpolation. At Bell Labs and elsewhere, human computers always followed a "plan of calculation," using (when one was available) a desk calculator to simplify their work. In thinking through how to specify the sequence of arithmetic operations required for interpolation,

Stibitz decided, like Aiken, to place the instructions on a strip of paper tape—the same tape used to hold the data for the flight profiles.[72] With this design, the Relay Interpolator could solve quite a broad range of problems. Stibitz, a mathematician, was quick to recognize this. He saw that his machine could solve differential equations, perform harmonic analysis, and extract the complex roots of polynomials—all familiar problems at Bell Labs.

Although formally Stibitz remained with Division 7 after the NDRC was reorganized, he continued to work closely with Weaver. In return, Weaver encouraged Stibitz to circulate his findings in a series of Applied Mathematics Panel reports. Thus, in spite of the explicit pressure for the NDRC to define its work in terms of what would be useful for the war effort, this work allowed Stibitz to extend his general interests in computing. And as others grew cognizant of his work, in no small part because of his circulation as a technical aide, Stibitz gained the opportunity to build machines with even greater computational capabilities.[73]

One other major wartime computer-development project unfolded in the context of cryptography. This work proceeded mostly outside the framework of the NDRC. Bush and his students' Comparator proved unreliable. While Bush tried to revive the project using NDRC funds, the Navy's Bureau of Ships stripped the project away from him when it grew disheartened with the slow progress. While the Bureau of Ships and the Navy's Division of Naval Communications issued many contracts to NCR and other firms, most of the machines built under this arrangement were electromechanical systems with extremely limited electronics capabilities.[74]

Cryptographers might have continued to shun electronics had it not been for the increasing casualties that Allied shipping was experiencing. In early 1942, the German Navy introduced a fourth encoding wheel to its famous Enigma cipher. Allied forces were unable to track the maneuvers of German submarines for the next nine months.[75]

While Britain turned to code-breakers (mathematicians trained to identify weaknesses in the enemy's code through stolen code books, cribs, and other means), the Division of Naval Communications pursued a more "brute-force" strategy for cracking the four-wheel code using electronics. The DNC was already familiar with the Bombe, a British electromechanical machine that was being used to crack the three-wheel Enigma code.

They also knew of a British plan to use electronics in a "super-bombe" to break the four-wheel code. Discontented with British progress, the DNC enlisted Joseph Desch and set up a new Naval Computing Machine Laboratory inside NCR. The American bombe, completed in September 1943, continued to use electromechanical commutators to test different wheel combinations. But the fastest commutator spun at a rate of 2,000 rpm. With 26 letters to a wheel, this meant that it had to perform more than 850 comparisons per second. At this rate, it was necessary to use electronic coincidence circuits. Desch went considerably further, using about 1,500 vacuum tubes for machine control and cryptanalysis.[76]

All this is to say that technical innovation was integral to the art of cryptography. Though there were different strategies for breaking a cipher, casualties made it imperative to embrace any innovation that would facilitate the work. But precisely to the extent to which this race was honed to a specific challenge, there was little room for open exploration. A general-purpose computer did not emerge in such a context.

Other Wartime Initiatives

Cryptography was not the only field in which computing innovations occurred outside of the National Defense Research Committee. Because it had mobilized mathematicians during World War I, the Army's Aberdeen Proving Ground was well positioned to do so again. In 1935, the diverse mathematical operations at Aberdeen had already been assembled into a new Research Division. This unit was rechristened the Ballistic Research Laboratory in 1938. Institutional memory, embodied in one of the BRL's civilian associate directors and in Oswald Veblen, now the BRL's chief scientist, helped facilitate rapid mobilization. The BRL began to assemble research mathematicians from various academic institutions. It also accelerated its work on artillery range tables. Production escalated to over six tables a day amidst rapid improvements in military ordnance. As a consequence, the BRL's Executive Officer, Paul Gillon, authorized the computing group to purchase IBM machinery so they could employ the methods used at Columbia University. When this proved insufficient, Gillon approached the Moore School, exercising a contractual clause that had specified that Aberdeen could take over that school's differential analyzer in a national emergency. And Gillon made other arrangements. Since it was easier to hire people with mathematical skills—female college graduates, mostly—in Philadelphia than in Aberdeen, Gillon asked the Moore School to assemble a human computing unit. Gillon, in other words, made the Moore School an auxiliary to the BRL's computing section. This was the arrangement that allowed the Moore School to work on electronic computers during World War II.[77]

However, the BRL's work during World War II was not limited to ballistic tables. For this war, the BRL envisioned a much broader research program. Its new mathematicians began to study a wide range of topics, including explosives, shock waves, and the ballistic trajectories of rockets. The general expansion of the BRL's research agenda mirrored the increasing sophistication of military ordnance. Assisted by a scientific advisory committee of prominent mathematicians and scientists, the BRL functioned as a military counterpart to Weaver's Applied Mathematics Panel, though with a narrower focus on military ordnance.[78]

Meanwhile, the Naval Proving Ground at Dahlgren, Virginia served as the naval counterpart to the BRL. Dahlgren produced the artillery range tables for the Navy's guns, including the lighter ship-board anti-aircraft guns that became the basis for the NDRC's later work on fire-control systems. Dahlgren made use of MIT's differential analyzer. It had MIT assemble a team of human computers to compute ballistics trajectories. Dahlgren also set out late into the war to create its own mechanized computing facility based on Stibitz's relay computer technology.[79]

The Harvard Mark I, meanwhile, was taken over by the Navy's Bureau of Ships. Howard Aiken, who had been a reserve officer, was called to active duty in early 1941. Although the construction of the Harvard Mark I had just begun, Aiken spent most of the war as a senior instructor at the Naval Mine Warfare School in Yorktown, Virginia. But when the Mark I was delivered to Harvard, Aiken convinced his superiors to reassign him. In May 1944, Aiken literally took command of his computer, retaining the rank of commander in the Naval Reserve. This allowed Aiken to use the Navy's recruiting machinery to draw in a staff of mathematically talented commissioned officers. The Mark I was used to solve some naval engineering problems, but soon Dahlgren appropriated much of its computing time.[80]

Los Alamos was one other place where a good deal of computing occurred during World War II. The early work on the atomic bomb was carried out by the NDRC, but in 1942 all work was transferred to the Army's Manhattan Engineering District. One of Warren Weaver's computing groups—presumably the Astronomical Computing Bureau—performed some calculations for the early gaseous diffusion experiments at Columbia University. However, the extreme secrecy that came to envelop the Manhattan Project ensured that Los Alamos would develop an independent capacity for computing and applied mathematics. Indeed, Los Alamos became a vortex for all knowledge that seemed even remotely relevant to the detonation of an atomic bomb. This included knowledge essential to the implosion mechanism of the plutonium bomb and early wartime studies of the hydrogen bomb. The physicist Donald Flanders, Richard Courant's colleague at New York University, brought with him the requisite knowledge about partial differential equations and its relevance to the study of shock waves. The associated non-linear equations required intense numerical computation. Los Alamos assembled its own group of human computers, along with a separate machine computing group that made extensive use of IBM machinery.[81]

Many other groups computed during World War II. Human computers calculated the anticipated performance of radar antennae. Groups working on operations research made considerable use of IBM machines to determine optimal military tactics and strategies. Wallace Eckert personally brought machine computation to the Nautical Almanac Office at the U.S. Naval Observatory, where he served as the office's wartime director. Military logistics also demanded clerical workers, and tabulating machines and other office machinery.

In spite of the NDRC's organized efforts, the wartime mobilization of computing and applied mathematics remained divided. Tensions between disciplines, organizational divisions among military agencies, and the decentralized structure of the NDRC raised real questions about how the new computing expertise should be organized. Stated in more pragmatic terms, the diverse aims of the war effort and the pressure to find a technical solution that would be available in time for the war meant that most of the new initiatives in computing continued to serve specific ends.

Yet although the war did not produce an entirely unified field, it produced some individuals who came to occupy an organizational vantage point from which they could see the breadth of wartime computing developments. When Dahlgren made a decision to create its own computing facility, it turned to Warren Weaver and his Applied Mathematics Panel for advice. In December 1943, in a study officially commissioned by Dahlgren as AMP Study NO-92, Weaver turned the task over to Stibitz. This study allowed Stibitz to survey the various analyzers and computing machines that had been built, and to continue to place his own work in a wider context.[82]

Clearly, some of the work, including those in cryptography and nuclear physics, remained outside of the purview of the Applied Mathematics Panel. Nevertheless, several individuals other than Weaver and Stibitz who were affiliated with the AMP—most notable among them Mina Rees and John von Neumann—emerged from the war as authorities on computing and applied mathematics. Outside the NDRC, Courant's team at New York University, the machine computing groups at Columbia University, and the computing unit at Los Alamos became broadly aware of the different uses for high-speed computation. And by the end of World War II, IBM equipment had been installed in a variety of esoteric scientific and engineering facilities. In spite of all the specialized work that occurred during the war, the overall ecology of knowledge had changed to support a view of computing as a more coherent discipline.

Conclusion

The period between 1900 and the end of World War II saw the tremendous circulation of knowledge about computing. Knowledge, techniques, and artifacts developed in one context were often adopted and extended in another to serve new and different purposes. These innovations were driven both by broad historical currents and by individual initiative. Industrial expansion, progressive reforms, the Great Depression, and World War II gave structure to the innovations, as mediated through the intermediate layers of an ecology of knowledge. At the same time, the actions taken by the various individuals described in this chapter make it clear that technical innovation was never entirely determined by a sustaining context or by technical and institutional precedents. In the same way in which the basic tenets of progressivism could be interpreted in so many different ways, how computing was refashioned to serve progressive agendas, economic efficiency, or military interest was worked out differently at different sites.

Peter Galison has advanced the notion of a "trading zone" to provide a way of thinking about how esoteric knowledge circulates across disciplinary communities.[83] This socially grounded metaphor offers a useful way to think about the technical exchanges that occur within a new interdisciplinary field, and the notion applies quite well to *postwar* developments in computing. Yet, although the concept has some relevance to early developments in computing, computing as a whole was not yet a trading zone as described by Galison. The metaphor of a trade relation works best when describing a sustained pattern of exchange that emerges between well-defined communities. In contrast, many of the innovations described in this chapter resulted from far more ephemeral encounters, where innovation occurred through the direct appropriation, recombination and extension of knowledge. Here, interdisciplinary and institutional networks did not always persist beyond the initial act of appropriation. New mathematical techniques traveled across disciplines and continents. Devices such as vacuum tubes and torque amplifiers were taken from very different fields and assumed a new meaning in the context of computing and cryptography. Even entire organizational arrangements could be borrowed from one context and reassembled in another. If the particular arrangements Lowan made for the Mathematical Tables Project was not alien to such work, the disciplined and unemployed New York clerical workforce at least eased his staffing difficulties.

A more direct understanding of why computing was such an innovative field during the first half of the twentieth century can be gained by recognizing that computing is an infrastructural technology. Like the steam engines and machine tools that drove the industrial revolution, computing devices and techniques were something that could be modified to serve an endless array of purposes. Yet, bound by the common language of mathematics, many of the different approaches to computing were, at least formally, equivalent. As the mathematicians at the Ballistic Research Laboratory discovered, it was possible to accomplish the same task using human computers, IBM machinery, or a differential analyzer. Different implementations simply offered different tradeoffs of cost, speed, and precision. Clearly, from the standpoint of a specific application, these tradeoffs mattered. Aiken chose precision; cryptography required speed; Lowan was bound by external constraints that defined the approach he could take.

Put differently, different institutional demands produced different innovations. Yet there was sufficient similarity in the underlying mathematics and methods so that the knowledge developed at one site was often useful to others. It was then in extending and reworking borrowed knowledge that each site produced new techniques and machinery, which were then placed back into the currents of exchange. It was like the finches of the Galapagos Islands. The overall ecology of knowledge had precisely the pattern of similarity and difference that made computing such a productive site for its own reconstruction. This contributed to the seemingly infinite malleability of computing even before the end of World War II.

But this history of incessant technical exchange did not leave the overall ecology unaltered. Those who came to occupy a special vantage point by the end of World War II, and the broad diffusion of certain objects and applied mathematics knowledge across many sites and disciplines, ensured that the persistent exchanges produced new interdisciplinary relationships and institutional affiliations. If rapid innovation had the centrifugal effect of producing greater diversity, this was now offset by a network of associations that made it more difficult to keep the knowledge about computing isolated in separate quarters. Thus, even with all the military secrecy and decentralized decision making that accompanied science mobilization during World War II, those involved with computing began to rebuild themselves into a coherent research community. With this coherence, computing came to assume the character of a trading zone as described by Galison.

This new ecology deserves one final comment. As is appropriate to the social metaphor of a trading zone, this new coherence was driven not just by a shared disciplinary identity, but also by new and fairly stable institutional associations that emerged to uphold this identify. Thus, in mimicking the broader "insaturation" that would drive postwar research as a whole, the relationships forged among military, industrial, and academic organizations in computing laid important foundations for subsequent developments in the field. Before we turn to these postwar developments, it is necessary to revisit one other wartime innovation: the development of the ENIAC.

Figure 2.1
John Mauchly with some of the ENIAC project's sponsors and staff members. From left to right: J. Presper Eckert, John Brainerd, Samuel Feltman, Herman Goldstine, John Mauchly, Dean Harold Pender, General G. M. Barnes, Colonel Paul Gillon. U.S. Army photo.

2 Biography and the Circulation of Knowledge: John Mauchly and the Origins of Electronic Computing

In June 1941, John William Mauchly stood at a crucial juncture in his career. Having had some minor successes in the field of statistical meteorology, Mauchly returned from a trip to Iowa where he saw an electronic computer built by his colleague, John Atanasoff. Fatigued by his teaching load at Ursinus College, Mauchly decided to enroll in a defense training program in electronics offered by the Moore School of Electrical Engineering. The program was really intended for college physics and mathematics graduates who were retooling themselves for the war. Mauchly learned of the program because he had received a letter as the head of the physics department. He had an offer to teach in a similar program at Penn State organized for high school graduates. However, faced with the opportunity to receive intensive training in a field that might support his budding research aspirations, Mauchly decided to become a student rather than an instructor.

Mauchly is widely recognized as one of the two principal inventors of the Electronic Numerical Integrator And Computer (ENIAC), a machine many consider to be the first electronic computer. Whether or not the ENIAC really was the first, Mauchly also stood at an important moment in the history of computing. Displaced from his studies of molecular physics by circumstances related to the Great Depression, Mauchly was forced to traverse through different institutions in a desperate attempt to sustain a productive research career. In the process, Mauchly came across many of the different approaches to computing described in the previous chapter. He was exposed to computational studies of radio propagation at the Carnegie Institution of Washington, to statistical approaches to meteorology as practiced at the Blue Hill Observatory in Massachusetts, to electronic scintillation counters at Harvard and Swarthmore, and to knowledge about "best practices" in electronics engineering at the University of Pennsylvania. For Mauchly, the circulation of knowledge was accomplished through the means of his own biography. Work on the ENIAC at the University of Pennsylvania emerged as a unique synthesis of the various interests and perspectives, and techniques and artifacts that Mauchly assembled during the course of his early career.

Earlier accounts of the ENIAC's prominence in the history of computing can be found in Nancy Stern's *From ENIAC to UNIVAC* (1981), in Alice Burks and Arthur Burks's *The First Electronic Computer* (1988), and in Scott McCartney's *ENIAC* (1999). These are all valuable accounts that describe different facets of this machine's history. Nevertheless, both these and other accounts of the ENIAC have been skewed by the effects of an acrimonious patent dispute, which has tended to set the governing questions and to polarize historical memory in a way that has intervened with a careful, historical understanding of the process of innovation. A close study of the circulation of knowledge, as described in relation to the technical and historical contexts described in chapter 1, should help resolve some familiar disputes that persist within the historiography.

Specifically, this chapter traces the knowledge Mauchly derived from earlier efforts (including that of John Atanasoff), addresses whether the ENIAC was a general-purpose computer or one dedicated to ballistics research, and asks, more generally, what was and what was not innovative about the work that unfolded at the Moore School of Electrical Engineering. It should be evident from chapter 1 that electronic computers, in one form or another, would have appeared regardless of the work on the ENIAC. A somewhat more interesting question that lurks beneath the current historiographic disputes has to do more with why this development took place at a relatively marginal institution rather than in the large-scale science mobilization effort conducted by the National Defense Research Committee. Here I draw and expound upon an answer provided by David Mindell, as already described in part in chapter 1.

This chapter should also be interesting for demonstrating what a careful biography can contribute to the understanding of the processes of innovation. In the introduction, I quoted Charles Rosenberg, who suggested how a biography can be a valuable means of identifying the "intellectual and institutional options that faced particular individuals or groups of individuals."[1] Historical accounts that focus on the actors, their career choices and disciplinary identity, and the particular approach to technical problem solving they take can provide insights into the "structure of integration" between esoteric knowledge and institutional contexts. In fact, Rosenberg recommends that we begin historical inquiries by reconstructing the detailed choices, assumptions and organizing ideas that governed an individual's conduct, even if the goal in the end is to describe how these individuals were "unknowing integers in a larger calculus."[2] On the other hand, in this chapter, I depart somewhat from Rosenberg's approach to biography. I turn to biography not as a means of describing disciplinary and institutional formations, but to trace how Mauchly's early training and socialization as a scientist brought him specifically to transcend institutional boundaries, and thus define a rather particular trajectory for technological innovation.

Mauchly's life confirms what has been observed in the literature about a historical shift from an era of independent inventors to one of institutionalized innovation. Although Mauchly assembled much of the knowledge required to build an electronic computer, he found himself displaced from his own invention after arriving at the Moore School. The disciplinary hierarchies and rules of conduct within an engineering school relegated Mauchly to a relatively marginal role in the actual construction of the ENIAC, with possibly detrimental effects to the scope of the innovation that initially emerged from the work. After the war, Mauchly found new opportunities resulting from the social capital bestowed upon "inventors." Nevertheless, Mauchly would find himself relegated back into the margins by the institutional contexts for computing research that emerged after World War II.

Science and Socialization

John William Mauchly was born on 30 August 1907 to Sebastian J. and Rachel E. Mauchly in the town of Hartwell, Ohio. Sebastian J. Mauchly was a physicist at the prestigious Carnegie Institution of Washington. Rachel Mauchly, an active woman who gained her own presence in the polite Washington suburb of Chevy Chase, Maryland. In the simplest version of the story, it could be said that John followed in the footsteps of his father. Enamored by his father's studies of terrestrial magnetism, John enrolled in a PhD program at Johns Hopkins University. He decided to study molecular physics. However, while growing up in a complex, industrial society, John was subject to other influences, not the least of which was the heady materialism of the 1920s. He also found himself drawn to a culture of independent inventors, which still remained quite vibrant during this period. Both parents, moreover, were divided about the merits of a scientific career, not least because of its association with the father's failing health, and the difficulties this presented for the family's social aspirations. Although there was nothing particularly extraordinary about Mauchly's upbringing, his early experiences and socialization gave Mauchly the values and commitments, and the knowledge and experiences that shaped his career trajectory.[3]

S. J. Mauchly began his career as a high-school science teacher in Cincinnati. During the early years of the twentieth century, the rapid development of wireless telegraphy generated interest in atmospheric electricity because magnetic variation affected radio propagation. This raised interesting questions for physicists on the terrestrial and extraterrestrial causes of this variation. Drawn to this topic, S. J. Mauchly turned to the study of physics at the University of Cincinnati and received his PhD in 1913. He was also drawn in by the culture of invention. His thesis was therefore built around an instrument he devised for measuring the vertical component of the Earth's magnetic field. This work was consistent with the emphasis on instrumentation that dominated U.S. physics programs, as described in chapter 1. It also earned S. J. Mauchly a position at the Carnegie Institution of Washington and its new Department of Terrestrial Magnetism (DTM). S. J. Mauchly then secured his scientific reputation by discovering the diurnal (daily) variations in the Earth's magnetic field.[4]

S. J. Mauchly's investigations took him out on ocean voyages where he gathered data pertaining to the more remote regions of the ionosphere. Separation was therefore an important theme in John's early childhood. On the eve of one of his departures aboard the *Carnegie*, S. J. Mauchly wrote to his eight-year-old son, encouraging him to take care of his sister and to look after his mother. John, in turn, was busy building a ship using his Meccano set. Yet even when separated by distance, S. J. Mauchly kept close tabs on his son's education, writing approvingly of John's exam papers in mathematics.[5]

Rachel Mauchly was a strong woman who enjoyed the gay style of the 1920s. If S. J. Mauchly took on more of the responsibility for his son's education, she took it upon herself to raise her family into a cultivated way of life. She did so mostly by example. Rachel Mauchly attended the regular meetings of the local Women's Club, organized "hen parties," and hosted occasional luncheons. If rules of class prohibited her from having an independent career, she nevertheless turned to such other sociable activities. Rachel Mauchly also worked directly to cultivate her children's tastes. She arranged their piano lessons, chided John for his penmanship, and made sure that the family took their summer vacations out at the Jersey Shore. The family sometimes partook in an occasional game of Mah Jong and other mildly exotic pleasures typical to the 1920s. In the United States, scientists were still a nascent community, and its members drew their identity through the styles and habits of a broader, socially ascendant middle class.[6]

Influenced by his father's interests, as well as the wider technical culture around him, John developed an early interest in "things technical." Mechanical devices captured his early imagination, but this soon gave way to the various electrical gadgets that accompanied the spread of electrification during the 1910s and the 1920s. In grade school, John spent the first dollar of his allowance on a buzzer and a battery. He soon devised a block signaling system for his model railroad set. By the time he was in high school, John had a minor business laying household lines and repairing appliances for his neighbors.[7]

John attended McKinley Technical High School, a school affiliated with the master mechanics at the Navy Yard in Washington. Although the school continued to offer some courses in the mechanical arts, by the early 1920s it had modernized its curriculum and served as a strong college preparatory school. John received high marks in a broad range of subjects including physics and mathematics.[8]

It is important, on the other hand, not to overplay John's technical inclinations. He did quite well in other subjects, and developed an interest in writing and literature. John was the chief editor of his high school newspaper. He also earned the rank of Commandant in the Washington-area High School Cadet Corps, which was a vestige of the United States' recent engagement in World War I. Outside of school John enjoyed playing tennis and pursued other leisure activities befitting a middle-class household. It is only in such items as his "Daily Record," a diary in which John kept a meticulous record of his daily activities, where it appears that John was socialized into a particular, technical way of life. The meticulous charts John produced in his diary were quite similar to the ones his father used to record diurnal variations in the Earth's magnetic field.[9]

Upon entering college, John's technical interests became the decisive factor in his choice of studies. In 1925, upon entering Johns Hopkins University, John chose to study electrical engineering. This decision was reinforced by an engineering scholarship he received from the State of Maryland, which had been created because of the rapid expansion of the state's engineering industries. But John had already internalized this interest. In an early English Composition paper in which he explained his choice of vocation, John wrote: "I speak what I believe to be the truth when I say the decision was based mostly on a natural desire to be engaged in such work."[10]

Rachel Mauchly's determination to oversee her son's upbringing was not diminished by his departure for college. The mother continued to look after her son's clothing and manner of dress. A woman of the 1920s, Rachel Mauchly had no qualms about encouraging her son's interest in girls. She encouraged John to go to the freshman dance, promising "If I had a car I'd take some girls over for you." Privately, she worried that her son would become too studious, like his father. "Don't you lose something of the college life by cutting some of those things," she asked. Chiding her son for failing to be "high brow," she added: "You are not queer, unless you let yourself grow so."[11]

Still, the difficulty of many mothers is that just when their children enter a stage of life where their tastes are coming into fruition, the children grow more distant from the family. This was especially true after a child's departure for college, since colleges were an institution specifically designed to provide a liminal period of adolescence when a child is exposed to the liberal influences of a broader society. John in fact had very little difficulty discovering his own pleasures. He and some friends drove to the Jersey Shore to attend a religious gathering. Though he preferred not to attend an organized dance, he had his girlfriends. Increasingly conscious of his appearance, John wrote home asking for money with which to buy new clothes.

John's life in college might have been uneventful, had it not been for his father's death. At about the time that John entered college, his father had fallen prey to a mysterious illness, later diagnosed to be bacterial encephalitis. Severe bouts of illness periodically incapacitated the father.

The family worried that the illness resulted from overwork. It also left the family in a difficult financial situation.

Rachel Mauchly did what she could to maintain the family's image. At first there were only polite references to the illness in her letters to her son. Still, financial constraints forced John to forgo a summer tour in Europe, and he spent his first college summer instead as part of an electrical work crew in North Carolina. John had never labored through a ten-hour day, let alone under the grueling heat of the midsummer sun. In a letter to her son, Rachel expressed hope that he wouldn't "over do" and added "I don't want any more break downs to look after."[12]

It was during one of S. J. Mauchly's more serious bouts with the illness that John decided to switch from electrical engineering to physics. At least on the surface, S. J. Mauchly supported, even encouraged his son's choice of a more practical career. Yet even from his sickbed, S. J. Mauchly exerted a subtle influence, supporting, for instance, the analytical approach taken in his son's General Engineering class. He also approved of John's interactions with Merle Tuve, a younger physics colleague at Johns Hopkins. In other letters, S. J. Mauchly wrote of other colleagues who had "made good" with their research. Perhaps most important, S. J. Mauchly continued to applaud his son for his academic accomplishments.[13]

John's outlook toward learning changed accordingly. He later recalled that at some point in his sophomore year "it looked to me that this engineering is just a bunch of cookbook stuff.... The problem was that my friends were not just garden variety engineers. My friends, and my father's friends, were scientists. They were over at Johns Hopkins taking PhD courses."[14] Whether or not this recollection is accurate in all details, it hints at the elaborate construction of meanings by which John began to distinguish science from engineering. During the following year, and against his mother's advice, John decided to apply for a scholarship that would allow him to transfer directly into Johns Hopkins University's PhD program in physics.[15]

S. J. Mauchly never recovered. He suffered another serious bout of the illness during the spring of 1928, after which his condition deteriorated. John tried to assuage his fears by delving into his studies. On the back of an envelope that had held a rather alarming letter from his mother, he scribbled "Mean free path of light thru gas should be a function of mean free path for atom, and perhaps density...."[16] Yet these thoughts could not shut out his mother's words:

Dear Bill: Your letter came yesterday & to say we are in bad is small talk. Dad had to have another spinal puncture Monday night. Had a bad time Monday from 11:30 on. He is having two days & nights of not much sleep & talks a good deal. Had Betty Dorsey again yesterday but she wasn't much good & stayed only till four. Had has it in for me & I had such a time to keep him quiet.[17]

S. J. Mauchly died on Christmas Eve, 1928.

Throughout John Mauchly's youth and adolescence, things took on greater significance through each event in his life. Education, health, family, and wealth meant only so much in and of themselves, but each came to be suspended within an evermore intricate web of meanings. A vacation in Idlewyld, New Jersey was no longer just about a family holiday, but about a father, his hopes for convalescence, and the well-being of the family. Work was as important as the family's well-being, but it was important not to "over do." Even the weather came to have its connotations, as John pictured his mother brooding over a gathering storm while his father lay ranting in his bed. Meanwhile, physics, for John, was not just a form of knowledge. It was something that encapsulated the memories of a father and a tenuous means of supporting a family. Mauchly's identity as a scientist blossomed in such a context.

Still, it should be reemphasized that there was nothing especially extraordinary about Mauchly's upbringing. Aside from his father's death, Mauchly's childhood was quite similar to those of other scientists who grew up in the United States in the 1920s. But the point of this opening passage is precisely that it is these mundane acts of socialization that give scientists their social and intellectual identity. Mauchly's upbringing introduced him to a vibrant culture of invention, as well as specific values and techniques within the sciences. By choosing to study physics, he chose to augment the latter through continued academic training. But both the social and technical commitments Mauchly gained early in his life would continue to shape his career. Shortly before her husband's death, Rachel Mauchly had written to her son, saying, "I'm glad you are getting along nicely,—pretty soon you can take care of us."[18]

Life in the Margins

John Mauchly chose to study molecular physics. During the 1920s, molecular physics still rivaled subatomic physics as the most promising area of research. The most exciting developments in subatomic physics were still around the corner, and there were those who doubted whether practical knowledge could ever be gained by delving further into minute particles of matter. In contrast, molecular physics offered real promise. The ionization of gases in the upper atmosphere had demonstrable effects on radio propagation, and physicists were well situated to study such effects. There was also the promise that a formal study of aggregate matter would allow physicists to challenge the chemists' expertise.

John's research was based on studying the energy levels of carbon dioxide molecules using mass spectroscopy. This line of inquiry drew on the work of his advisor, Gerhard Dieke, a Dutch émigré physicist who had just arrived at Johns Hopkins. Spectroscopy emerged during the 1920s as a promising means by which to study molecules, and Dieke brought with him the physical models and analytical techniques necessary to pursue this work. Despite this European intellectual lineage, the routine work of analyzing energy levels was quite consistent with the experimental tradition in U.S. physics. It was also the highly repetitive nature of this work that introduced Mauchly to the rigors of calculation. Mass spectroscopy produced photographic traces as intermediate data, which made it necessary to work backwards through the physical models to calculate molecular energy levels. John and his fellow graduate students became quite skilled at analyzing spectroscopic data using mechanical calculators similar to the ones John had seen in his father's laboratory. His work was good enough to earn him a PhD. It failed to earn him a job.[19]

The year 1932 was simply a bad time to graduate. Subatomic physics had eclipsed molecular physics. The first wave of scientific émigrés from Europe, who had been trained differently, were arriving. Above all, the Great Depression made it difficult for any newly minted academic to find a good appointment. Mauchly applied for several different positions, including one at the Department of Terrestrial Magnetism, but was rejected by all of them. Mauchly stayed on at Johns Hopkins, calculating molecular energy levels for his advisor for 50 cents an hour.[20]

During the following year, Mauchly found a position at Ursinus College, a small liberal arts college outside Philadelphia. Unlike the research institutions to which he had applied the previous year, Ursinus was extremely impressed by Mauchly's credentials. During the 1910s and the 1920s, Ursinus made a transition from a teacher's college to a respectable school where students could pursue preparatory work for law and medicine. Consistent with the progressive reform of medical education, Ursinus placed considerable emphasis on building a new science curriculum. Pleased to receive an application from a graduate of Johns Hopkins, the President of Ursinus himself welcomed Mauchly into the faculty. Mauchly started his career as a tenured faculty member and as the head of a one-person physics department.[21]

The appointment at Ursinus allowed Mauchly to earn a living, but not to pursue research. It came with a heavy teaching load, which included both an introductory and advanced course in physics, and responsibility for the students' teaching laboratory. At the same time, physics, as a field, was changing. The leading laboratories in the country were equipped with apparatus well beyond the reach of many state universities, let alone an individual physicist working at a liberal arts college. During the first couple years, Mauchly devoted his energies to teaching. He revised Ursinus's physics curriculum. He also made improvements to the library's holdings, figuring that this might serve his research interests. Mauchly did continue to dabble in his chosen field of research, using borrowed data from his former advisor. He even purchased a Marchant calculator. Yet Mauchly grew painfully aware of the limits of what he could accomplish on his own.[22]

Insofar as his own identity remained inextricably tied to that of being an active researcher, Mauchly set out to identify new strategies for research. He began by delving into statistics, reasoning that a deeper understanding of statistics might allow him to extract new findings from the data generated by others. This led Mauchly back to the Carnegie Institution of Washington during the summer of 1936, albeit as a Temporary Assistant Physicist and Computer. He worked under his father's former supervisor, John Fleming. Though he was hired as, more or less, a glorified human computer, this position at least gave Mauchly access to the Department of Terrestrial Magnetism's vast pool of data, with the added hope that his new statistical knowledge would allow him to "get farther in analysis of that data than anybody had before." Mauchly was also befriended by another émigré physicist, Julius Bartels, not least because Bartels's work was based on extending Mauchly's father's observations. Bartels, who had studied at Göttingen, brought with him a statistical approach to geophysical morphology, which coincided with Mauchly's new interest. Prior to Mauchly's arrival, Bartels had already massaged the DTM data in identifying a separate, 27-day cycle in the Earth's magnetic field. Because this corresponded to the sun's period of rotation, it established an important connection between sun spots, cosmic rays, and the Earth's magnetic field.[23]

Encouraged by this affirmation that statistical techniques could generate new findings, Mauchly began working intensively with the DTM data in search of other regularities. Much of this work fell outside of his formal assignment. By the end of the summer, Mauchly felt that he had demonstrated a diurnal depression in midday ion densities, based on a correlation between the angle of incidence and the intensity a solar effect. The results were complicated, however, by the apparent unreliability of the data collected from the observatory at Huancayo, Peru, where meteorological conditions seemed to also affect ion densities. This prompted Mauchly to develop a new technique for bivariate statistics, since he had to isolate the two independent causes.

Mauchly pursued this work over the course of three summers at the DTM. It helped that he could write up his results while back at Ursinus. After the third summer, Mauchly submitted an article to the *Journal of Terrestrial Magnetism and Atmospheric Electricity,* published by the DTM and edited by Fleming. But after an internal review, Fleming questioned whether Mauchly had sufficient theoretical understanding of the ionosphere to "draw general conclusions regarding the causes of variation in the F_2 region." Fleming also criticized the paper for relying on too short a period of analysis. Perturbed, moreover, that Mauchly was working on this article rather than writing up his formal assignment, Fleming sent an admonition: "As you realize, it must play an important part in your record here."[24]

Fleming's initial rejection of the paper led Mauchly in two directions. First, it caused him to shift his interest to meteorology. Mauchly reasoned, perhaps correctly, that part of Fleming's apprehensions was about allowing someone outside the DTM to publish independently using the laboratory's data. If the DTM's atmospheric data had to be regarded as private, the daily weather maps and precipitation data generated by the Weather Bureau were entirely in the public domain. This brought about a technical turn in his reasoning. Instead of treating weather as an independent variable that could induce changes in atmospheric electricity, he postulated that meteorological conditions and terrestrial magnetism were both dependent variables affected by the same extraterrestrial cause. If sunspots could generate disturbances in the upper atmosphere, it was theoretically plausible that a similar effect could be discerned in the lower atmosphere.[25]

Mauchly also met his supervisor's challenge more directly through two complementary strategies. The first was to delve even further into the study of statistics, since a stronger statistical argument could legitimate the inferences he made from a limited pool of data. Mauchly knew that Fleming and the other researchers at the DTM were intrigued by his use of multivariate statistics, since the technique could be applied to other work at the DTM. However, upon examining the literature, Mauchly found that other statisticians had already given formal treatment to the subject, albeit in the context of educational testing and the social sciences. Undaunted, Mauchly approached some of these statisticians, including Samuel Wilks at Princeton University and Harold Hotelling at Columbia University. Both in turn were drawn to the fact that Mauchly was using their statistical techniques in the physical sciences.[26]

Mauchly soon acquired a working knowledge of multivariate statistics, to positive effect. First, Fleming accepted Mauchly's article for publication, albeit as rewritten to emphasize the statistical technique instead of his inferences about midday ion densities.[27] Second, Mauchly was able to publish a separate piece in the *Annals of Mathematical Statistics,* a leading statistical journal. Although his work was derivative, Mauchly's introduced a new "sphericity test" that offered a new way of drawing inferences using multivariate statistics. Perhaps as important, his new knowledge of statistics opened up opportunities in his newly chosen field of statistical meteorology. He earned enough credibility as a statistician to secure a position during the summer of 1940 as an advisor to the Blue Hill Observatory in Massachusetts.[28]

Mauchly's other strategy was to extend his capacity for data analysis at Ursinus. Quantity could overcome what he could not achieve through theoretical understanding or statistical methods alone. Mauchly began hiring his students using funds provided by the National Youth Administration, a program similar to the Works Projects Administration. However, Mauchly's frustration with human computers—a frustration exacerbated by the fact that many of his students were not accustomed to the rigors of calculation—brought him to consider whether he could automate or simplify computing through new computing instruments.[29]

Mauchly had previously returned to his childhood interest in electrical invention out of a need to maintain a teaching laboratory. Given the limited departmental budget, Mauchly built the power supplies his students used. Then at around the time he began spending his summers at the DTM, Mauchly began to dabble with electrical inventions that had little to do with his formal research agenda. This included an electronic thermostat that took into account the first derivative (rate of change) of the room's temperature. It was expensive to heat a home during the Depression. Going one step further, Mauchly designed a radio-frequency control system that would enable a thermostat to send control signals to the furnace using the household electrical lines. It was also in conjunction with this work that Mauchly first considered the use of the physicist's scaling circuits, which he discovered in the pages of the *Review of Scientific Instruments*. Reading a 1937 article by E. C. Stephenson and I. A. Getting in the context of his new interest, Mauchly reasoned that such a device could be used in industrial control applications where it might be useful to have an "absolute," or in modern parlance, digital frequency meter that could function at a rate of up to 50,000 cycles per second. Mauchly pursued this idea during his 1937 Christmas holiday. Desperate to find a successful career, Mauchly was willing to go well outside of his disciplinary interests in search of success.[30]

The possibility of an invention that would be useful to statistical meteorology pulled Mauchly's inventive efforts back in line with his research interests. At some point in the late 1930s, Mauchly came across a description of an electrical harmonic analyzer that operated on the same basic principle as the mechanical devices built by William Thomson and Albert Michaelson. Having just designed an electrical device for computing a mathematical derivative, Mauchly found it easy to follow the basic operations of a harmonic analyzer. Mauchly recognized that a harmonic analyzer could be used to identify statistical regularities. His statistical methods, as well as his earlier work on molecular spectroscopy, were both quite similar, in practice, to Fourier analysis. Sometime between 1939 and 1940, Mauchly built an electrical 12-ordinate harmonic analyzer. That unit was best suited for studying diurnal variations. Mauchly then began to design a 27-ordinate analyzer for studying the effects of solar flares.[31]

Insofar as Mauchly continued to supervise National Youth Administration students, he also tried to think of better ways to mechanize numerical computation. He was aware of the scientific uses of IBM equipment (possibly because of his statistical connections with Columbia), but the cost of IBM equipment precluded serious interest on his part. His tenure as a statistical consultant at the Blue Hill Observatory led him to another possibility. As a consultant, Mauchly found official sanction to explore new interests, and this brought him to the 1940 meeting of the American Mathematical Society at Dartmouth University. This was the meeting at which George Stibitz demonstrated his Complex Number Calculator, and Mauchly saw it. Although the records are somewhat vague, it was most likely just after this demonstration that Mauchly began to consider how a physicist's scaling circuit could be used for calculation. Mauchly built an electronic decade (scale of ten) counter using gas-filled triodes—a cheaper but somewhat slower substitute for vacuum tubes—based on a description he found in the literature.[32] He then reported to H. Helm Clayton, his former employer and now colleague at the Blue Hill Observatory: ". . . we are now considering the construction of an electrical computing machine to obtain sums of squares and cross-products as rapidly as the numbers can be punched into the machine. The machine would perform its operations in about 1/200 second, using vacuum-tube relays, and yielding mathematically exact, not approximate results."[33] Mauchly also shared his idea with one of his former NYA students, advising him to keep the information in the dark since he would "like to 'be the first.' "[34]

Still, at this point Mauchly remained committed to meteorology. Like others before him, Mauchly set out to build an instrument to support a specific line of research. His correspondence with Clayton continued to be primarily about meteorological statistics. His letter also contained a discussion about barometric pressures and tides, a statistical technique for manipulating barometric data, and a report about the persistence of recent peaks in cosmic ray activity.

In December 1940, Mauchly redirected his energies to a paper he was scheduled to present that month before the Physical Society of the American Association for the Advancement of Science. This was one of his earliest opportunities to present his new work on statistical meteorology. Mauchly was prodding his students to complete the data analysis just days before the talk.[35]

It was during his AAAS presentation that Mauchly met John Atanasoff. Atanasoff expressed interest in Mauchly's use of harmonic analysis, and Mauchly was immediately drawn to Atanasoff's work on electronic computing. Both were somewhat cautious. Atanasoff was being supported by an organization that encouraged its grantees to obtain patents. Mauchly had just written that he wanted to be the first. Still, both were molecular physicists with an uncannily similar interest in computing. Both remained committed to the norms of scientific exchange. After briefly describing his work, Atanasoff invited Mauchly out for a visit.[36]

This trip transformed Mauchly's career. In June 1941 he drove to Iowa, where he spent several days with Atanasoff and his assistant. Even though the machine was not yet fully operational, it was the largest electronic device designed to perform useful computation to date, and it was designed to compute a rather elaborate mathematical problem with broad application to the sciences and engineering. Mauchly did wonder why Atanasoff chose not to use electronic counters, and he spent some time describing his own efforts. However, this discussion was generally moot, since an all-electronic computer designed to implement even Atanasoff's limited algorithm would have required at least 3,000 vacuum tubes to store the coefficients. This was something neither Mauchly nor Atanasoff could afford to do.[37]

At the time of this meeting, it was Atanasoff and Berry who clearly knew more about electronic computing, and the two of them spent considerable time conferring their knowledge to Mauchly. Impressed by what he saw, Mauchly wrote to another of his former students: "The calculator in Iowa really deserved the prefix 'blitz,' for although it is not finished it is easy already to see that it will outcalculate anything so far constructed (and cost a lot less than any previous mathematical wonder, too)."[38] Mauchly redoubled his efforts in electronics.

The immediate consequence of this visit was that Mauchly decided to commit himself to the war training program in electronics at the University of Pennsylvania. This was a somewhat weighty decision, in view of the circumstances within his household. Mauchly married his first wife, Mary, shortly after graduating from Johns Hopkins, and they now had two young children, Jimmy and Sidney. There was little leeway in the Mauchly household finances. Even during the summers when he worked at the DTM, Mauchly wrote embarrassing letters to companies apologizing for his unpaid bills. It was difficult to uphold a middle-class lifestyle on a meager academic salary. Mauchly began once more to record his expenditures, as he had while at college—telephone services $9.50; meat $15; Jimmy's tuition $60; milk for the children $18, and so on.[39]

Mauchly's commitment to research now stood in a rather complex juxtaposition with his obligations to his family. On the one hand, a successful research career seemed the most obvious way to improve his family's well-being. On the other hand, even the ten dollars spent on a set of gas-filled triodes weighed against the immediate needs of his family. By 1941, Mauchly was convinced that he had to look outside of a traditional academic career in order to meet his family's demands and expectations. Although his meteorological research showed some promise, the path ahead seemed torturous. In contrast, war mobilization was creating a renewed demand for physicists willing to undertake engineering work.[40]

Still, Mauchly decision to delve into electronics was difficult. He wrote to his mother:

[Mary] particularly thought that it would be better for me to break the following news: I am studying engineering at the University of Penna. You know from her letters, I think, that I was offered one of the defense teaching jobs—one at Hazelton, Pa. The money looked big. But when I considered that an opportunity to get a compressed course in electrical engineering might never present itself again, I decided to be student rather than teacher.

Tomorrow Mary is moving herself and kids over to the Carter's farm (near Evansburg). I stay in Phila. During the week at 3806 Spruce St. . . . Are you surprised off your feet? Mary was.[41]

Working at the Moore School

In late June, when Mauchly arrived at the University of Pennsylvania's Moore School of Electrical Engineering, the school was just beginning to gear up for war. It had its new agreement with the Ballistic Research Laboratory. It had some new research contracts with the National Defense Research Committee, and it was pursuing more of them. The program in which Mauchly enrolled was yet another component of the Moore School's mobilization.[42]

Mauchly found the formal curriculum of the summer program rather mundane. However, the program and its laboratory facilities gave him time and space to think about electronic computers. Most important, the program placed Mauchly in touch with J. Presper Eckert.[43] At the time, Eckert was a stellar Master's student who was in charge of the summer program's teaching laboratory. Eckert was intrigued by Mauchly's ideas about electronic computing. With his advice and encouragement, Mauchly began designing elementary circuits for an electronic calculator. During this period, Mauchly also wrote up his first specifications for an "ideal" computing device. This remained a simple device. It contained several registers and supported the four basic arithmetic operations. Although Mauchly considered the possibility of automatic operation, the control panel he sketched out was designed to support manually keyed operation, just like a mechanical calculator.[44]

After the summer program, Mauchly, along with the one other student in his class who already had a PhD, was offered a teaching position. Several of the Moore School's faculty members had been reassigned to wartime projects. The dean, Harold Pender, was anxious therefore to fill the vacancies, especially because of the prospect of having to accelerate his undergraduate program. Pender could only offer a temporary position. Accepting it meant forfeiting his tenured position at Ursinus. Nevertheless, the Moore School was a better place for his new interest in electronics, and the excitement over war research generated a sense of job security absent since the beginning of the Depression. Although the salary Pender offered was only marginally better than what he already earned, the lower teaching load— 11 hours a week, rather than 30 plus—promised to give him the time to pursue his research.[45]

Although Mauchly continued to collect meteorological data, his primary research interest shifted to electronic computers. Over the summer, he had continued to weigh different designs and approaches. By the middle of August he had become convinced that the cost of electronic devices made electronic computing justifiable only for "more involved, more lengthy, or more specialized jobs." This was something he might have inferred from the designs of Atanasoff's and Stibitz's machines. By the time he accepted Pender's offer, Mauchly was exploring the possibility of building an electronic, numerical version of the Moore School's differential analyzer.[46]

Mauchly apparently learned of this possibility through Irven Travis, the Moore School faculty member who designed and supervised construction of the Moore School and Aberdeen differential analyzers. Travis had become something of an expert on mathematical instruments and their use in engineering analysis. During the late 1930s, General Electric employed Travis as a consultant to deal with a minor transmission line crisis on the West Coast. Travis solved the problem using the differential analyzer, which impressed GE managers. These managers asked Travis to assess whether the firm could benefit from such a machine, and then to carry out a design study. In treating this as an academic investigation, Travis considered several design alternatives. In two reports, he described an analog electronic differential analyzer and an entirely numerical alternative to the differential analyzer. For the latter, Travis envisioned a computing system made up of a series of ganged adding machines. He recognized that such a machine could offer greater precision, but noted that electronic circuits might be required to deliver the requisite speed.[47]

Mauchly denied that he had seen these reports. However, he did meet with Travis on several occasions prior to his trip to Iowa. The two had spoken extensively about computing devices and their applications. This included some new work Travis was doing on GE's behalf for the NDRC's anti-aircraft fire-control section; the work would eventually make Travis the principal contracts administrator for the Navy Bureau of Ordnance's work on fire-control systems. Travis left the Moore School just before Mauchly arrived for the summer program. In fact, Mauchly sat in Travis's office for the duration of the war.[48]

In the fall of 1941, Mauchly wrote Atanasoff asking whether he would object if he, Mauchly, built a computer that incorporated some of the features of his machine, or if the Moore School ultimately decided to build an "'Atanasoff Calculator' (a'la Bush analyzer)." Atanasoff was already aware of this technical option, having himself been asked by the NDRC to evaluate such a possibility. But he continued to be constrained by the policies of the Research Corporation, and therefore sent a friendly but negative reply: "Our attorney has emphasized the need of being careful about the dissemination of information about our device until a patent application is filed." For the moment, Mauchly respected Atanasoff's wishes. However, this blocked his most promising avenue of research. Disheartened, Mauchly wrote to one of his former NYA students: ". . . if you know of any jobs that look like they were my meat, I don't have to stay here as long as I did at Ursinus."[49]

The war did bring new opportunities. After helping with another of the Moore School's war training programs, Mauchly was given the chance to supervise a task within a new Signal Corps contract. The contract was for evaluating new parabolic radar antennas, and Mauchly was assigned to calculate the electromagnetic patterns associated with these new antennas. While Mauchly's knowledge of physics was valuable here, it was his experience supervising human computers that made him the ideal candidate. While this meant returning to the drudgery of supervising human computers, the work presented some interesting challenges. Because of the inaccuracies of the traditional method for computing electromagnetic patterns at extreme angles, Mauchly had to resort to the method of numerical integration. Significantly, numerical integration was also the means human computers used to solve the non-linear differential equations associated with exterior ballistics equations.

In August 1942, the extensive use of human computers at the Moore School and the growing backlog of unfinished ballistic tables inspired Mauchly to draft a memorandum titled "The Use of High Speed Vacuum Tubes for Calculating."[50] The memo was sent to John Brainerd, the Moore School's overall supervisor of war projects. Although this document is sometimes referred to as an early version of the ENIAC proposal, the memo is as interesting for its limitations as for what it describes. Mauchly set out in the memo to describe an "electronic computor," which he suggested was well suited for solving any iterative and hence computationally intensive problem such as numerical integration. Most of the memo was based on a close analogy between mechanical and electronic calculators. This was useful for explaining the basic principles of operation of an electronic computing device. However, the comparison had another intentional aspect to it, for although Mauchly could have used a combination of registers and arithmetic circuits as he had suggested in his earlier design studies, this might have required circuits quite similar to those employed by Atanasoff. Mauchly therefore chose to stay quite close to Travis's idea of using ganged adding machines, which although unoriginal, did at least "belong" to the Moore School. This also made it easier to draw a comparison to the differential analyzer and its intensive use for military ballistics.

Yet although the memo makes several references to the differential analyzer, the machine's basic mode of operation remained firmly rooted in the procedures employed by human computers. Significantly, although

Mauchly proposed to interconnect 20–30 electronic calculators, he still proposed doing all the calculations sequentially as if a human computer was working on a single mechanical calculator. Despite the opportunity for parallelism, Mauchly spoke of a central "program device" that would coordinate a sequence of operations that would "yield a step by step solution of any difference equation within its scope."[51]

Significantly, this central program device was also the least developed aspect of the proposal. This lack of description of a concrete programming device points to the invisibility of human computing work and all the skills associated with setting up a complex mathematical operation. The Moore School would soon learn about the challenges of conferring such knowledge when they set out to expand their human computing operations, only to discover that they had exhausted their local supply of willing, women mathematics graduates. In any event, Mauchly's memo failed to draw much interest. Although Brainerd read the memo, he suggested only that "it is easily conceivable that labor shortage may justify development work on this line in the not too distant future."[52]

Mauchly's fortunes changed with the arrival of Herman Goldstine. Goldstine was an Army lieutenant from the Ballistic Research Laboratory, and a young research mathematician with a PhD from the University of Chicago. In the fall of 1942, the BRL assigned Goldstine to the Moore School specifically to oversee its accelerated computing operations and its apparent growing pains. Mauchly had continued to study the differential analyzer in operation. He may have also gained greater familiarity with manual methods for computing ballistics trajectories after his wife—a mathematics graduate—was recruited for this work. It was the following spring when Goldstine finally heard about Mauchly's interests and memo. Unlike Brainerd, Goldstine could immediately assess the mathematical advantages of electronic computing speeds. And unlike Brainerd, he lacked the electronics engineering experience that might have given him pause. Moreover, Goldstine, specifically charged to deal with the backlog of ballistic tables, had every reason to be enthusiastic about Mauchly's memo. Goldstine convinced both Brainerd and the Army Ordnance Department to entertain a research proposal.[53]

The formal proposal, titled "Report of an Electronic Difference Analyzer," was submitted to the Army on 8 April 1943. As the war projects supervisor, Brainerd wrote the administrative parts of the proposal. The main body of the proposal, including its technical appendices, was written by Mauchly and Eckert. By this point Eckert had completed his Master's degree, and had been designated a special consultant to all war projects because of his engineering skills and inventive abilities.[54]

In many respects, this proposal continued to be quite similar to the earlier memorandum. It was still primarily an analyzer built using ganged adding machines. However, there was a difference in its manner of presentation. The proposal was now explicitly built around a direct analogy between an electronic computer and a differential analyzer. It in fact opened with a preamble that described the BRL's reliance on the Moore School's analyzer, and proceeded to explain how an electronic "difference" analyzer would perform the same task, but with greater speed and accuracy.

While this was done first and foremost as a selling strategy, drawing a strong analogy to the differential analyzer had consequences for the proposed machine's design.[55] This is made quite evident in the makeshift setup diagram assembled for the proposal. Compare figures 2.2 and 2.3, which consist of a differential analyzer setup diagram and the one devised for the ENIAC proposal. Across the top of both diagrams are the computational elements. The horizontal lines in the new setup diagram also look like fixed physical entities, just like the mechanical shafts used to interconnect the mechanical integrators in a differential analyzer. Figure 2.3 can itself be said to have been derived from the differential analyzer's setup diagrams.

However, close inspection also reveals a difference. Whereas the mechanical integrators of a differential analyzer represented a mathematical operation, with the rotating shafts representing mathematical variables and computed values, this arrangement was inverted in the case of Mauchly's setup diagram. In the new diagram, it was the adding machines, or more technically "accumulators" (see below), that held the variables and mathematical constants. The horizontal lines, then, were not just an unordered column of numerical values, but an implied sequence of mathematical operations. Such a sequence of operations, or "program" as referred to in the proposal, was based quite directly on the familiar "plan of calculation" used by human computers.

The influence of the two different approaches to computing can also be read off of the machine's proposed organization. The proposal, in fact, remained quite divided as to the machine's basic design. In their formal

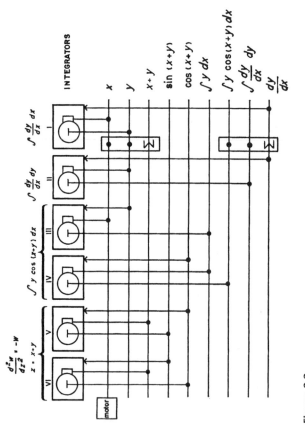

Figure 2.2

Setup diagram for differential analyzer. From Hartree 1949. Courtesy of University of Illinois Press and Richard Hartree.

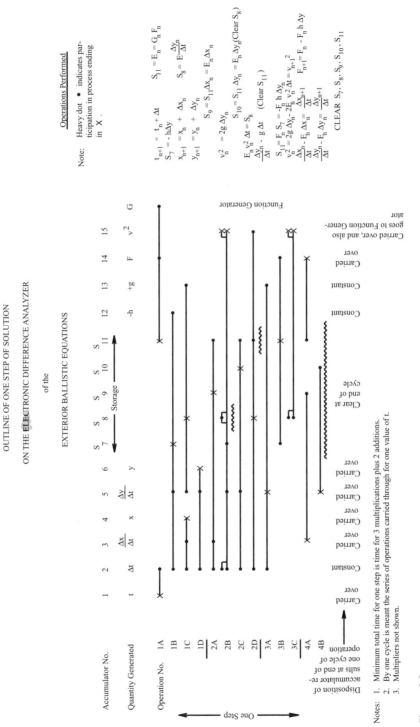

Figure 2.3
Setup diagram for the ENIAC, from the 8 April 1943 proposal. From collections of University of Pennsylvania Archives.

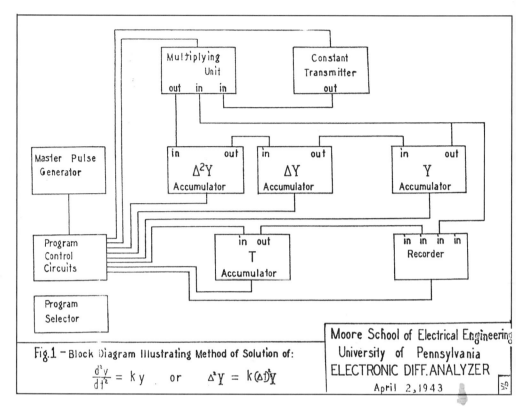

Figure 2.4
Block diagram of the ENIAC. From Collections of the University of Pennsylvania Archives.

definition of a "Program Control Unit," Mauchly and Eckert continued to emphasize the kind of centrally controlled, sequential operation that had its origins in human computing work. However, in the other major diagram submitted with the proposal, Mauchly and Eckert described a very different approach to programming, which was based on using direct interconnections between the various units of the machine just like the differential analyzer. (See figure 2.4.) This was the particular amalgam between the differential analyzer and the knowledge of human computers that became embedded into the ENIAC's proposed design.[56]

  Technically, it is possible to reconcile the two different approaches to programming as suggested in the proposal. So long as the setup diagrams were seen only to be a representation of specific programs set up by a central program control unit, the setup diagram and the idea of a central programming device were not incommensurable. But more to the point, the programming mechanism was hardly more developed in the 8 April proposal than in Mauchly's earlier memorandum. Although the proposal mentioned the possibility of a "program selector" that would load a control program from IBM punched cards, there was no technical description of how the program control unit would control the larger machine. While it is meaningless to say that the ENIAC was not yet invented in 1943, this was the state of the machine design upon which the project gained the Army's approval in June 1943.

As others have noted, Mauchly and Eckert's proposal was criticized by various NDRC members before the Army approved it. Hazen, who by then was heading the NDRC's fire-control division, learned of the Moore School proposal through the BRL's chief physicist, Thomas Johnson. Johnson had participated in the Moore School's first presentation at Aberdeen (9 April 1943), and he discussed the proposal with Hazen. Hazen proceeded to conduct an informal NDRC review, apparently mistaking Johnson's inquiry as an unofficial solicitation for support. As pointed out by Mindell, there was nothing surprising about the adverse reaction of Hazen, Caldwell and Stibitz. What little Mauchly and Eckert submitted in the way of circuit diagrams was far inferior to the work done by NCR and RCA. (They had to produce the proposal in just several days.) These diagrams would not have appeared credible to any electrical engineer familiar with the latest work in electronic computing. Moreover, during the review of Zworykin's work at RCA, Weaver had made a distinction between systems and components, and between fundamental research and developmental work. He specifically decided that the only work in the field of electronic computing the NDRC should support was work on fundamental components. Both Stibitz and Hazen embraced both distinctions. Stibitz made this clear in his subsequent written exchanges with the Army Ordnance Department, Hazen by deciding to terminate even the fundamental components research on fire-control systems. Hazen's comment "How's for some bright ideas as to how to get the Moore School profitably occupied?" must be read within this context.[57]

Lieutenant Colonel Paul Gillon, who had built up the relationship between the BRL and the Moore School, had moved on to become the Deputy Chief of the Ordnance Department's Research and Materials

Division. For his part, Gillon regarded the NDRC's evaluation as "imperti-
nent interference." He also had the authority to approve the project. Gillon
viewed ballistic tables production in concrete, strategic terms. For instance,
the construction of a new gun director for the sidewise firing of anti-aircraft rounds from aircraft had recently been delayed for several months
because of a backlog at the Moore School that prevented the BRL from
delivering the necessary firing tables. This, presumably, would have been
for the ball turret of a heavy bomber such as the B-17, which was suffering heavy losses from their operations in Europe. It is also significant that
rapid wartime advances in applied mathematics, which would not have
been as apparent to all of Hazen's crowd, made electronic computing
speeds especially attractive to the BRL's research mathematicians. Meanwhile, as a permanent unit of the military, neither the BRL nor the
Ordnance Department had quite the same interest in limiting its activities
to wartime research as the NDRC.[58]

By moving to the Moore School, Mauchly amassed not only the remaining knowledge, but the institutional reasons and infrastructure with which
to build an electronic computer. Mauchly clearly instigated what became
the ENIAC—Project PX, in its wartime designation. Yet although Mauchly
drew together the necessary resources, this also marked the point where
he himself became sidelined from his own invention. Brainerd, the war
projects supervisor, would not allow a temporary faculty member with so
little experience in electrical engineering or research administration to
supervise such a large undertaking. Brainerd assumed the title of project
supervisor. Eckert was made chief engineer. Mauchly was given the title
Principal Consultant. In a memo to Dean Pender, Brainerd suggested that
Mauchly and the other member of the staff who continued to teach in the
undergraduate program "could carry their normal teaching loads."[59]

Project PX

The ENIAC may have been a specific implementation of Travis's idea to
build a numerical equivalent of the differential analyzer. However, the
work of finding this implementation was no trivial matter. Getting a device
with more than 17,000 vacuum tubes to operate in an unfamiliar numerical domain required a large number of well-integrated innovations. The
technical achievements of the ENIAC have in fact been glossed over in

much of the existing literature. Closer examination of the different forms of knowledge involved in the project, and of the scientific and engineering practices that supported the project, makes it possible to document much more precisely the specific synthesis of ideas and experience that was coterminous with the invention of the ENIAC. This also makes is possible to provide a more fair assessment of what exactly was innovative and what was conservative about the machine's design.

It is useful to begin at the end, with a technical description of the ENIAC.[60] Physically, the ENIAC consisted of 40 frames arranged in a U-shaped layout with a total length of 80 feet. These frames housed 30 semi-autonomous units, 20 of which were called *accumulators*. Each accumulator was like an adding machine, in that it could add, subtract and store a 10-digit number and its sign (+ or –). There were two other arithmetic units, namely a high-speed multiplier and a divider square-rooter. The ENIAC also had three function tables and a large array of manual switches whose

Figure 2.5
The ENIAC at the University of Pennsylvania's Moore School of Electrical Engineering. U.S. Army photo.

values could be read at electronic speeds. These function tables were needed to represent the non-linear, incalculable component of the exterior ballistics equation that was used to represent air resistance (specifically, the drag coefficient). The ENIAC was also designed to receive (input) and record (output) data using a pair of specially built electronic units that interfaced with a modified IBM punched-card reader and an IBM card punch machine. These units were known, respectively, as the constant transmitter and printer.

Internally, the ENIAC processed electronic signals at a rate of 100,000 pulses per second. Each of the decade ring counters could advance one *stage*, namely to shift, say, from the number 6 to the number 7, during each pulse. As such, it took a pre-programmed sequence of control pulses that lasted exactly 20 *pulse times* to get an accumulator to perform a single addition (+) or subtraction (–), with the necessary carry propagation and other bookkeeping operations. The Cycling Unit produced nine different streams of control pulses that controlled the basic operation of the accumulators and all other units of the machine. Because it took 20 pulse times (one *addition time*) to perform an addition, each accumulator could perform 5,000 additions per second. The Multiplier could perform approximately 350 multiplications a second. Both division and the extraction of a square root were performed using the relatively slow method of repeated subtraction, so that the Divider Square-Rooter proceeded at a slower rate of approximately one hundred operations per second.[61]

If the specific function of each unit in the ENIAC was fairly straightforward, setting up the machine to solve a problem was less so. Significantly, the ENIAC's accumulators did not have all the functions of an adding machine. Mechanical adding machines perform addition by storing two numbers, x and y, in mechanical form before carrying out an addition $(x + y)$. However, an accumulator was designed to store only one number so that every addition had to occur by precisely coordinating the action of two units so that the contents of one accumulator would be transmitted to another unit to advance the latter's electronic counters. This coordination was achieved by sending a separate control, or *program pulse* simultaneously to both units. Programming the ENIAC meant coordinating the actions of its various units at the scale of the entire machine.[62]

To fully understand the ENIAC's programming mechanism, it is useful to keep in mind that all numerical calculations have to be carried out through an exact sequence of operations.[63] In the case of the ENIAC, calculation began with the Initiating Unit, which sent the first program pulse that triggered the overall sequence of operations. All the other units were designed to receive a program pulse, and to issue a new pulse once it finished its designated operation. This pulse was then routed to the units involved in the next step of the calculation through a direct, wired connection. There was one unit known as the Master Programmer, which was used to cycle the machine through a sequence of operations for a preset number of iterations. The Master Programmer's mode of operation was quite simple. It simply counted the iterations, issuing a program pulse to one set of units during the iterations, and to the unit or units representing the next operation when done. The exact sequence of operations within an iteration still had to be set up through precise interconnections between units, with the program pulses being passed from unit to unit during this sequence. Most of the units also had what the project engineers referred to as a *local programming* capability. The accumulators, for instance, could be set up to add, subtract or increment its own value, and to perform this task from one to nine times. To allow each unit to be used more than once, each unit had multiple sets of local program control circuits, each with its own pair of sockets for receiving and transmitting a program pulse. Quite esoteric in its design, this was the final realization of the programming mechanism that was left unresolved in the 1943 proposal.[64]

Mauchly presided over the project's early meetings, in June 1943. Despite his official role as consultant, all the engineers recognized that Project PX was his idea and deferred, at least initially, to his authority. Working on the presumption that their first task was to build an "electronic version of [an] ordinary computing machine," Mauchly proceeded to convey his ideas about how electronic circuits could be built to perform the function of a mechanical calculator. Eckert followed with some notes from his early design experiments on ring counters. After these initial discussions, each of the engineers assigned to the project—T. Kite Sharpless, Robert Shaw, Hyman James, and others—returned to the bench in a common effort to develop a working knowledge of vacuum tubes, pulse generators, ring counters, and other devices.[65]

Despite Mauchly and Eckert's joint leadership over the project, Project PX remained, in many respect, a collaboration among the faculty and research staff of the Moore School. Nearly all of the engineers took con-siderable initiative in determining their own contributions to the project. For instance, Sharpless, a senior project engineer, strove to see whether any of his substantial knowledge of radio electronics could be transferred to the domain of electronic computing. Vacuum tubes were already being used in FM broadcasts and early studies of television broadcasts, both of which operated in the megahertz range. The Moore School was a reposi-tory of the latest thinking about high-frequency vacuum-tube engineering in this domain.

In laboratory notebooks, Sharpless examined his knowledge of high-fre-quency filters and amplifiers in regard to their possible application to the design of more responsive computing circuitry. Especially important was any information he could glean about relevant design tradeoffs in the high-frequency domain, and about the source of parasitic oscillations that might cause erratic behavior. Although numerical calculation offered the promise of arbitrary accuracy and precision, the work of supporting the electrical abstractions that enabled numerical (in latter-day parlance, digital) computation required a good deal of effort in high-frequency analog design. Sharpless contributed substantially to the ENIAC's basic cir-cuitry. He also designed the Cycling Unit, which had to deliver precise, 2.5-microsecond square pulses throughout the machine with adequate signal capacity.[66]

Robert Shaw was an engineer whose broad interests complemented Sharpless's narrower focus. Shaw and his assistant Harry Gail designed a "function generator" that would have performed interpolation using a mathematical technique known as the Gregory Newton Method. Although this device was never built (building the three function tables proved to be a better work-around), the work nevertheless indicated the comfort Shaw had with the mathematical aspects of the machine.[67] Shaw also worked on the constant transmitter and printer. Then, once the general design of the ENIAC was completed, Shaw shifted his focus to a "megacy-cle" counter, which could be used in subsequent improvements to the machine. Others, meanwhile, persisted with the arduous task of building the machine. In most cases, Eckert and Mauchly simply formalized the division of labor that emerged through collective discussions among the project engineers.[68]

While most project members focused on the engineering aspects of the machine, Mauchly turned his attention to other problems. For instance, during an early meeting with the BRL, it became apparent that the validity of the project rested on an assertion he made in the 1943 proposal that it was possible to use a somewhat simpler algorithm to solve the exterior ballistics equation, where a smaller interval of calculation would compensate for the greater accuracy of the current algorithm employed by human computers. After delving into the literature on numerical integration, as he had done earlier with statistics, Mauchly realized that the validity of this assertion rested on a formal analysis of rounding and truncation errors. This time Mauchly chose not to pursue the work himself, but made an arrangement with Hans Rademacher, a mathematician at the University of Pennsylvania.[69]

Eckert's principal responsibility, meanwhile, lay with the overall engineering effort. Eckert did make quite a few detailed technical contributions. He designed the ENIAC's ring counters, and he contributed to the accumulator, the master programmer, the cycling unit, and the self-test circuitry built into the initiating unit. However, Eckert's main responsibility was effective management of the complex project. Even as individual initiative contributed to the project's success, Eckert had to coordinate these interests to serve a common goal.[70]

The productive tension between individual initiatives and project coordination, along with evidence of the project's reliance on external knowledge, can be seen by following the ENIAC's development from the perspective of one of the staff members, Arthur Burks. Although Burks is often portrayed as one of the project engineers, his degree was in philosophy. He was the other PhD who had enrolled in the summer program in electronics. That summer, and the following year, Burks and Mauchly were housemates. When Mauchly invited Burks to join his project, Burks allowed curiosity to guide his career.

Burks, whose knowledge of electronics was rudimentary, was first assigned to conduct routine electrical tests on the project's experimental ring counters. This work allowed Burks to become familiar with these devices, and their modes of failure. Because Eckert was occupied with other matters, he let Burks write up the project's early work on ring counters. Eckert was never fond of writing. Burks had written a dissertation in philosophy. This was the first substantial set of reports issued by Project PX.[71]

These reports, along with Burks's laboratory notebook, help document the extent to which the Moore School relied on outside knowledge. Eckert maintained that he never used the circuit designs supplied to him by

NDRC officials. Technically, this may have been true. The fact that the Moore School was not within the inner circle of the NDRC meant that information had to be secured through official channels, and this may have added to Eckert's frustrations and the decision to shun existing solutions. But Burks's notes also make it clear that Eckert's original ring counters could not be made to operate much above 40 kilohertz. It was only after Eckert secured a series of reports on NCR's high-speed thyratron counters, along with other work at RCA that he came up with a design that could meet the ENIAC's targeted design speed of 180 kHz. The ring counters used in the ENIAC may have been sufficiently original to warrant a subsequent patent claim. Nevertheless, exposure to the broad array of wartime digital electronics developments, and the fundamental circuit design ideas that they embodied, contributed to the project's success.[72]

Once he had completed his work on the ring counter, Burks proceeded to help with the accumulator. By late 1943, Eckert felt that the group had made sufficient progress on the ring counter to shift to this larger unit. This work still remained mostly beyond Burks's expertise. Eckert, along with Frank Mural and several other project engineers made most of the design decisions regarding the accumulator. Burks, however, did make one significant contribution. Unlike a mechanical ring counter, electronic ring counters were designed only to advance in the forward direction. This meant that an accumulator could subtract a number only by adding its complement. The mathematical details of how to do this were somewhat esoteric, though not beyond the ability of the engineers. Still, with his background in philosophy, which at the time would have included formal training in logic, Burks could treat this as a routine mathematical exercise. Burks was also asked to write the report describing the accumulator.[73]

In October and November 1943, Eckert, Mauchly and the other project engineers worked out the ENIAC's programming system. The most pressing issue for them was that of demonstrating the feasibility of their project, which they were technically required to do by the end of the year. The engineers felt that the best proof of concept was a pair of accumulators—again, a single accumulator could perform no useful operation—that could be tested at the full design speed. This meant, however, that they had to have some kind of programming system. As a matter of expediency, Eckert and Mauchly decided, basically, to implement the decentralized programming system that was already implied by the diagrams they submitted with the April 1943 proposal. (See figures 2.3 and 2.4.) Though it was possible to design a central program control mechanism, it was easier to conduct the two-accumulator test by using local switches and physical wires to interconnect the two units. Designing the entire control system now was too much work, especially under the deadline; designing a dedicated control system for this single test would be wasted effort. Moreover, for the engineers who had just designed the arithmetic control circuits for the accumulator, extending these circuits to encompass local programming seemed a much easier task.[74]

Burks followed these discussions closely. In the process, he secured for himself an even bigger role, namely that of assembling Project PX's first semi-annual progress report, issued in December 1943. While other project engineers focused on the remaining technical problems, Burks, who was responsible for the report, became adept at visualizing the entire machine. This led him to try his hand at producing a diagrammatic mockup of the entire machine, as set up to perform an exterior ballistics calculation using the proposed second-order Heun method. This served as an important means of fleshing out the ENIAC's programming system in determining the necessary scope of the local program controls, and in deciding just how many accumulators and other units they had to have.[75]

The work during the second half-year of the project centered on three major tasks consisting of the remaining design work, the two-accumulator test, and settling on the machine's final configuration. Each of these efforts presented their own challenges. However, the most interesting work from the standpoint of how the project continued to coordinate its work with outside knowledge was the work on the machine configuration. The original contract did not specify a final configuration, since there was insufficient knowledge about the computational requirements of the proposed algorithm, or about the ENIAC's actual speed. But by early 1944, the Moore School had in hand Rademacher's analysis and a precise specification for each of the ENIAC's units. They also had Burks's implementation of the exterior ballistics equation.[76]

However, Eckert and Brainerd chose to leave the final design decision to Leland Cunningham, the head of the BRL's machine computation group. The mathematics of military ballistics was the BRL's expertise. Moreover, placing the decision in the hands of the BRL gave the decision official sanction, thus limiting the Moore School's liability. After several joint meetings, Cunningham concluded that the ENIAC had to have at least enough units to implement the algorithm Burks had employed. However, Cunningham also requested several additional units that would augment the ENIAC's ability to handle other problems that were of interest to BRL mathematicians. This choice points to the expanded program in applied mathematics at the BRL, as well as to the laboratory's tendency to look beyond wartime research. This was the social process through which the ENIAC was determined to have the "proper proportion of the various kinds of units."[77]

While the remaining work on the ENIAC was fairly straightforward, there was one other aspect of the work, not related to Burks, that is worth considering. In the literature, Eckert is generally credited with having discovered the need for a burn-in period for vacuum tubes, without which the ENIAC could not have sustained useful operations for very long. However, this is exactly the kind of "asymmetric" explanation rejected in the science-studies literature, where the end result is used to explain an innovation. Such explanations tend to produce linear and highly functionalist accounts of technological innovation. In contrast, a sustained focus on engineering practice makes it possible to provide a more contingent—and compelling—account of technological innovation.

Working directly from the laboratory notebooks, it was around January 1944 when the project engineers began purchasing significant lots of vacuum tubes in order to build the first pair of accumulators. At this point the ENIAC was designed using nine standard types of tubes, and the first lots that had arrived demonstrated substantial variability in their operating characteristics. Eckert assigned Sharpless and Hyman James, who was a relatively junior engineer, to study this difficulty. Drawing on the conventional industrial practice of testing incoming parts, Sharpless and James designed a meticulous testing regimen whereby they proceeded to test every incoming lot of vacuum tubes to make sure that the devices operated according to the specifications they set for the tubes. This work also involved the design of a specialized testing assembly. James then began to analyze the particular aspects of a rejected tube that seemed to contribute to its poor performance.[78]

James had tagged all the tubes that had undergone the incoming test, which is to say all of the tubes in the ENIAC. Thus, as operational testing of the ENIAC's units began in June 1944, it was not too much of a stretch to track the tubes that failed during operation, and to record the particular mode of failure. By early 1945, this had expanded into a meticulous program for tracking the life history of all the tubes, and the particular aspect of the ENIAC's operations, or features of an incoming lot, that seemed to contribute to tube failure. The common modes of failure became immediately apparent from this data. Tubes clearly tended to fail during power-up and during power-down. Moreover, tube failures tended to occur early in a tube's life span. Whether it was Eckert, James or someone else who first did so, all that remained was to plot out the time-dependent distribution of failures to determine what an appropriate burn-in period would be for each category of vacuum tubes.[79]

On the other hand, it is important to note that analyzing the failure of vacuum tubes was only one component of a much broader program for reliability engineering put in place by Eckert. Quite early in the project, Eckert had instituted a system of "design principles" (now called "design rules") that were imposed across the project. Some of the more simple principles dictated that all vacuum tubes had to be operated at less that 50 percent of their rated plate voltage (E_p) and 25 percent of their rated plate current (I_p). A more complex principle held that none of the program pulses could be routed from unit to unit since this could result in progressive signal degradation. More broadly, whether in the mechanical design of the electrical lines, the use of modular plug-in units, or the systematic reuse of standard circuit design elements, the ENIAC was built using a rather robust set of "best practices" for building reliable electronic systems. It is significant in this respect that at the start of the work Eckert was a recent graduate who had just been indoctrinated into these practices. If this was not always as practiced in industry, Eckert's rigid adherence to the principles of reliability engineering proved essential to the ENIAC's success.[80]

It is worth thinking more generally about what the Moore School, as an institution, brought to the project. The Moore School was not MIT or Stanford, both of which had built their reputation on the formal and experimental analysis of electrical networks. On the other hand, especially in view of its proximity to the work on radio transmission at RCA and at Bell Labs, the Moore School was a major regional center for radio electronics and engineering. This did not include the work on the fundamental physical properties of a vacuum tube, as was pursued at these industrial laboratories.[81] Nor did the Moore School have a strong orientation toward mathematics, as Bell Labs and MIT did. Nevertheless, the Moore School remained a major repository of knowledge about best practices in electronic circuit design. This is what the Moore School engineers brought to the development of the electronic computer.

Rebuilding a Career

What happened to John Mauchly during the course of the ENIAC's development? When the engineering work began in earnest, the relationship between Mauchly and Eckert became "Eckert and Mauchly." Once Mauchly conveyed his ideas about electronic computing to the Moore School engineers, much of the remaining knowledge needed to complete Project PX was within the electrical engineer's domain of expertise. This is not to say that Mauchly had no role in the project. Nevertheless, from the disciplinary perspective of electrical engineers, as well as the bustle surrounding war work, much of Mauchly's, as well as Burks's efforts became invisible. Or, they were at least easier to erase during subsequent reconstructions of historical memory, especially as backed by a patent system whose emphasis lay with physical artifacts.

The remaining story about Mauchly's career is told better elsewhere, notably in Nancy Stern's *From ENIAC to UNIVAC* (1981) and in Scott McCartney's *ENIAC: Triumphs and Tragedies of the World's First Computer* (1999). After working with the mathematician John von Neumann in designing a new machine, the Electronic Discrete Variable Automatic Computer (EDVAC), Mauchly and Eckert left the University of Pennsylvania to create the Electronic Controls Company. The firm was formally incorporated as the Eckert-Mauchly Computer Corporation in 1948. This company proceeded to build the first widely available commercially manufactured computer, the Univac I. Financial difficulties forced Eckert and Mauchly to sell their firm to Remington Rand two years later, although both were given senior management posts. Some elements of this story are picked up in subsequent chapters, though they are told mostly from the perspectives of other individuals and organizations.

Mauchly's early career was hardly predictable. Unfettered by any single institution, Mauchly drew as much from a culture of invention as from a culture of science and from the multiple resources and opportunities that

World War II created. Different academic disciplines and organizations shaped his career: physics, statistics, meteorology, electrical engineering, the specific organizations where he studied and worked. All this was structured, in turn, by the commitment to a successful career that he had acquired during his youth and his adolescence. Even negative institutional constraints served a useful purpose. It was the disciplinary and institutional constraints presented by the Department of Terrestrial Magnetism, by Ursinus College, and by the interwar scientific labor market in general that propelled Mauchly into a peripatetic career.

The specific technical trajectory of Mauchly's efforts after he arrived at the Moore School suggests that it may be worth revisiting the claim that the ENIAC's design was simply a matter of wartime expediency. Expediency did matter, both from the standpoint of the war effort and from the standpoint of the work schedule laid out in a contract. Nevertheless, from the standpoint of engineering practice, what is most evident is the incremental nature of the design innovations that led to the ENIAC. From Travis to Mauchly to Eckert, all the design ideas were based on taking known methods in analog and human computation and finding an automated, numerical means of implementing them in mechanical and then electronic form.[82] The focus on flexibility, incrementalism and technical practice places this study in line with Andrew Pickering's (1995) observations about the "mangle of practice." Pickering holds to one of the strongest positions on the contingencies of research and on the opportunities that researchers like Mauchly create for themselves not by realigning their knowledge but re-defining the goals of their research. But whereas Pickering explicitly avoids using the notion of constraint, I find it useful to think about how disciplinary practices as well as the knowledge embedded in artifacts do as much to constrain innovation as to enable it, generally in the very same instance.[83] This is not to say that disciplinary boundaries are absolute. Clearly, Mauchly transcended all kinds of boundaries during the early course of his career. Yet the knowledge he acquired was deliberate and was guided by his prior choices. He was also influenced by the differential analyzer and by the practiced skills of human computers.

Moreover, once Mauchly's ideas were brought into the disciplinary and organizational context of the Moore School, work on the ENIAC proceeded according to a fixed notion of the problem, with little attention to other possibilities. This is especially evident when the work is contrasted against that of Howard Aiken and George Stibitz. Although Aiken and Stibitz both worked entirely within the electromechanical domain, they separated the program control unit from the arithmetic unit, and the arithmetic unit from the registers in making more efficient use of circuitry. I do not consider this to be the result of any particular "brilliance" on their part. Unencumbered by the design of the differential analyzer, both Aiken and Stibitz set out simply to automate the work of human computers. Notably, human computers laid out their calculations assuming that they were working with a single calculating machine (it was easier to move numbers than to move bodies). Thus, whereas in principle researchers may always be free to alter their perspectives, it makes little sense to ignore the technical and institutional contexts that bind researchers to certain traditions without which the notion of discipline carries little weight.

There is in fact a stark contrast between the first and second half of this chapter. The first half of Mauchly's early career was marked by wide and undisciplined exploration, during which he was able (and forced) to move from one institutional context to another. This gave him the freedom to change his ideas and priorities (even as his actions were constrained by a more basic commitment to research). But upon his arrival at the Moore School, Mauchly's work took a specific technical turn and proceeded along a more predictable path. So long as Mauchly chose to accept his position as a consultant, and to shape his work according to the institutional expectations and disciplinary standards of the Moore School (not to mention a contract), his subsequent work followed a specific "trajectory," precisely as defined by Donald MacKenzie (1990) in describing Charles Stark Draper's work at the MIT Instrumentation Laboratory.

Moreover, institutional constraints became encoded in the project's organization. Once the basic commitment was made to go with a series of ganged adding machines, most of the project engineers honed in on specific tasks to which they felt they could apply their expertise. No one was specifically assigned to review the overall design to consider whether there were significant design alternatives. Put more simply, Project PX was, from the outset, an engineering project and was organized along such lines. Thus, despite all the contributions that Mauchly (and Burks) made to the project, so long as he accepted the underlying premise for a project that he himself had put forward, he could not amass the authority or even the perspective from which to pursue more radical changes to the technology. That this need not have been the case is made amply evident in the next chapter.

Upon leaving the Moore School, Mauchly resumed the peripatetic strategies that marked his earlier career. At the Electronic Controls Company, Mauchly retrained himself to be an entrepreneur. His prior knowledge of statistics proved useful in securing the company's first contract with the Census Bureau. Mauchly focused his energies around a study of computer applications and markets. But when he and Eckert's firm was absorbed by Remington Rand, his self-taught marketing skills became eclipsed by those of a professionally trained marketing staff.[84]

Mauchly never mastered the institutional contexts within which he hoped to build his career. In this sense, he never became a "system builder" (to use a term coined by Thomas Hughes in describing, mostly, an earlier era characterized by inventor-entrepreneurs).[85] In 1959, Mauchly left Remington Rand to establish Mauchly Associates, an independent consulting firm. In the late 1960s, he started another company (Dynatrend). Mauchly had been designated a consultant at the Blue Hill Observatory, and then for Project PX. He now found a more permanent niche as a consultant within the emerging computing industry, where he helped to introduce quantitative project management techniques to U.S. construction firms and other businesses.[86]

Mauchly had a successful career. Whatever the various turns in his life, he helped to design the machine that many continue to see as the first electronic computer. He also created a startup firm that built the first commercially successful computer in the United States. Late in life, Mauchly returned to the problem of the weather, which had fascinated him long ago. He retired to a quiet suburb outside Philadelphia. He died on 9 January 1980, at the age of 72.[87]

Figure 3.1
John von Neumann and the IAS Electronic Computer. Courtesy of Archives of the
Institute for Advanced Study, Princeton.

3 From Ecology to Network: John von Neumann and Postwar Negotiations over Computer-Development Research

The years immediately after World War II were crucial for the future of computing. A substantial consensus formed around general-purpose, digital electronic computers built around a particular design that would become known as the von Neumann architecture. These years also produced intense negotiations over U.S. postwar research policy. This chapter aims to demonstrate that the two were intertwined. The dominance of the so-called von Neumann architecture cannot be understood except in relation to the extensive negotiations about the proper course of postwar computing and computer-development research.

The origin of the von Neumann architecture is a major topic in the history of computing. Not surprisingly, John von Neumann has attracted the attention of many historians and biographers. The most recognized works include Norman Macrae's *John von Neumann* (1992), Steve Heims's *John von Neumann and Norbert Wiener* (1980), and William Aspray's *John von Neumann and the Origins of Modern Computing* (1990). Aspray's book, in particular, is a solid academic history documenting von Neumann's accomplishments.

However, von Neumann's prominence and the narrative conventions of the biographic genre have generated accounts that have skewed the history somewhat from the point of view of the "symmetry postulate" in science studies—the proposition, embraced by most constructivist scholars, that technical developments should be described according to the viewpoints of each of the major actors.[1] There can be valid reasons for rejecting the symmetry postulate, often having to do with retrospective assessments of the value of an innovation. Nevertheless, so long as the aim of this book remains that of describing the *process* by which new knowledge and institutions came into being, it is important to retain a focus on the detailed circumstances under which this occurred. Even Aspray's book, with its effort to assess von Neumann's many contributions to computing, has a different purpose than documenting the process of innovation. What remains necessary is a study that considers von Neumann's work in light of other postwar computing initiatives and the emerging social organization of postwar research.

Although I will begin this chapter with some important biographic details from von Neumann's early life, the chapter's main purpose is to pick up where I left things at the end of chapter 1. I will attempt to trace how computing and computer-development work went from a loosely configured ecology to a "trading zone" to a well-ordered network assembled around von Neumann and his ideas. Whatever controversy surrounds the origins of the stored-program concept (a cornerstone of the von Neumann architecture), historically von Neumann and the Institute for Advanced Study emerged as an obligatory point of passage for doing digital computers research. It literally became a "center of calculation," to draw once again from Latour. In fact, this chapter draws quite extensively from constructivist and post-constructivist analyses of the social organization of knowledge, even as it historically situates the competing metaphors of "ecology," "trading zone," and "network" through an assessment of their relevance to different phases in an interdisciplinary field's origins.

In view of the substantial discussion of ecological perspectives in the introduction, and the general familiarity of sociotechnical networks in science and technology studies, it may be useful to also provide a more precise definition for the notion of a trading zone.[2] The concept originates with Peter Galison's essay "Computer Simulations and the Trading Zone" (1996). As Warwick Anderson (2002) has suggested, the concept of a trading zone can be seen as part of a broader movement in technoscience studies that aims to offer a more decentered account of knowledge production running counter to an earlier generation of laboratory studies. As I suggested in chapter 1, this makes Galison's work similar to earlier symbolic-interactionist studies of science, which also focuses on disciplinary boundaries, exchanges networks, and the circulation of knowledge. However, more specific to Galison's notion of a trading zone is an explicit interest in questions of epistemology. The main aim of his work is to document how the validity of knowledge is upheld through social relations of trust forged between collaborators not entirely conversant with all the constitutive bodies of expertise. Such collaboration, and trust, can foster the rapid development of a new interdisciplinary field, even as the participants continue to pursue goals defined by different disciplinary agendas. Galison also draws on the linguistic notions of pidgin and creole to describe the increased coherence of the conversations that emerge within interdisciplinary communities. This produces a bounded semantic space that eventually becomes occupied by a new group of specialists who lay claim to the newly emergent field.

This is what makes the "trading zone" metaphor appropriate for depicting the intermediate stage between a loosely coordinated ecology and a network built around a narrowly defined notion of computing. During the postwar years, digital computers generated considerable interest and interdisciplinary exchange, even as different actors continued to pursue this work through their respective interests in applied mathematics, physics, electrical engineering, and other contributing fields. A sense of common interest helped to build up the nascent field in attracting the attention not only of potential sponsors but new researchers. Although I focus less on questions of epistemology than Galison does, I place the same emphasis on social trust as a basis for creating a new interdisciplinary field. Meanwhile, I extract from the "trading zone" metaphor a more explicit interest in questions of power and influence. This facilitates the segue into an initial discussion about how the early technical exchanges gave way to a more coherent network built around von Neumann's ideas.

Von Neumann's Early Life and Work

John von Neumann was born Janos Lajos Neumann in Budapest on 28 December 1903. His father, Margittai Miksa Neumann, earned the "von" for his work as a director and chief counsel of a major Hungarian bank. Aspiring to join Budapest's upper class, Miksa and his wife, Margit, cultivated their son's taste for high culture and learning. Aided by private tutoring, Janos entered a renowned Lutheran *gymnasium* at age ten. Quickly recognized for his mathematical abilities, Janos began taking lessons from faculty members at Budapest University. He then proceeded directly into the doctoral program there in 1921. Concerned that his son ought to also consider a more practical career, Miksa convinced him to also study chemistry at the University of Berlin. He then transferred in 1923 to the famous chemical engineering program at the Eidgenossische Technische Hochschule (ETH) in Zürich. However, by this point Janos had already begun pursuing serious inquiries into abstract mathematics. Especially while in Berlin, he became part of an increasingly prominent circle of Hungarian scientists who had taken residence there—a group that included Denis Gabor, Leo Szilard, and Janos's childhood friend Eugene Wigner.[3]

Janos received a PhD from Budapest University in 1926 (as well as a degree from the ETH), whereupon he chose to stay in Germany so that he could remain at the heart of the mathematical research activities that were unfolding at Berlin and Göttingen. Around the turn of the twentieth century, David Hilbert and his cohort at Göttingen made a number of fundamental advances in mathematical logic, number theory, set theory, and abstract geometry. During the 1910s and the 1920s, Hilbert turned his attention to constructing a logical foundation for mathematics. He enlisted many mathematicians, including young Janos, to assist him with this work. This formalist program foundered on Kurt Gödel's famous 1931 proof that no formal system powerful enough to formulate arithmetic could be both complete and consistent. Still, Janos's affiliation with Hilbert enabled him to publish a paper on the formalist program, along with six other papers in the general realm of mathematical logic.[4]

What earned Janos Neumann his broader reputation was his extension of work in abstract geometry by David Hilbert and Richard Courant—an extension that provided a rigorous mathematical formulation for quantum mechanics. Building on early speculations by Hilbert and his physics assistant, Lothar Nordheim, Janos was able to advance an axiomatic system that offered an abstract unification of the wave and particle theories of subatomic behavior advanced by Schrödinger and Heisenberg. While this system was of limited use to physicists, it helped to silence critics inside mathematical physics who doubted whether such unification was possible. These findings appeared in a series of papers published between 1927 and 1929, which were compiled as *Mathematische Grundlagen der Quantenmechanik* (*Mathematical Foundations of Quantum Mechanics*) in 1932. Among his other major accomplishments from this period was a theory of measurement, which offered a persuasive mathematical foundation for describing the indeterminate and probabilistic nature of quantum theory, defined in terms of the interactions between observer and the observed.[5]

From 1933 on, Janos Neumann (John von Neumann, as he styled the name in his immigration papers) would make the Institute for Advanced Study in Princeton his intellectual home. The IAS was established through an endowment created by Louis Bamberger and Caroline Bamberger Fuld, whose family had made its fortune in the department store business. Bamberger and Fuld had approached Abraham Flexner, known for his work on the reform of medical education, with the idea of creating a medical school in northern New Jersey that would be open to Jews. But Flexner, by then, had shifted his attention to creating in the United States an institution of higher learning that could rival the best research centers in Europe. Having secured Bamberger and Fuld's support, Flexner agreed to serve as the institute's first director. Drawn to abstract knowledge, Flexner hoped to organize the IAS initially around a powerful school of mathematics. His vision became real when he enticed Albert Einstein into accepting an appointment. This enabled Flexner to draw together other notable figures in mathematics. Having already secured Oswald Veblen's interest and help, he enticed James Alexander from Princeton's mathematics department. He also secured a promise from Hermann Weyl, who had succeeded Hilbert as the director of the mathematical institute at Göttingen.[6]

Von Neumann was a serious contender for one of the initial appointments. Flexner had assembled a faculty with well encrusted reputations, and there were reasons for drawing in a younger scholar. Moreover, Oswald Veblen pushed for von Neumann's appointment. Veblen, by then an eminent figure in mathematics in the United States, had helped to build up Princeton's mathematics department during the 1920s. Veblen contributed directly to Flexner's plans for the IAS, and would come to tussle for authority with Flexner. Von Neumann had already drawn Veblen's attention several years earlier, and as a result he had been lecturing at Princeton University since 1930. But Flexner opposed the appointment, in part because the funds were limited and in part because he desired to placate Princeton, from which he had already extracted two faculty members. However, when Weyl changed his mind and decided to remain in Göttingen, von Neumann emerged as the obvious candidate. It is historically significant that the aging IAS faculty always thought of von Neumann as "young Johnny."[7]

Another feature of existing historical accounts of von Neumann is that they tend, collectively, to offer an "American" interpretation rooted in postwar perspectives. This has led to subtle erasures of the nationalistic and anti-Semitic impulses that were integral to U.S. academic institutions in the late interwar years. This, along with the anachronistic reading of von Neumann's accomplishments, has created some misunderstandings about von Neumann's personal ambitions and intellectual trajectory. Of considerable importance in this respect is the strength of the émigré culture that emerged in physics and mathematics with the rise of the National Socialists in Germany. As anti-Semitism spread across Eastern European academic institutions, the Rockefeller Foundation and other American and European institutions worked to create opportunities for those affected by the scientific diaspora. Yet, although universities in the United States and Canada saw themselves as helpful, many émigré scientists and mathematicians remained frustrated by the opportunities extended to them. By necessity, many American universities continued to protect their tenured and tenure-track faculty and the graduate programs they worked so hard to cultivate. As such, they insisted on a two-tier system of placement that excluded many European mathematicians from positions they felt more qualified to take.

Among von Neumann's cohort, this frustration is most visible in the correspondence he received from Stanislaw Ulam, a friend and colleague from his days in Europe. Ulam began his work in Lwów, Poland. In 1936, after a brief stay at the IAS, he received a temporary position at Harvard. Quite confident of his abilities, Ulam wrote: "The 'job' situation is of perennial sadness. Some people here, especially Langer seem to be very much anti-foreign & I do not expect any promotions here, in spite of the fact that there will be several vacancies here."[8] Disappointed, he took an assistant professorship at the University of Wisconsin, where Warren Weaver had taught some years earlier. However, Ulam became distraught with what in the United States was the normal teaching load at a state university. Hoping to escape his intellectual isolation, Ulam wrote von Neumann asking whether he would consider spending part of the summer in Madison. Von Neumann replied only that he had enjoyed visiting Wisconsin before and was "hoping for the best."[9]

Differential treatment and isolation increased the social solidarity among émigré mathematicians. As against the nationalistic rivalry in mathematics that existed back in Europe, differences were quickly brushed aside to create a pan-European identity of the sort typical of expatriate communities. By virtue of his intellectual reputation and institutional location, von Neumann quickly came to sit at the heart of this new network. Not unlike the social gatherings of fin-de-siècle Vienna, or more proximately those of the belle époque in Budapest, von Neumann and his wife opened their home to gatherings of intellectuals, often organized around a visit by some famous mathematician. Von Neumann's "parties" have been reinterpreted in an American vein: Good time Johnny, always ready with a joke. No doubt, von Neumann partook in assimilation strategies—strategies which were all too familiar because of his Jewish background. Jokes, in particular, have long been known to signal attempts to be "American." Nevertheless, the emphasis von Neumann and his first wife placed on social gatherings should also be read as an effort to reconstruct a distinctive social milieu marked by European sensibilities. Notably, mathematics in Europe sat squarely within a rich tradition in the arts and letters, not the instrumentalism or the hurried escape from it that characterized mathematics in the United States.[10]

More important to the constitution of this émigré community was a distinct culture of mathematical correspondence, made all the more important by the scientific diaspora. The intense correspondence among émigré mathematicians of this era is documented in any of a number of archives.[11] These letters make it clear that personal correspondence among mathematicians were an exceptional mechanism of intellectual exchange. They provided a means for working out problems, communicating ideas, initiating collaborations, and building consensus. These letters and the intellectual games they displayed also sustained the social bonds essential for defining new research agendas in mathematics. The relevance of such networks to postwar developments in computing is considered below. For the moment, it is interesting that von Neumann's habit of closing his letters with the phrase "from our house to yours" was adopted by his European cohort and by his junior American colleagues.[12]

As I noted in chapter 1, World War II gave new articulation to the boundary between pure and applied mathematics. Von Neumann inhabited this changing boundary. The mathematical research at Göttingen played a major part in this shift, more so than described in chapter 1. It was Hilbert and Courant's extension of Felix Kline's speculations about quantum theory that allowed von Neumann to develop a formal foundation for quantum mechanics. And although much has been made of Courant, Friedrichs, and Lewy's 1928 paper on the use of finite difference methods for solving partial differential equations—and its contribution to the implosion calculations of the atomic bomb—less historical attention has been paid to the broader military implications of this paper. Basically, the parabolic, elliptic, and hyperbolic partial differential equations that characterized Hilbert spaces mapped directly onto an extremely wide range of physical systems—shock waves, aerodynamics, hydrodynamics, blast effects, and the like. Courant, Friedrichs, and Lewy's paper made it theoretically possible to compute a solution to all such non-linear phenomena. To reiterate, wartime priorities, joined by new computational machinery, made it increasingly feasible to do so.[13]

Von Neumann's involvement with this line of research grew directly out of his ties to Göttingen, as well as his physical and social proximity to the Army's ballistics work at Aberdeen. Contemporary work on exterior ballistics benefited from interwar advances in applied mathematics, especially the work on the calculus of variations pursued by Gilbert Bliss at the University of Chicago. Nevertheless, Veblen, who remained an advisor to Aberdeen, recognized the potential relevance of the newer developments in abstract mathematics. It was through Veblen's introduction that von Neumann began to consult for Aberdeen in 1937. In 1940, when Aberdeen—by then the Ballistic Research Laboratory—constituted a Scientific Advisory Committee, von Neumann was made a member. The fact that several of the committee members, Theodore von Karman, I. I. Rabi, and Harold Urey among them, were émigrés was not a trivial issue. Von Neumann himself had become a naturalized citizen only in 1937. Yet once the relevance of their knowledge was demonstrated through the Scientific Advisory Committee's regular meetings,[14] von Neumann and his fellow advisors all found broad circulation in and beyond the Army despite their cultural standing as foreigners.[15]

By September 1941, von Neumann was a consultant to the National Defense Research Committee's Division 8 (explosives). He worked specifically with George Kistiakowsky, E. Bright Wilson, and G. J. Kirkwood on the theory of detonation and shock waves. From September 1942 to July 1943, von Neumann shifted his attention to the quasi-mathematical realm of operations research and began working for the Mine Warfare Section of the Navy Bureau of Ordnance. This work took von Neumann to England during the first half of 1943. Combining his interest in shockwaves with naval mine warfare, he then formulated an Applied Mathematics Panel project for studying the hydrodynamics of underwater explosions. This was carried out under his direction at IAS. Throughout late 1943 and early 1944, von Neumann also spent a good deal of his time collaborating with von Karman on a formal study of blast effects for the NDRC's Division 2 (structures). Because of this work and the growing appreciation of his knowledge, von Neumann was made an official member of this division.[16]

In September 1943, von Neumann also became an official consultant to Los Alamos. J. Robert Oppenheimer invited him to Los Alamos because of his knowledge of hydrodynamics, which was crucial to the early stages of the detonation of a plutonium-based atomic bomb. The Manhattan Project had recently redirected some of its attention from uranium to plutonium amid concerns about U-235 production. However, the physical characteristics of plutonium made it necessary to have a much more complex detonation mechanism to ensure that the material would reach the critical mass necessary for a sustained nuclear reaction. Von Neumann was instrumental in convincing Oppenheimer and other senior Los Alamos scientists, based on his knowledge of shaped charges and shock waves, that the implosion mechanisms proposed by physicist Seth Neddermeyer was indeed feasible.[17]

Von Neumann first developed an interest in machine computation while visiting Britain in conjunction with his mine warfare studies. He had paid a visit to the Nautical Almanac Office, studying its staff and its equipment quite carefully. However, it was the implosion calculations at Los Alamos that really propelled von Neumann to pursue this interest. Again, it was Donald Flanders who brought the knowledge of Courant, Friedrichs, and Lewy's paper to Los Alamos. Meanwhile, it was another physicist, Dana Mitchell, who brought Columbia's IBM methods to the hand computing group there. Still, von Neumann stood at an interesting confluence of different forms of knowledge, organization, and machinery. As a "consultant" at a time when most other scientists were required to maintain a permanent residence in Los Alamos, von Neumann remained free to circulate among wartime military projects and organizations in an effort to secure greater knowledge about applied mathematics and computing machinery.[18]

Drawing on his connections with the Applied Mathematics Panel, von Neumann submitted an inquiry to Weaver in January of 1944 asking what computational facilities might be available for complex hydrodynamic calculations. He followed this up with a visit, during which he revealed that this was for Los Alamos, not his work at the IAS. Weaver immediately placed von Neumann in touch with Howard Aiken, George Stibitz, and Jan Schilt, who in Wallace Eckert's absence remained responsible for the Astronomical Computing Bureau at Columbia University. After conferring with Courant on the feasibility of the proposed work, Weaver did what he could to ensure that von Neumann gained access to the equipment he needed. Since there was a direct technical relationship between this work and von Neumann's own work on explosives, machine computation also became a part of the AMP project at the Institute for Advanced Study.[19]

Those who criticize von Neumann for his involvement with applied mathematics and war work forget the patriotic fervor with which émigré scientists set out to recapture their own countries from the Nazi regime. All the same, von Neumann did clearly enjoy his new "obscene interest" in computing machinery, along with the military applications of his mathematical knowledge.[20]

Von Neumann and the EDVAC Design

In more popular accounts, historians have tended to attribute von Neumann's encounter with the ENIAC to a fortuitous meeting with Herman Goldstine, who again was serving as the Ballistic Research Laboratory's official liaison to the Moore School of Electrical Engineering. However, in light of his affiliation with the Ballistic Research Laboratory and Los Alamos, and the computational demands of his work, it was really only a matter of time before von Neumann would discover the ENIAC.[21]

Von Neumann's contributions to the Moore School have been the source of considerable controversy, centered on the origins of the "stored-program" concept. This was basically an idea for treating computer programs as a form of data held in high-speed memory that could be interpreted by a central program control unit. A computer built by this design would need only one arithmetic unit, a unified memory system and other associated subsystems, all of which were controlled by a central device. This presented a very different design as compared to the ENIAC.

Aside from the relative ease by which a program could be set up on such a system, the major aim of this design was to reduce the amount of hardware by severely limiting the machine's parallelism. As Paul Ceruzzi has noted, "the *First Draft of a Report on the EDVAC*, written in 1945, described a machine that economized hardware by doing everything, including addition, one bit at a time (serially)."[22] The origins of this idea lay in a series of discussions among von Neumann, J. Presper Eckert, Mauchly, Burks, and Goldstine, occasionally joined by other members of the Moore School. This was itself an important early "trading zone." By August 1944, when the group began its discussions, it is clear that both von Neumann and those at the Moore School were familiar with the electromechanical computers that Howard Aiken and George Stibitz had built. Both the Harvard Mark I and Stibitz's various relay computers represented programs as a sequence of coded instructions placed on a strip of paper tape, which was interpreted by a central control unit. Meanwhile, the ENIAC's function tables offered an important technical precedent whereby numerical data loaded into electrical relays could be read at electronic speeds. (Technically, this was an auxiliary unit, known as the *constant transmitter*, which read and transmitted the data contained in the manually set switches of the function table.) Though this may seem like a spurious connection, within two weeks of von Neumann's first visit Goldstine was describing a scheme in which electrical relays could be used to implement, in effect, the automatic "program selector" that John Mauchly mentioned in the original 1943 ENIAC proposal. More important, by early September, Goldstine reported that the group was proposing to use "a centralized programming device in which the program routine is stored in coded form in the same type of [mercury delay line] storage devices." In many respects, the stored-program concept was the least difficult part of the new machine's design.[23]

In acknowledging von Neumann's interest and influence, the Army Ordnance Department decided in October to support further studies into what all the collaborators recognized would be radical improvements to the ENIAC design. Specifically, the Moore School was tasked to lay out a "comprehensive plan" for an Electronic Discrete Variable Automatic Computer, with the advice and consultation of von Neumann. Joint discussions continued through the following spring. Deferring to von Neumann's reputation, S. Reid Warren, the senior Moore School faculty member assigned to supervise the project, honored von Neumann's request that he be allowed to produce "a summary of these analyses of the logical control of the EDVAC."[24]

At this point, there is again an odd disparity in the historiography. Although von Neumann's report is often cited as the source of the stored-program concept, or alternatively, as a "logical" description of modern computers, the document was not a formal treatise in mathematical logic. Nor was it principally about the stored-program concept. Formal logic was directly relevant to only a small part of von Neumann's analysis; by his own account, he set out to describe instead the general "structure" of a computing machine.[25]

Indeed, the overall coherence of the report was sustained not through formal logic, but a neurophysiological metaphor. During late 1944, von Neumann began participating in a group known as the Teleological Society, organized by the MIT mathematician Norbert Wiener. Wiener's reputation in cybernetics parallels and overlaps von Neumann's in computing in interesting ways. Wiener was involved with the NDRC's work on fire-control systems. After being relieved of this responsibility because of his attachment to less-practical solutions, Wiener set out to develop a formal theory of human-machine interactions, which, as David Mindell notes, was already integral to this work.[26] Shortly before the war, Claude Shannon, then a master's student in MIT's electrical engineering program, demonstrated that there was a formal equivalence between relay switching circuits and Boolean algebra. For Wiener, this raised the bigger prospect that the various advances in computing, communication, and cognition could give way to a unified mathematical representation, much in the way that the unified theory of electromagnetism, quantum mechanics, and (more elusive) general relativity had (or might). Wiener, a self-conscious former child prodigy, assembled a group of mathematicians, philosophers, and neurophysiologists for just such a purpose. This group was called the Teleological Society because they (Weiner, at least) felt that their investigations might shed light on the intentional nature of human action—and eventually of intelligent machines.[27]

Von Neumann remained skeptical about Wiener's ambitions. He built his early career around just such a search for unified representations, and by his reading, computing and cognition did not yet reveal the kind of regularities that might give way to such unity. Nevertheless, these discussions, which took place in parallel with the EDVAC design meetings, gave von Neumann the neurophysiological metaphor that made him refer to the various parts of the computer as "organs."

More important, the underlying organic metaphor helped von Neumann maintain a systematic focus on the highly interrelated tradeoffs that contributed to machine efficiency. The report described, for instance, how it was necessary to employ both short and long delay lines for the registers and the main memory so that the length of the former was matched to the behavior of the central arithmetic organ. The report also called for a fully bit-wise serial operation, where all calculation was done one binary digit at a time. Although framed in terms of a thought experiment, von Neumann suggested that such a radically simple design would enhance component reliability, permitting a higher circuit speed that would substantially compensate for the lack of parallelism. Von Neumann also offered a tentative instruction set ("order codes" in his parlance) that weighed the tradeoffs between engineering costs and computational efficiency. He did this not in the abstract, but in specific relation to different kinds of problems—problems that were defined by his mathematical interests and his military advising work.[28]

In view of the report's multiple references to engineering concerns, there is little doubt that many of the specific ideas emerged through joint discussions with members of the Moore School.[29] Yet whatever individual contributions the Moore School members made, von Neumann's analytical training, the discursive style of mathematics, and the borrowed language of neurophysiology allowed him to write an elegant and comprehensive description of a machine that went beyond any of these earlier discussions. The draft report was an eminently practical document. By situating the report precisely halfway between mathematical abstractions and detailed electronic designs, von Neumann created a document that many could regard as the blueprint for postwar computer development.[30]

Technically, the stored-program concept, as embodied by the draft report, conforms to a formal theory of automata pertaining to machines capable of modifying their own programs. But when von Neumann wrote his report there was not yet a formal distinction between an order (i.e., an instruction) and an address (i.e., a number representing a specific location in the memory). It was only in the *Preliminary Discussion of the Logical Design of an Electronic Computing Instrument, Part I, Volume 1,* as written by

Burks, Goldstine, and von Neumann (1947), that a specific order was created to perform what was referred to therein as the "partial substitution" of a program. This was an order that would replace an address contained in another order stored elsewhere in memory. This was one way of implementing a "variable address," in which a fixed sequence of operations, generally referred to as a subroutine, could be made to act upon data contained in different parts of the memory. A more effective solution, developed some years later at Manchester University, involved storing a variable address in a separate index register.[31]

With an index register, there was neither a formal need nor a practical need to modify an order by treating it as a form of data, at least for the classes of problems von Neumann had identified in his draft report. It would have been quite sufficient, for instance, to load a program into a bank of mechanical relays. The general interest in treating programs as data had to do instead with self-modifying programs and their relationship to recursive sequential functions, a formal mathematical understanding of which was not established until 1964. In fact, the historiographic interest in the stored-program concept really has its origins with the emergence of theoretical computer science, a development that was 20 years in the future. At the time of the report, the stored-program concept entailed only the practical consideration that storing the program along with the data would obviate the need for designing two separate memory systems.[32]

To understand the historical disagreements from the viewpoint of von Neumann and his contemporaries, it is important to consider that a disagreement erupted because the authorship of the draft report was credited solely to von Neumann. The wide distribution of the report has been attributed, properly I believe, to a difference in the practices of disclosure among mathematicians and engineers.[33] Whereas mathematicians tend to be quite open with their ideas, engineers tend to be more closed as necessitated by a system of patents. Stated more concretely in terms of situated practice, von Neumann was actively circulating knowledge about computing in his role as a wartime consultant. In contrast, J. Presper Eckert and John Mauchly, during the early part of 1945, were enmeshed in a difficult negotiation with the dean of the Moore School in an attempt to retain their patents rights for the ENIAC and their contributions to the EDVAC.[34]

If Goldstine's later recollections are correct, it is possible to be even more specific about the set of practices that contributed to the wide distribution of the report, namely the culture of mathematical correspondence. Von Neumann sent an early draft of his manuscript to Goldstine sometime in late April. He then sent some corrections and clarifications in a long letter, which was necessitated by his return to Los Alamos. Von Neumann worked with Goldstine, not only because he was the official BRL liaison, but because he alone possessed the knowledge—a PhD in mathematics from the University of Chicago—necessary to handle this letter. Goldstine then accepted the task of compiling the final manuscript. Goldstine felt no need to work through issues of authorship, since the document was intended for internal distribution. Nor did Goldstine think to secure military classification for the report. However, Goldstine was a young mathematician who was enamored with von Neumann. Seeing a brilliant synthesis of discussions to which he himself contributed, Goldstine allowed the report to have wider distribution. He apparently did so without concern for *von Neumann's* rights as an author. The report circulated among Goldstine's colleagues within the BRL, and other trusted individuals in and beyond the NDRC who knew about the Moore School work. This circulation impinged on Eckert and Mauchly's plans to patent some of the ideas contained in the draft report. This dispute, which grew more acrimonious over the years, was about the entire body of ideas contained in the draft report.[35]

Still, this attention to intellectual property obscures the fact that there was no prima facie reason why von Neumann's description of the EDVAC should have been superior to an engineering implementation of the machine. Public demonstrations rather than written descriptions are often the most effective means of disseminating new technologies. This has been true for electrical lighting, telephones, and even the World Wide Web.

Recognition accrued with von Neumann for several historically contingent reasons. First, so long as the work at the Moore School remained classified, an unclassified report had privileged circulation. Second, an abstract design offered enough of a description to garner the interest of an applied mathematician who hoped to use—or to fund the development of—an electronic computer. Even for those who set out to build a computer, von Neumann's report functioned as an attractive "blueprint," specifically by allowing them to choose their implementation based on their technical expertise and institutional capabilities. Perhaps most important, the immense network assembled because of wartime applied mathematics work offered a rapid path for the dissemination of new ideas about

high-speed computation. Here, the mathematical culture of correspondence played an even greater role, as word about von Neumann's report
spread quickly through mathematical circles desperate to have a machine
that could solve the problems that wartime research had identified.

Postwar Negotiations as a Trading Zone

At the end of World War II, many groups stood poised to enter the exciting new arena of digital computers research. The field received its vitality
not only because of new computing machinery, but because of extensive
wartime developments in applied mathematics. In transcending traditional
distinctions between context and content, the social arrangements constructed to pursue this work determined the very boundaries of a new interdisciplinary field. This section should also make it clear how postwar
negotiations over computing and computer-development research could
give precise articulation to evolving institutional conversations about
postwar research policy.

A Postwar Applied Mathematics Panel

As might be expected, serious discussion about a peacetime applied mathematics program originated not with von Neumann, but with the Applied
Mathematics Panel. As head of the AMP, Warren Weaver paid close attention to Vannevar Bush's plans for demobilizing the Office of Scientific
Research and Development. Bush and Weaver were close colleagues, and
Weaver was a foundation officer with a keen interest in this discussion.
By all indications, Weaver supported Bush's interest in dismantling the
wartime research organization and creating a new federal agency—Bush
referred to it then as the national research foundation—that would administer all fundamental scientific and engineering research, including work
of military significance. Yet as early as May 1944 Weaver and the other
members of the AMP were evaluating the prospects of having something
like the AMP persist after the war. Among their concerns was the planned
dissolution of Arnold Lowan's Mathematical Tables Project, which had
received its support under a now-defunct Depression-era program. Although the members of the AMP stopped short of "setting policy" by
recommending a peacetime version of their panel, they wrote their final
report in such a way as to suggest the possibility.[36]

The prospects for a peacetime version of the AMP became more real when Britain established a new Mathematics Division within the National Physical Laboratory, its premier national laboratory for research in the physical sciences. The 2 April 1945 meeting of the AMP was devoted to whether a similar arrangement could be made in the United States via the Research Board for National Security, an interim organization set up to administer fundamental research paid for by the military services.[37]

Two historical contingencies interrupted these plans. The first was the collapse of the RBNS. Irreconcilable differences between Bush and other federal bureaucrats blocked the work of this organization. The other was an inner-ear infection that incapacitated Weaver. This left the work of charting the AMP's demobilization to Mina Rees, the technical aide to the panel.[38]

Rees, who held a PhD in mathematics from the University of Chicago, had been assigned to the AMP at the recommendation of Richard Courant. She was relatively passive at first, having been assigned to take minutes. It was a full year before her own voice began to appear in the proceedings. As Rees grew familiar with the activities of the AMP, she began performing many of the evaluations upon which the panel rendered its decisions. Her thoughts and evaluations became a regular part of the group's deliberations. Although Thornton Fry replaced Warren Weaver as the AMP's acting head, it was Rees who really served as Weaver's substitute.[39]

Still, the sudden collapse of the RBNS left Rees with few options. The Navy had made substantial use of the AMP, and so Rees approached Admiral Julius Furer, the director of the Navy's Office of Coordinator for Research and Development. Contrary to Bush's plans, Furer (more accurately, his staff of eager young officers) was setting out to create a central naval office for peacetime research. By May, this had led to the creation of a new Office of Research and Inventions, which absorbed several naval laboratories and offices, including Furer's. Furer would have been pleased to support a peacetime program in applied mathematics. His limited budget kept him from doing anything more than sustaining the Mathematical Tables Project. Still, this arrangement laid an important foundation for further negotiations that unfolded after the end of hostilities. In the process, Rees secured herself a job. In August 1946, with the formal establishment of the ORI's successor, the Office of Naval Research, she went from being an administrator of a civilian research organization to a civilian program manager charged with overseeing a military research organization's interests in mathematical research.[40]

MIT and the National Research Council

Next to act were MIT and the National Research Council's Committee on Mathematical Tables and Other Aids to Computation. These two organizations stepped forward to jointly organize the first major computing conference, in October 1945. Like most other units of the NRC, the Committee on Mathematical Tables was displaced by Vannevar Bush's wartime organization, and its members were now eager to reclaim their historic role. Samuel Caldwell, who was in the process of regaining his position as the head of MIT's Center of Analysis, was acutely aware of the extent to which MIT had lost its lead in computing as a result of wartime developments. Caldwell was a member of the NRC committee. He became the vice-chairman of the subcommittee that was set up to organize the conference.[41]

MIT had made an even earlier move in its effort to recapture its reputation in computing. It had made an offer to John von Neumann. Although this fact is duly noted in the literature, what is not so well recognized is the extent to which it was tied to MIT's broader aspirations in nuclear physics. Despite the immense wartime expenditures at MIT, the Institute missed the boat on the new developments in nuclear physics. (MIT's Radiation Laboratory worked on radar, not nuclear radiation.) As the end of the war neared, senior MIT administrators, including President Karl Compton and Dean of Science George Harrison, set out to recruit prominent physicists from the Manhattan Project. Von Neumann's name came up in this context. It was Norbert Wiener who then arranged the necessary conversations, because of their mutual interests. Wiener knew that von Neumann was looking for a place where he could pursue research in digital computers. But Harrison was mostly eager to make von Neumann an offer more because von Neumann's work in applied mathematics would firmly align MIT's mathematics department with an expanded program in nuclear physics. Harrison offered von Neumann the chair of the department.[42]

The initial plans for the computing conference occurred while MIT was trying to recruit von Neumann. It is therefore entirely possible that MIT administrators agreed to host the conference because of their desire to assess the broad relevance of this work before embarking on a major machine-development project. In August 1945, von Neumann rejected Harrison's offer, choosing to stay at the IAS. This led the MIT administration to place their trust in local talent to restore MIT's reputation in the field. After the conference, Caldwell was given the opportunity to remake the Center of Analysis through a coordinated effort involving himself, Wiener, and the new Research Laboratory for Electronics (the postwar incarnation of the Radiation Laboratory). Weaver agreed to sponsor Caldwell's project, acting as an officer of the Rockefeller Foundation.[43]

Like others, Samuel Caldwell set out to build a general-purpose computer. Having come out of the analog tradition, Caldwell continued to hope that analog devices might find a place within an optimal design. In view of wartime developments in analog electronics, this hope was not unjustified. However, in retrospect, the decision to leave the overall design of the machine open was a mistake. Wiener's interest stemmed from a more abstract understanding of computers, and not one based on practical trade-offs between analog and digital computation. The RLE, meanwhile, could find no compelling interest in the project so long as its underlying use of electronics remained unresolved. There were, in any event, other pressing opportunities that stemmed from their established work on microwave radar. No doubt the overall assessment that Caldwell was overtaken by Jay Forrester and his work on Project Whirlwind—an assessment offered by the historian Larry Owens—is quite correct. (See chapter 5.) Nevertheless, it should be noted that Caldwell failed to construct within the confines of MIT a research program where mutual interests and interdependence constituted a viable trading zone.[44]

Von Neumann and the IAS

Meanwhile, von Neumann began laying his own plans. Von Neumann had amassed a considerable reputation through his wartime work, including the EDVAC report. The conversations initiated by the AMP and by the MIT conference also generated considerable interest, much of it focused on his contributions. Still, it remained necessary for von Neumann to convert this reputation into tangible assets, especially if he hoped to pursue computer-development research.

William Aspray, in describing the negotiations at the IAS, suggests that the IAS's reluctance to break with tradition represented a serious challenge for von Neumann. Von Neumann was genuinely concerned about whether the IAS faculty and its trustees would allow him to undertake a computer-development project. At the same time, he was quite aware of the effort with which research universities in the United States were courting those affiliated with the Manhattan Project—he himself was involved with an effort to bring Enrico Fermi to Princeton. At a minimum, von Neumann was willing to entertain the thought that he could find an institution better suited to his evolving research interests, which had in fact strayed from the IAS's emphasis on abstract mathematics and theoretical physics. Von Neumann secured an offer from the University of Chicago as well as the one from MIT, and Harvard and Columbia also expressed serious interest. Though von Neumann was clearly aware of the value of these competing offers, it seems unlikely that he pursued them simply as a means of forcing his colleagues to accept a computer-development project.[45]

In fact, the historical records of the IAS reveal relatively little direct opposition to von Neumann's wishes. In 1939, Frank Aydelotte, formerly president of Swarthmore College, replaced Abraham Flexner as the director of the IAS. At least from that point forward, governance at the IAS remained divided between the IAS's director, the faculty as dominated by the School of Mathematics, and the IAS Board of Trustees (which included Veblen as a faculty representative). At the end of World War II, the major policy issue for the IAS was (not surprisingly) whether the IAS should create new appointments in experimental physics. This was a contentious issue, with von Neumann and Veblen in favor of the new appointments (Veblen always supported the practical application of mathematics) and with other faculty members arguing that experimental physics and the facilities necessary to pursue such work would harm the purity of their institution. Einstein was of the latter opinion. Taken out of context, his protest that experimental physics might fuel ideas of "preventative wars" may seem quite out of place. Yet amid the backdrop of the Holocaust, the aging Einstein would not have been alone in fearing the effects of a powerful weapon placed in the hands of a totalitarian regime, or even a military-minded democratic state. It was precisely at the end of this heated discussion that von Neumann expressed his desire to undertake a project to build a digital computer at the IAS.[46]

In April 1945, the director of the IAS, Frank Aydelotte, issued a directive calling for the faculty to "curtail or give up entirely their war work." Yet the immense success of wartime research—in which Aydelotte himself had played a minor part—had an influence on the institution. In October 1945, Aydelotte issued a report documenting the various wartime contributions of the IAS faculty.[47] Moreover, he had made a specific exception for John von Neumann in his earlier directive: "It may happen, furthermore, that some men like von Neumann will be doing work of such public and scientific importance as to make it necessary for them to continue after the end of the war. I shall report such cases to the Trustees at future meetings and ask their approval of whatever arrangements may seem to be advisable."[48]

Building on his rapport with Aydelotte, von Neumann set out to convince him, and then the IAS trustees, that an electronic computer with 1,000 times the computational power of any machine then in existence could revolutionize science. He suggested how computers would pave the way for fundamental advances not only in quantum mechanics, hydrodynamics, and other fields made computable through partial differential equations but also in more traditional subjects, including celestial mechanics, dynamic meteorology, statistics, and economics. Von Neumann argued that by acting now the IAS could lead the field. It was important, moreover, for an academic institution to assume the lead so that it could "exercise a directing influence on all future developments." Von Neumann mentioned MIT's offer, which Aydelotte took seriously. Aydelotte also chose to trust von Neumann on a subject with which he was unfamiliar, and assumed the responsibility of presenting von Neumann's wishes before the trustees. Whether for the trustees' benefit or his own, Aydelotte accepted as "soberly true" the idea that an electronic computer would open up vast new domains of knowledge. He drew an analogy to the 200-inch telescope at Mount Palomar—the biggest, foundation-backed interwar project undertaken in the name of fundamental advances in knowledge. The board approved a request for a guarantee of $100,000 from the endowment with which to underwrite von Neumann's project.[49]

The faculty of the school of mathematics discussed this arrangement several months later, albeit in a displaced conversation about von Neumann's subsequent plan to initiate a program in dynamic, computational meteorology with support from the Navy's Office of Research and Inventions. Speaking really about computing in general, the faculty discussed the effects that such activities would have "upon the progress of mathematics and upon the general atmosphere of the Institute." Opinions varied from that of "Professor Siegel who, in principle, prefers to compute a logarithm [by hand] . . . through that of Professor Morse who considers this project inevitable but far from optimum, to that of Professor Veblen who simple-mindedly welcomes the advances of science regardless of the direction in which they seem to be carrying us."[50]

Of these responses, Marston Morse's response was probably the most indicative of the attitude that allowed von Neumann to remain at the Institute for Advanced Study. While some faculty members opposed new institutional policies that would embrace experimental science, neither were they willing to challenge a colleague's right to choose a research topic. Because the faculty was even more committed to academic freedom, it was indeed "inevitable" that it would allow von Neumann to pursue the work he chose to pursue. Had the $100,000 promised by the trustees come at the expense of other programs, the issue might have been contentious. However, the recent deaths of Louis Bamberger and Caroline Bamberger Fuld had left the IAS awash with money just as the institution was setting out to define its postwar mission. In any event, in the eyes of the senior faculty von Neumann remained a young, amicable faculty member whose reputation had grown substantially during the war. They knew they would lose him should they constrain his work. Thus, despite some faculty members' recent opposition to the experimental sciences, no one stepped forward to oppose von Neumann's efforts to bring not only a laboratory but an engineering facility into the fold of the IAS.[51]

Von Neumann's mathematical contemporaries lamented his departure from abstract mathematics. Yet whatever the outcome, von Neumann's postwar research agenda was a product of intellectual ambition, not of intellectual retreat. In the context of wartime developments, traditional questions in abstract mathematics seemed quite staid. Especially in view of the defeat of the formalist program on which von Neumann had cut his teeth, the new frontiers of mathematics seemed to lie in interdisciplinary areas outside the current agenda of the mathematical research community. Von Neumann's contributions to economics, to game theory, and to stochastic processes, in addition to his contributions to computing and to the theory of automata, suggest how far afield he was willing to go to make a fundamental contribution to knowledge.[52]

For the immediate task of building an electronic computer, which was instrumental to some of this work, von Neumann leveraged the IAS's support to secure comparable commitments from the Army Ordnance Department and from RCA. Aydelotte intended the IAS's contribution to be used just for such a purpose. RCA provided support in kind, agreeing to develop a high-speed memory system based on the cathode-ray tube. The Rockefeller Foundation and the Navy's Office of Research and Inventions took part in the initial conversations. Weaver backed out because the project was entangled with industrial interests. Meanwhile, negotiations with the ORI foundered over the issue of whether the IAS could retain title to the computer and thereby control the choice of the scientific problems it would handle. This prompted von Neumann to broker a deal in which the Navy was asked instead to support the work on dynamic meteorology. This work had clear strategic implications for the Navy, in terms of weather prediction. At the same time, the complexity of atmospheric phenomena made dynamic meteorology an attractive research topic, since the underlying hydrodynamic equations could foster new developments in applied mathematics, and perhaps even abstract mathematics. Far from being an instance where military interests simply shaped academic agendas, this provided an occasion for von Neumann to press the IAS and the ORI in a direction he himself wished to go.[53]

The Moore School and the Ballistic Research Laboratory

In approaching the Army, von Neumann was careful to consider the respective responsibilities of the IAS and the Moore School, given the tensions generated by the EDVAC report. Yet although tensions continued to mount between von Neumann and Eckert and Mauchly, the Moore School and the BRL remained interested in an arrangement that would allow them to remain within von Neumann's fold. The Moore School did not oppose Goldstine's move to the IAS. In fact, the Moore School tentatively accepted an arrangement under which "Pres" Eckert would have served as the chief engineer for both the Moore School and IAS computer projects. More important, it was agreed through mutual consent that the IAS would focus on fundamental design studies, while the Moore School would focus on engineering aspects while working from the original EDVAC design. The BRL, meanwhile, would focus on applied mathematics and the use of digital computers in military ordnance applications. These arrangements suggest that the IAS, the Moore School, and the BRL were all content to extend the roles they adopted during their wartime collaboration.[54]

This is not to say that this trust in a smooth division of labor was sustainable. The Moore School, in particular, faced numerous difficulties during the postwar transition. Foremost among these difficulties was Eckert and Mauchly's ambivalence about staying at the university. But the Moore School, and the BRL, was also hampered by security regulations. This inability to speak publicly was also compounded by the fact that Eckert and Mauchly were not especially eager to publicize their work, at least until their patents were filed.

As it turns out, it was the Army itself, and the BRL in particular, that pushed for the rapid declassification of the work on the ENIAC. The BRL had a good technical reason for continuing to support computer-development work: electronic computers opened new horizons for applied mathematics research. But equally important for the BRL was the ENIAC's publicity value. The BRL mathematicians' esoteric wartime achievements were difficult to document. In contrast, the ENIAC was a highly visible artifact that rivaled the NDRC's best accomplishments. Viewed in the context of demobilization, public disclosure of this work could strengthen the claim that the BRL was an institutional asset that ought to be preserved in peace.[55]

These sentiments were quite evident during the public dedication of the ENIAC, held on 16 February 1946. Substantial declassification enabled the BRL and the Moore School to publicize their accomplishments through a carefully staged event organized with the assistance of the War Department's Bureau of Public Relations. Although a number of press releases focused on the ENIAC and its inventors, considerable attention was given to the wartime activities of the BRL. Most telling was a press release titled "High-Speed, General Purpose Computing Machines Needed," which was really about the significance of applied mathematics in the postwar era: "It took an atomic bomb to make the average man conscious of how far the physical sciences have advanced. No longer is the physicist thought of as a long-haired, wild-eyed, impractical dreamer whose work is of no immediate value. With the release of atomic energy, the public became conscious of the physicist's role in society. . . . It is hoped that the announcement of the ENIAC will bring the same recognition for the mathematician."[56]

Within a month of the public dedication, tensions flared between the Moore School and Eckert and Mauchly, leading to the latter's departure. A year earlier, Eckert and Mauchly successfully secured an exemption from the university's president, George McClelland, which allowed them to retain the patent rights for their work.[57] However, in early 1946, Irven Travis, who again was serving as a contracts administrator at the Navy Bureau of Ordnance during the war, returned to the Moore School to become the new supervisor of research. It is not clear whether Travis felt that Mauchly and Eckert appropriated his ideas, whether he regarded their commercial interests to be incompatible with academic norms, or whether he simply hoped to institute policies consistent with the best practices he saw as a contracts officer. Determined in any case to place the Moore School's patent policies on a uniform basis, Travis demanded that Eckert and Mauchly sign a standard release that would require them to transfer any future patent rights over to the university. This was the specific incident that convinced Eckert and Mauchly to try their hand at entrepreneurship.[58]

Eckert and Mauchly's departure wrecked havoc to the Moore School's computer-development program. Other defections followed, as individual engineers placed their trust in other organizations that seemed better situated to do the most exciting work. Several joined Eckert and Mauchly. Defections to the IAS renewed the tension between the two laboratories. The Moore School tried hiring Samuel Williams, Stibitz's chief engineer for the Bell Labs machines, to serve as the EDVAC's technical director. However, Williams's age and unfamiliarity with electronics prevented him from exercising effective leadership. Williams soon left, leaving the project directly in Travis's hands. Travis eventually brought the project to completion. Still, this work failed to gain much recognition as compared to other efforts.

The National Applied Mathematics Laboratories

During the spring and summer of 1946, negotiations continued over the postwar incarnation of the Applied Mathematics Panel. The immediate issue at hand was the fact that the Naval Research Laboratory, the Bureau of Ships and Dahlgren Proving Ground were all in the process of setting up new computing facilities. Given the ORI's mandate to coordinate naval research, this office convened a Naval Computing Advisory Panel. Among those on the panel were Rees and Stibitz, Stibitz having been the one who advised Dahlgren through the AMP study mentioned in chapter 1. Lieutenant Commander John Curtiss, an applied mathematician stationed with Aiken's group at Harvard, represented the Bureau of Ships. Mathematicians continued to dominate the panel. Owing to his reputation, von Neumann was asked to be a member.[59]

Once this group was assembled, it quickly interjected the AMP's earlier discussion about a postwar program in applied mathematics. Indeed, the basic recommendation that emerged from this panel was to follow through on the earlier plan to establish a national laboratory for applied mathematics. The panel advised that the ORI director approach the National Bureau of Standards (the U.S. counterpart of the British National Physical Laboratory) with such a plan. Fortunately, the director of the NBS, a well regarded nuclear physicist by the name of Edward Condon, was busy laying his own plans for the postwar reorientation of the NBS's research activities. Condon, who was appointed director shortly after the war, inherited a research organization that had expanded immensely through wartime military funds transfers. Condon made it his mission to use this wartime expansion as a means of remaking the NBS into a powerful, peacetime federal research laboratory. The prospect of securing new funds transfers to support applied mathematics research fit neatly in his plans. At this point, Curtiss moved from the Bureau of Ships to the NBS to facilitate further planning.[60]

Opportunism continued. After a basic deal was struck between the ORI and the NBS, further negotiations were opened up to a series of "public discussions." When funds transfers became a contentious political issue several years later, Curtiss would refer to these early discussions in claiming a public mandate for a national laboratory. At the time, however, the principal aim of these discussions was to leverage ORI's support by securing similar commitments from other, principally military agencies.[61]

The early plans for the National Applied Mathematics Laboratories were both a continuation of and a departure from the AMP. On the one hand, two of the units in the proposed organization were a direct outgrowth of the AMP. The NAML's Computation Laboratory would continue the work of the Mathematical Tables Project with personnel who were willing to move from New York to Washington. The Statistical Engineering Laboratory was created to pursue the kind of work performed during the war in Columbia University's Statistical Bureau.[62]

However, the research mathematicians assembled for this new round of conversations also convinced military sponsors that one of the NAML's principal missions ought to be fundamental advances in applied mathematics. Von Neumann drew quite directly on the rhetoric constructed during his negotiations with IAS trustees in arguing that fundamental advances in applied mathematics remained necessary before anyone could take full advantage of electronic computers. The third major recommendation was therefore to create a new institute for research on numerical analysis, with the stipulation that such an institute should be set up at a major research university to facilitate recruitment and to ensure academic integrity. But then in turning this last argument on its head, von Neumann and his colleagues also convinced the military representatives that the NAML would need an electronic computer to remain at the vanguard of research. Trusting the mathematicians' technical judgment, ORI agreed to support a separate machine-development project at the NAML. By no accident, those most involved with wartime applied mathematics work wound up endorsing the same basic arrangement that von Neumann set up at the IAS, namely a program that tied machine-development work to research in applied mathematics and its practical applications. While it took some time to complete all the necessary arrangements, Curtiss proceeded to build up the NBS's applied mathematics and computing activities along these lines until the NAML's official dedication in January 1947.[63]

Programs on the Periphery

Among those most skeptical about the new developments in electronic computing were those at Harvard and Bell Laboratories. Despite his background, Aiken failed to appreciate the full scope of the wartime advances in mathematical physics. Continuing to see applied mathematics from the perspective of an earlier mathematical tables tradition, Aiken saw no need to personally embrace the radical changes in computing machinery. Given, moreover, that his principal sponsor, Dahlgren, wanted to have a machine that could immediately handle their routine ordnance calculations, Aiken deemed electronics to be too unreliable. Aiken proceeded to work entirely with electromechanical devices in building the Harvard Mark II.[64] Meanwhile Bell Laboratories found no compelling reason to sustain its wartime efforts in computing. Driven by the logic of demobilization, Stibitz had left Bell Labs to accept an academic appointment at the University of Vermont. As mentioned previously, Bell Labs did build several additional relay computers based on contracts it accepted during or shortly after World War II. Nevertheless, this did not turn into a program for electronic computers work.[65]

IBM and National Cash Register were driven even more explicitly by the dictates of the marketplace. Office equipment manufacturers belonged to a highly competitive industry, and this created a strong incentive to focus on recapturing market share during the postwar reconversion period. NCR felt it had little choice but to terminate military contract work that did not contribute directly to a sustainable product line. Still, this allowed those who were involved with NCR's wartime cryptography effort to establish Engineering Research Associates. Although ERA would eventually embrace electronic computers, their initial strategy would be based on building highly specialized equipment for the Navy.[66]

IBM's president, Thomas J. Watson, also instructed his engineers to focus their attention on commercial sales. However, Watson paid greater attention to the ascending wartime reputation of scientists, and the possible impact of the wartime work in electronics. Watson allowed his product-development groups to study how electronic devices could be integrated into the firm's existing product line. (See chapter 6.) In March 1945. influenced by the emerging (if misguided) consensus that science was the wellspring of new technologies, Watson, through a special arrangement with Columbia University, created a Department of Pure Science. This department operated as the Watson Scientific Computing Laboratory at Columbia. Watson hired Wallace Eckert to serve as its first director. Eckert was given the resources to build a large-scale computer. However, Eckert and his staff began this work before von Neumann's EDVAC report. And he and the IBM engineers at Endicott froze the design of the Selective Sequence Electronic Computer before the widespread postwar discussion about electronic computer designs. As a consequence, the SSEC was a hybrid, where only the internal arithmetic circuits and registers were driven by electronics.[67]

In view of the computational demands of designing nuclear weapons, Los Alamos was another site that might have made an aggressive entry into electronic computing. However, Los Alamos faced considerable instability in the period after World War II, as most of its scientists returned to academic positions. Moreover, because of the political debate over whether the United States should develop the hydrogen bomb, the core mission of the laboratory remained unresolved. Under these circumstances, Los Alamos chose to trust von Neumann and his work at the IAS, even as the IAS proceeded to amass an extensive collection of older IBM computing equipment. This eventually became a formal agreement. Los Alamos's postwar director, Norris Bradbury, agreed to defer work on an electronic computer until the IAS had demonstrated its feasibility by designing a "prototype" suitable for replication.[68]

The NBS, Eckert and Mauchly, and the Census Bureau

Some of the early arrangements were unstable. For example, the overall plans for the NAML changed as a result of Eckert and Mauchly's entrepreneurial efforts. As was mentioned in chapter 2, the first contract Eckert and Mauchly got was with the Census Bureau. J. C. Capt, the bureau's director, and James McPherson, the administrator responsible for the bureau's tabulation operations, liked the idea of using an electronic computer to process the 1950 decennial census. Moreover, that idea was potentially even more important to some of the bureau's other routine tabulations. Unwilling, however, to draw from existing programs, Capt appealed directly to Congress for a new appropriation. Having contributed to Herman Hollerith's efforts in the 1880s, the Census Bureau saw itself as a technologically forward-looking organization. A small appropriation would allow it to resume leading the United States in its use of a revolutionary administrative tool. In the wake of hearings about a supplemental authorization for the Census Bureau during the summer of 1946, Congress authorized the bureau to spend $500,000 on the procurement and experimental use of an electronic computer.[69]

Having cast itself as a technologically innovative organization, the Census Bureau had little more than a machine shop to service its aging tabulating machines. It had no experience with electronics, nor did it have the legislative authority to issue a research and development contract. However, the NBS's charter allowed it to perform technical services for other federal agencies, and it had the authority to oversee research contracts. At the Census Bureau's request, the NBS agreed to serve as a contract monitor for a procurement contract that called for the development of an electronic computer. The Census Bureau allowed the NBS to extract a significant overhead from the $300,000 transferred to the NBS in order to carry out whatever technical evaluations were necessary to supervise the contract. The NBS decided, meanwhile, that the NAML's own computer could also be built by Eckert and Mauchly's Electronic Controls Company.[70]

The NBS's decision, especially on the latter arrangement, requires closer scrutiny. The plans for the NAML called for the NBS to have its own machine-development laboratory. But Curtiss was a research mathematician with no particular attachment to electronics research. His own experience at Harvard, along with the planning meetings for the NAML, made it clear that building an electronic computer would be anything but a simple undertaking. It was at this point that the Census Bureau arrived with the funds for a second computer *and* a contractor that happened to be the inventors of the first electronic computer. Curtiss reasoned that by issuing two contracts to a single manufacturer, he could reduce the risks of research and development. Curtiss trusted Eckert and Mauchly and accepted that the NBS's primary task would be that of supervising work in computer development.

Other historians have suggested that this change in the NAML's mission may have also resulted from the direct intervention of the NBS's director, Edward Condon. Condon had a rather complex relationship with the military despite his expansionist plans that required a good working relationship. After the war, Condon joined Leo Szilard and other scientists in their efforts to secure the civilian control of atomic energy. He spent much of his time in late 1945 testifying before the Senate Select Committee on Atomic Energy. These statements drew the ire of certain factions within the military establishment. But there were other groups within the military that remained eager to work with the NBS. Condon, for his part, remained committed to the idea of military funds transfers. Though not entirely consistent with Bush's postwar plans, these transfers at least placed military money back into civilian hands. By the same token, Condon welcomed the opportunity to channel these funds out to private industry when the opportunity arose. This was consistent with the NBS's obligations as a unit of the Department of Commerce. Although it is not clear to what extent Condon intervened directly, both he and the general culture of the NBS created an environment that favored the NAML's shift to an implicit policy favoring the privatization of research and development. The NAML's now-misnamed Machine Development Laboratory became, in effect, a federal procurement service for electronic computers.[71]

All these negotiations turned Vannevar Bush's policies on their head. Columbia University agreed to play host to a private manufacturer's basic research facility. The IAS and Harvard University, despite their stated intent to curtail military research, allowed their faculty members to continue to embrace military development work. At the same time, military institutions and industrialists found themselves struggling to uphold a rhetoric of fundamental research. The Army's support for von Neumann's work was carefully crafted along these lines. Although the Navy's Office of Research and Inventions initially expressed reservations about such an arrangement, that office's successor, the Office of Naval Research, wound up supporting basic research in applied mathematics through a civilian, governmental research facility located at a public university. The Office of Research and Inventions also agreed to fund machine-development work at the NBS. But then, through a curious turn of events, the NBS permitted these military funds to be re-directed to private enterprises. This was hardly the centralized planning process envisioned by Bush. Indeed, there emerged no clear demarcation between military and civilians, or between public and private interests. Postwar arrangements for research in computing and applied mathematics were as complex as the negotiations concluded at each juncture.

Still, these negotiations reconstituted computing as a whole into a trading zone—albeit one rife with institutional politics and intersubjective evaluations that put certain groups at the center and others on the periphery. At the center, there were already the substantial beginnings of a coherent network, with many groups clearly aligned to von Neumann and his work at the IAS. This included the Moore School, the BRL, RCA, and the Army Ordnance Department, along with Los Alamos and a substantial part of the work at the NBS. On the periphery were the groups at Harvard, MIT, Dahlgren, IBM, and Columbia, in addition to Eckert and Mauchly's startup. Each of these organization defined its separation from von Neumann in a different way. Still, even the groups furthest out on the periphery could no longer deny that they were engaged in a common endeavor. No group could afford to ignore the work undertaken by other groups. On the other hand, different disciplinary agendas and divergent interests continued to prevail even among the groups most closely aligned to von Neumann, as anticipated by Galison's notion of a trading zone.

Social relations of trust also sustained this emerging network. Both the ORI and IBM placed their trust in the value of fundamental research, even as the faculty and trustees at the IAS trusted von Neumann to pursue inquiries that would uphold their institution's commitment to fundamental knowledge. And although the arrangement was more ephemeral, the IAS, the Moore School, and the Ballistic Research Laboratory agreed to partition their respective responsibilities in terms of design, engineering implementation and ordnance-related research. More broadly, those at Dahlgren, Los Alamos, and the NBS trusted others to build electronic computers, even as the mathematicians who headed machine-development programs—von Neumann and Curtiss among them—trusted engineers to deliver the equipment they needed to perform their research. all this made computing a very lively trading zone, one that defined the boundaries of a new interdisciplinary field.

From Trading Zone to Coherent Network

Subsequent events transformed the trading zone into a coherent network built around von Neumann's ideas. Describing *how* this occurred requires a more detailed, process-oriented account of local choices made under very different circumstances, something which I do in chapters 4–6. In those chapters, I provide a detailed case study of two "new" entrants at the NBS and MIT, and of the ongoing work at IBM. These chapters examine the nuances of how different groups came both to uphold and challenge von Neumann's assumptions about digital computers research. Here, I merely set the stage by providing a series of static or "synchronic" descriptions of the overall field between 1947 and 1949. This will also bring this chapter on von Neumann to at least a contingent closure.

Fortunately, it is easy to document the growing dominance of von Neumann vision for computer-development research. Because of the enthusiasm surrounding digital computers, there was a continuous series of computing conferences between the end of World War II and the early 1950s. Supplementing the historical record is also a series of more critical studies and investigations, which became more common after unforeseen technical challenges caused significant delays in nearly all U.S. computer-development projects. In speaking about the early computer conferences, William Aspray notes in his introduction to a reprint edition of the 1947 Harvard Symposium on Large-Scale Digital Calculating Machinery proceedings that such events offer a valuable snapshot of an emerging field.[72]

The choice of the specific conference and studies examined here are somewhat arbitrary. They consist of the 1947 Harvard symposium, a study produced by Engineering Research Associates between 1947 and 1949, and a study released by the Department of Defense's Research and Development Board in late 1949. It is necessary to acknowledge that these are all problematic sources. Even academic conferences are carefully staged events, as seen in the case of the MIT gathering. Yet, while it is important to consider the perspectives of those organizing a particular conference or study, the increasing prominence of von Neumann's ideas can be discerned quite clearly in spite of the apparent perspectives and biases. Again, I do so in very simple terms below, leaving the more subtle analyses to the following chapters.

The 1947 Harvard Symposium

The Harvard symposium was held on the occasion of the dedication of the new Computation Laboratory at Harvard, held at a time when Aiken was about to commit himself to the design of yet another machine for Dahlgren. This time Aiken was willing to consider electronics, but he followed Wallace Eckert's approach in choosing a hybrid design for the Harvard Mark III. Dahlgren's Commanding Officer, C .T. Joy, opened the symposium with the statement that he wanted Aiken to propose a machine of "any particular type, mechanical or electronic."[73] Nevertheless, those at Dahlgren were quite concerned about Aiken's technical conservatism. Dahlgren's decision to co-sponsor—indeed insist upon—a symposium was a means by which the Navy could hold Aiken accountable for decisions that seemed at odds with projects elsewhere.

Aiken attempted to skew the program as indicated by the very title of the symposium. It focused on "calculating machines," not computers. Moreover, the conference program was built around the machine designs, mathematical techniques and components that could support Dahlgren's immediate need for computing. This focus on immediacy allowed Aiken to restrict the session on machine design to "Existing Calculating Machines," thereby excluding all the work on stored-program computers. There was also an implicit disciplinary hierarchy to the program, with machine design and mathematical methods receiving greater emphasis, while electronic and non-electronic subsystems, aside from high-speed memory, were relegated to the second half of the symposium.

Yet, despite this effort to defuse the interest in electronic computers, the proceedings tell a different story. Many of the presentations on new memory systems and input-output devices, and even the papers on numerical analysis, were predicated on the desirability of all-electronic computers. The work on mercury delay lines, and the even faster CRT-based memories being developed by MIT and RCA would have been improperly matched if placed in a machine where instructions were delivered at electromechanical speeds. Over half of the presentations on numerical methods were about hydrodynamic equations or large systems of linear equations, where the complexity of the scientific problems that a machine could handle was strictly bounded by the size of the internal memory system designed to store intermediate results. Moreover, in looking at the list of attendees, it is clear that a majority of those attending the symposium did so because of their interest in electronic computers. With 336 people registered, as opposed to the much smaller group that met at MIT, it was also clear that digital computers were beginning to draw a large audience.[74]

The Engineering Research Associates Survey

To Aiken's credit, all the electronic computer projects in the United States ran up against technical difficulties and bureaucratic challenges that seriously hampered their progress. The difficulties with one of these projects, Jay Forrester's work at MIT (see chapter 5) prompted the Office of Naval Research to encourage Engineering Research Associates to produce a general survey of research on digital computers.[75]

Engineering Research Associates' director of research, Charles Tompkins, agreed to conduct this survey just as ERA itself was deciding to enter the arena. Technically, the ONR's task order had a much narrower purpose—a study of memory systems—but it gave Tompkins an open license to visit all government sponsored computer-development projects, including the private-sector efforts of Eckert and Mauchly, and Raytheon. This should have raised concerns about a conflict of interest—by August 1947, ERA already had a separate task order for the paper design of a general-purpose, stored-program computer.[76]

The resulting survey, *High-Speed Computing Devices* (1983/1950), is also a somewhat problematic source from the standpoint of the present analysis. Tompkins conducted his visits and assembled the information he needed during 1947 and 1948, but did not publish his report right away. Two different ERA staff members made significant revisions to the text during 1949, after Tompkins had left the firm. This raises complex issues of authorship, and questions about the exact timing of this window into digital computers research. Nevertheless, the broader context of the report—a private manufacturer's desire to make a competitive entry into digital computers research—lends a discipline to the text that makes it possible to recover a consistent treatment of von Neumann's ideas.[77]

Composed as a primer for those interested in high-speed computation, ERA's survey was quite ecumenical. It gave equal attention not only to Aiken and Wallace Eckert's efforts, but to the desk calculator and punched-card computing systems that remained integral to technical computing work. On the surface, the book was also descriptive, not proscriptive; at no point is the text written in a directly critical or evaluative tone.

Yet, as with the proceedings from the Harvard symposium, there remained an underlying order to the text. The central chapters on computing systems offered a steady progression from manual methods to large-scale digital computers, as organized around the principle of speed. The text also tied the choice of the "right" equipment to a careful study of the intended application, and placed considerable emphasis on problems such as compressible fluid flow and non-linear dynamics that would have favored electronic computing speeds. Moreover, in an implicit critique of Howard Aiken and Wallace Eckert's designs, this study explicitly stated that a high-speed memory system ought to be used along with electronic components, if the goal was to solve complex mathematical problems that required high-speed computation.[78]

In all these respects, ERA's study re-encoded many of von Neumann's ideas, ideas that could be traced back to the EDVAC report. This included the close integration of computer design and applications research, a systematic attention to design tradeoffs as evaluated against the intended application, and above all else a strong interest in advanced applied mathematics problems that demanded electronic computers. Not surprisingly, important passages of the ERA study were based on a direct, if not entirely careful reading of Burks, Goldstine, and von Neumann's *Preliminary Discussion of the Logical Design of an Electronic Computing Instrument* (1947). The ERA study cited this report, along with Goldstine and von Neumann's subsequent volumes on the planning and coding of problems,[79] which were the most significant technical reports produced by von Neumann's group. ERA's own entry into the field—Atlas[80]—would use a magnetic drum, as opposed to mercury delay lines or CRT memory for internal storage. But this made sense in view of the firm's technical capabilities and the difficulties other groups were experiencing, especially with CRT-based memories. As a new commercial entrant into a competitive market, ERA had no reason to remain wedded to old technologies. Nor could it afford to. It used the ONR-funded study to survey and settle on an efficient machine design, and they drew heavily on von Neumann's work.

Computer-development work proceeded through the entire period spanned by ERA's study. Yet, despite the lavish funding bestowed on U.S.-based efforts, it was two British academic laboratories, F. C. Williams's group at Manchester and Maurice Wilkes's at Cambridge, that built the first stored-program computers, in 1948 and 1949 respectively.[81] As we will see in chapter 4, it was not until June 1950 that the first stored-program computer entered useful operation in the United States, and this through the "new" entrant at the NBS that did not fully get its start until late 1948.

The Research and Development Board's Report

It was these delays that prompted the Research and Development Board to conduct a survey of the nation's computer-development programs in late 1949. Established initially as the Joint Research and Development Board in 1946 (it was a joint services initiative that replaced the RBNS), the RDB was officially transferred to the Department of Defense when the latter was set up in 1947. The RDB's official mission was the same as for the RBNS, namely to coordinate the national program in defense R&D. With the simultaneous entry of so many groups into the field of digital computers research, this presented an obvious candidate for review. The RDB was chaired by Vannevar Bush.[82]

Originally, the RDB created a Sub-Panel on Computing Devices within its Committee on the Basic Physical Sciences. In setting up this sub-panel, the CBPS's executive secretary, Martin Grabau, was instructed to approach the "more or less obvious" candidates. Grabau consulted directly with von Neumann, whereupon von Neumann basically appointed his lieutenant, Goldstine, to chair the sub-panel. This allowed von Neumann to deflect criticisms of his and other projects, specifically by directing the sub-panel to survey the national need for computers before conducting a critical study of U.S. efforts in computer development. Unfortunately for von Neumann, these plans were disrupted by former MIT president, Karl Compton. When Bush became ill in 1948, Compton left MIT to replace Bush as the interim chair of the RDB. One of his first actions was to request a direct assessment of U.S. computer-development projects carried out by an independent ad hoc panel. This too resulted from the difficulties that were unfolding at MIT. While I leave the detailed discussion of the origins of the ad hoc panel and its findings about Forrester to chapter 5, here I wish to focus on the general aspects of the study, including the response it elicited from the emerging computing community.[83]

One important thing to note about the ad hoc committee, and the RDB itself, was that they were both manifestations of the "management and organization" movement that spread through corporate and public administration circles in the decade after World War II. The CBPS specifically charged the ad hoc panel to survey the nation's digital electronic computer efforts "from the management point of view." For this reason, as well as for reasons of objectivity, the members of the ad hoc panel were not eminent figures in computing, but consulting engineers and engineering managers with substantial industrial experience. Grabau enlisted Gervais Trichel, a former Army Ordnance officer who was now an assistant to the General Manager at Chrysler; Lyman Fink, an electronics engineer at General Electric; and Harry Nyquist, a control-systems engineer at Bell Labs. Nyquist was also instrumental in bringing a separate, systems engineering perspective into this review of digital computers research. The "management and organization" perspective, meanwhile, was conveyed to the panel through the efforts of Grabau, who contributed directly to the work of the ad hoc panel.[84]

The panel's report was written in a familiar, investigative tone that sought to uncover duplication, waste, and inefficiency. But despite its criticisms of the individual projects, the report was quite mild in its overall recommendations. While suggesting that the national program in digital computers research was "far from optimal," the report suggested that it was best to sustain most of the projects already underway because of the sunken investments. Recognizing that they had only limited knowledge of digital computers, the panel constructed several abstract criteria upon which to judge the merits of different projects. Drawing on a perspective von Neumann himself promoted, namely that of tying computer design to computer applications, the panel constructed the criteria of "end use" to determine whether a project had a clearly defined application that steered its work. They also drew on von Neumann's (and as it turns out, Forrester's) notion of a "prototype," in arguing that no contract to build multiple copies of the same machine should be issued until a machine had entered into reliable operation.[85]

Yet, as reasonable as these claims might have seemed, the ad hoc panel's lack of technical knowledge about computers placed them in a bind. Their criteria for a "prototype," led to the conclusion that only the Harvard Mark III was suitable for replication, which they recommended that the services do. Many in the field, beginning with von Neumann, ridiculed this suggestion as a recommendation to replicate an obsolete machine. Meanwhile, those committed to the idea of pursuing basic research on machine designs criticized the panel's definition of "end-use," which left no room for such work.[86]

In fairness to the ad hoc panel, it should be noted that the panel simply assembled the critiques that researchers were beginning to levy against each other as perpetual delays created considerable pressure for each group to be more accountable for their work. However, in criticizing just about every project, and by implication, the good judgment of their military sponsors, the report brought the fledgling computing community into a unified defense. This once again placed Goldstine's CBPS sub-panel in a privileged position. Goldstine specifically moved to have the ad hoc panel's report referred down to them for consideration. The military representatives to the larger CBPS concurred, as did T. K. Sherwood, the dean of engineering at MIT, after insisting that the sub-panel be augmented to include representatives from other computer-development projects and sponsoring agencies. This maneuver was overturned by the incoming chair of the RDB, William Webster. However, Webster also instructed the ad hoc panel to convene a meeting at which it would "hear the views of representatives from the Departments and of the contractors."[87]

Ultimately, it mattered that the ad hoc panel seriously transgressed its authority. Driven by its desire to establish managerial efficiency, the panel's main recommendation was to institute a central system of oversight, in which all computer-development contracts over $25,000 would be subject to an annual review by a permanent panel established under the CBPS. However, this exceeded the RDB's statutory authority, which was to merely advise and coordinate the work of the military agencies. Instructively, Webster sent the ad hoc panel's final report to the RDB's military representatives with the instruction that military agencies should examine the report as "an instrument for program guidance."[88]

Still, the broader effect of the report was to galvanize the computing community around von Neumann's ideas, even as they continued to endorse diverse approaches to digital computers research. How this failed to bring complete closure around von Neumann's vision of computing research is considered more carefully in the subsequent chapters.

The IAS Electronic Computer Project

It is important to note the disjuncture between the prominence of von Neumann's ideas and the fate of his own machine-development project. Only by doing so is it possible to understand what was at the heart of von Neumann's network.

From the outset, the IAS's Electronic Computer Project was fraught with minor difficulties. When the project was initiated in 1946, the IAS had very little space, let alone experimental facilities, to offer. Most of the first year was spent on such mundane tasks as erecting a building and assembling the machine shop necessary to fabricate mechanical components—pieces of the input-output equipment, the overall housing for the computer and the like. Acquiring a competent staff also presented a significant challenge. Although von Neumann hoped to attract five or six engineers and a dozen wiremen, it proved difficult to draw an experienced technical staff to the intellectual oasis of Princeton, NJ. Von Neumann was able to gather a full complement of engineers by June 1946, including two engineers from the ENIAC project, Robert Shaw and John Davis. But by the end of the year, both Shaw and Davis had left to pursue better opportunities elsewhere.[89]

Shaw and Davis's departure was also tied to an implicit disciplinary hierarchy within the project. Because of the project's emphasis on fundamental design studies, and because of von Neumann's sense of the proper temporal order for the work, the design studies received the greatest attention during the early months.[90] As indicative of the higher status that even the engineers accorded to design, the annual reports describing the engineering aspects of the project were titled "Interim Progress Report on the Physical Realization of an Electronic Computing Instrument."[91] At the IAS, engineers clearly worked within the tradition of technicians charged with building a scientific instrument to specifications set by a scientist.

Nor was this hierarchy simply a matter of stature. It affected the disciplinary practice of engineering. Von Neumann wound up hiring Julian Bigelow to be his chief engineer after rescinding the offer he extended to J. Presper Eckert. Bigelow was an electrical engineer who had worked for Wiener under Wiener's NDRC contract, and as such, he was already accustomed to building instruments to a mathematician's specifications.

It remains difficult to pinpoint the exact consequences this had for engineering practice. Nevertheless, as compared to the wartime efforts of the Moore School, Bigelow tended to be more methodical and sequential in how he ordered his tasks. He and several engineers on the project placed considerable emphasis on the mechanical transport of the magnetic wire used for secondary storage, after determining that this was an important unsolved problem. This work was done at the expense of a more robust program on electronic circuit design. There are also indications that Bigelow took at face value the claim, found both in the draft EDVAC report and Burks, Goldstine, and von Neumann's *Preliminary Discussion* (1947) that a digital computer would be but a fairly straightforward combination of basic circuits such as an adder and shifting registers. Bigelow deferred much of the work on the control circuits, the power supply, and other elements of a complete system until January 1948. By then, others, especially Jay Forrester at MIT, were well on their way toward building an entire computer. While more a difference of degree than of kind, Bigelow saw his work explicitly as a "Program of Experimentation" where "experimental components work" was placed ahead of an "operable design" and the actual construction of the machine.[92]

At a more basic level, Bigelow was affected by the IAS's institutional emphasis on freedom of inquiry and fundamental advances in knowledge. Bigelow therefore gave his engineers considerable autonomy, without regularly drawing together and assessing the group's overall progress. He also began to push for independent investigations initiated by members of his group, and began to view certain aspects of the IAS computer as being within the purview of an electronic engineer's design expertise. He began to clash with Goldstine, who had stepped forward to act as the project's general manager amid von Neumann's frequent absences. All in all, the group's monthly progress reports suggest that the engineers at the IAS did not emerge as a cohesive team until the final phases of the project, when external pressures drove the project to completion. By 1951 the engineering leadership had also transferred to James Pomerene, after von Neumann arranged for Bigelow to receive a Guggenheim fellowship to pursue independent research at the IAS.[93]

The IAS, in the end, was not very conducive to engineering work and the feelings in this respect were mutual. However, it is important once again not to read history backwards. Certainly, there were early indications of awkwardness and tension, as when the faculty expressed its concern that the project would have an adverse effect on the "atmosphere" of the institute. Early on, even Aydelotte remarked innocently that the only way he could immediately "dispose" of von Neumann's "workers" would be to place them in the basement next to the men's lavatory. Nevertheless, the Electronic Computer Project was tolerated, and there was little in the way of hardened attitudes early on.[94]

One of the first serious issues to arise concerned hiring. Permanent appointments at the IAS were reserved for a very select group of faculty. The remaining "members" all held PhDs, and by custom such scholars were invited to be in residence for one or at most two years. The only group clearly exempt from this rule was the permanent staff of the institute, which consisted of secretaries, librarians, and custodians. This created a category problem: How should the permanent staff of the ECP be characterized?

The records of the IAS seem to indicate that many seemingly trivial things did as much to harden the attitudes of those within the institute as did overt discussions of policy. Preserved within Aydelotte's letters to von Neumann is a sense of his growing frustration with von Neumann's staff. The IAS maintained a quaint, British tradition of serving tea in the afternoon. Because of the distance that separated Fuld Hall (the main building at the IAS) from the building that housed the ECP, those associated with the ECP began serving tea in their own building. On one occasion, the ECP staff members ran out of sugar, and one of the meteorologists went to Fuld Hall to raid its sugar supply. Aydelotte drafted a letter to von Neumann complaining that the sugar had been carefully rationed, and moreover that he did not approve of ECP staff members helping themselves to multiple serving of tea and having the audacity to take their tea back to their desks. Aydelotte wrote that he was disturbed by the lack of order and discipline in the computer project and that the atmosphere of the institute was "a little bit injured by its proximity to the Institute."[95]

Still, the ultimate reason for the ECP's demise was von Neumann's own departure from the IAS. Von Neumann made many significant intellectual contributions after World War II—many mathematicians continue to regard his corpus of postwar work as rivaling that of his earlier contributions to abstract mathematics. Nevertheless, after the war, von Neumann found himself drawn increasingly to his consulting work. This included his continued involvement with the BRL, the Navy Bureau of Ordnance, and Los Alamos. He also held new consultancies with the RAND Corporation, the Weapon Systems Evaluation Group in Washington, the Central Intelligence Agency, and about a dozen other government agencies and corporations. These consultancies also included all the major nuclear weapons labs from Oak Ridge to Sandia. Von Neumann then served on the General Advisory Committee of the Atomic Energy Commission from 1952 to 1954, and on the AEC itself from 1954 to 1956. On top of this, von Neumann secured a consulting arrangement with IBM that earned him an annual income of $12,000 (his IAS salary was around $15,000), for no more than 30 days of work a year.[96]

Historiographically, this consulting work has been described as a kind of adornment on a prestigious career. Yet, from a social historical perspective, von Neumann's intellectual achievement and his consulting work must be viewed integrally. Coming out of World War II, the scientific émigrés in the United States emerged with a very different social (and technical) stature. The interwar reputation of abstract mathematics, rooted in a European tradition of arts and letters, faded along with the destruction of the old world. But, as von Neumann walked through the corridors of Washington and through the nation's new weapons laboratories, he could feel the same sense of power and prestige that drove him to the study of mathematics. This was not so much a corruption of his intellectual commitments, as it was an extension of his social ones. The military served as a surrogate "society" in which von Neumann could place his social aspirations.

In this respect, von Neumann's appointment to the Atomic Energy Commission represented the pinnacle of his career. Von Neumann's appointment log, as an AEC commissioner, documents the steady stream of renowned scientists, congressional representatives, military officials, and other important visitors who came to him for advice. They all asked him to apply his mind to vital issues of nuclear weapons and international security. Von Neumann enthusiastically attended the atomic tests conducted at the Bikini atoll back in 1946. In 1955 he was diagnosed with cancer. As he lay confined to a bed in the Army's Walter Reed Hospital, a steady stream of visitors continued to arrive, offering encouragement. For von Neumann, the greatest frustration in his life was that of having to bear an excruciating illness that slowly tore away at his mind.[97]

The IAS faculty did not wait until von Neumann's death to determine the fate of the ECP. With his departure to the AEC, the rest of the faculty felt free to discuss the future of the project. In von Neumann's absence, it was easier to raise the question of whether the ECP would continue to make—really, whether it had ever made—a fundamental contribution to knowledge. This question was complicated by the work in dynamic meteorology. While this work was clearly rooted in military and civilian applications, von Neumann had once again placed his hopes on the fact that the hydrodynamic complexities of the atmosphere might necessitate fundamental advances in applied mathematics and mathematical physics. However, it had become easier over time to view this work as having been driven by practical considerations. Meanwhile, the broad consensus surrounding von Neumann's ideas about computer design clearly reduced this work to an engineering effort.

Concerned nevertheless with due process, the IAS faculty appointed one of their own members, Freeman Dyson, to survey von Neumann's peers by posing the same question they had posed to themselves. Yet both the questions offered and the respondents chosen for the survey—Subramanyan Chandrasekhar in astrophysics, Edward Bullard in geophysics, and M. J. Lighthill and Geoffrey Taylor in meteorology—tended to ensure a foreordained result. The final dissolution of the ECP did occur only after von Neumann's death. Jules Charney, who headed the meteorological work, transferred to MIT in 1956 as the director of a new project there. The IAS agreed to keep operating von Neumann's computer only for a year, with the hope that Princeton University would then take over the machine to provide computing services to its faculty. By pushing for this outcome, the IAS faculty ensured that their institute would remain a bastion of abstract knowledge, this despite the broader context of the Cold War that drew many institutions in the opposite direction.[98]

Figure 4.1
The NBS's Standards Eastern Automatic Computer, which was sometimes known by the alternative name shown in the photograph. Courtesy of Charles Babbage Institute, University of Minnesota, Minneapolis.

4 Knowledge and Organization: Redefining Computer-Development Research at the National Bureau of Standards

In this chapter and the next, I take a look at two groups that got their start by aligning themselves closely with von Neumann, only to find themselves diverging from the work at the Institute for Advanced Study as their work progressed. This chapter is about the computer-development efforts of the National Bureau of Standards. The next one is about Jay Forrester's work at MIT. In both cases, these research programs would come to exhibit, to borrow Wiebe Bijker's phrase, a more "limited degree of inclusion" within the "technological frame" represented by von Neumann's vision for computer-development research.[1]

These chapters also draw attention to some of the specific dynamics that can unfold within an ecology of knowledge. This chapter is built around the question of how the structure of administrative organizations, inter-organizational commitments, and the politics of organizational decision making affect how research is done, and conversely, how research organizations become transformed during the course of successive judgments about its work. In the chapter on Forrester, I focus more closely on inter-organizational relations, and on the codetermination that occurs between esoteric knowledge, institutional objectives, and the commonplace rhetoric that researchers use to sustain the validity of their work. Neither of these chapters is built exclusively around these questions. I consider matters of historical context, disciplinary hierarchy, contractual relations, and other things as necessitated by an ecological view of knowledge. Nevertheless, both chapters aim to elicit specific observations about the process of technical and institutional change.

The question of how organizational structures influence the course of technological development has been addressed most directly by organizational sociologists. Robert Thomas's 1994 book *What Machines Can't Do:*

Politics and Technology in the Industrial Enterprise would probably be the work most familiar to historians of technology. Thomas draws explicitly from the science-studies literature, including the ethnographic approach of laboratory studies. Drawing also from broader symbolic-interactionist traditions, Thomas brings to his study of engineering organizations a keen eye for the politics that occur inside a workplace and how this affects the research-and-development process. For those familiar with Thomas's work, this chapter should present nothing too surprising about the internecine negotiations that come to define technological projects and organizational responsibilities.

This chapter differs from Thomas's work mainly in that it employs a historical time scale to make visible certain processes that remain inaccessible to accounts relying exclusively on ethnographic techniques. Also, even though I focus on the NBS, this account moves beyond a firm or an organization as the exclusive unit of analysis. It is necessary to reiterate that there are definite limits to what can be done with historical material. Nevertheless, by maintaining an ethnographic eye even in approaching historical sources, it is possible to consider how the politics of organizational decision making at the NBS, and external influences on these decisions, gave shape to the machine-development work that unfolded within one federal research laboratory. Equally important, I show how these negotiations transformed the organizational structure and conduct of the NBS, and its capacity for carrying out such work. In this respect, this chapter takes the kind of negotiations Thomas describes in his book out to a point of relative stability, or closure, to again draw from Bijker's (1995) theorization of sociotechnical change. Thus, while I build on Thomas's insights, this chapter should provide, in the end, a more extended and expansive view of organizational dynamics and technological change processes.[2]

The historical narrative for this chapter unfolds around an evolving tension between two groups at the NBS. The first was the Machine Development Laboratory, set up within the NBS's new National Applied Mathematics Laboratories. The second, a fledgling group of electronics engineers set up within the NBS's Ordnance Development Division. At first, it was clear that the MDL held greater authority given the prominence of applied mathematicians in postwar computing. The MDL also oversaw the NBS's popular computer procurement service. However, the contractual obligations that emerged from this work bound the NAML to certain responsibilities for which the relevant expertise began to reside increasingly with electronics engineers. This allowed Samuel Alexander, as the head of the Electronic Computers Group, to slowly usurp most of the NAML's authority over machine development. Moreover, Alexander used what amounted to a very opportune position to build the NBS's own stored-program computer. The Standards Eastern Automatic Computer was the first machine of this type to go into reliable operation in the United States. Those familiar with the history of computing will know that computer-development work quickly became the purview of industry, most notably of IBM. Nevertheless, the negotiations over computer-development work that unfolded at the NBS contributed directly to the definition of what a civilian, federal research laboratory could do in an era of unprecedented federal interest in research.

The Electronic Computers Group at the NBS

John Curtiss had already reoriented the NBS's responsibilities with regard to machine-development work. But these responsibilities shifted again as a result of a group of electronics engineers at the NBS. During the war, the National Defense Research Committee enlisted the aid of several electrical engineers in the NBS's Ordnance Development Division (ODD). They had worked on a variant of the proximity fuse recently developed at Johns Hopkins University. Despite the National Defense Research Committee's involvement, the work was financed by the Army Ordnance Department—the same agency that supported the work on the ENIAC. Thus, when the ENIAC was publicized in early 1946, these engineers combined a hobbyist's curiosity with their newfound electronics skills to build some model digital computing circuitry. This group came to be headed by Samuel Alexander, who had been elsewhere during the war but had returned to lead these engineers into the work on electronic computers.[3]

This group's initial interest became far more serious when it became obvious that the NBS, as a major federal research laboratory, possessed many different forms of expertise that could be relevant to computer-development work. In addition to those engineers who worked on proximity fuses, there were engineers within the ODD that had worked on high-frequency radio-frequency circuits, radar displays, magnetic materials, and dielectric substances, all of which seemed relevant to digital computers.[4]

It was von Neumann himself who helped to draw these groups into his fold. Von Neumann knew of the work at the ODD because of his involvement with the Naval Computing Advisory Panel. As indicated in the last chapter, both he and the Moore School were experiencing staffing difficulties during the early part of 1946. Because they shared an affiliation, von Neumann asked the Army Ordnance Department to convene a meeting to consider how the NBS could contribute to the work.

There was a clear hierarchy to the ensuing arrangements. Although his own group was engaged in fundamental computer design studies, von Neumann reasoned that for the purpose of the engineers at the NBS, computer systems design was now "reasonably well advanced." He therefore encouraged the NBS's engineers to work on computer components that could be utilized by the IAS and Moore School projects. While this clearly reinforced the prevailing hierarchy between mathematicians and engineers, NBS engineers welcomed the arrangement. Not only was this consistent with their current level of expertise, but it gave them a definite window through which to join the national computer-development effort. It also gave them privileged access to two of the most prominent postwar computer-development programs.[5]

This new arrangement ended the applied mathematicians' exclusive license over machine-development research at the NBS. Alexander and the other engineers within the ODD were able to secure a two-year contract from Army Ordnance, in the guise of a transfer of military funds, which allowed them to work on electronic components, computer memories, and input-output devices. The organizational jurisdictions that separated the ODD from the National Applied Mathematics Laboratories allowed Alexander to conduct this negotiation without the explicit approval of Curtiss's group. (The fact that the NAML did not technically exist yet also limited its ability to intervene.) On the other hand, the administrative ties that emerged between the two groups were somewhat more complex. From the outset, Curtiss had drawn Alexander into his planning meetings in the hopes of benefiting from his electronics expertise. Moreover, when the NAML was formally established in January 1947, and an Applied Mathematics Executive Committee set up to advise this body, Alexander's group was made to also report to this external committee of scientific advisors. These organizational entanglements would have specific consequences for the course of machine-development work at the NBS.

The Federal Government as Contract Monitor

During the early part of 1946, Curtiss was busy grappling with the question of how the NAML could serve as a technical monitor for the pending contract with Eckert and Mauchly's Electronic Controls Company. Truth be told, both he and his fledgling staff knew very little about electronic computers. At this point, Curtiss turned to George Stibitz instead of Alexander. Stibitz seemed very knowledgeable about computer design, and he was an important participant in the naval advisory panel meetings. Curtiss asked Stibitz to help him evaluate the proposal Eckert and Mauchly submitted based on the new arrangement with the Census Bureau.

Historical accounts have mostly treated Stibitz harshly, portraying him as a conservative naysayer to the NBS's work on digital computers. Stibitz was indeed critical of the Census Bureau and the NBS's decision to lend their support to Eckert and Mauchly. However, so long as the focus of this study is on the process of innovation, including the circulation of knowledge, it is helpful to consider what information Stibitz did convey to the NBS, especially on how to play the part of a monitor of technical contracts. Stibitz, after all, was a technical aide to the NDRC's fire-control division, and he had received direct tutelage from Warren Weaver on the art of managing complex research-and-development projects.[6]

It does remain important to consider how Stibitz's earlier evaluation of the ENIAC colored his interpretation of the proposed new work. In his exchanges with Army Ordnance, Stibitz had challenged the Ballistic Research Laboratory's exclusive interest in electronic computing speeds. Drawing on the same language that von Neumann later espoused for tying computer-development work to its applications, Stibitz had suggested that BRL pay greater attention to a machine's overall "load-bearing capacity." After the war, Stibitz had changed his mind about the overall value of electronic computers; he was among those who fully embraced electronic computers research for its potential. Nevertheless, Stibitz shared the concern, held by Aiken, Caldwell, and others that those seeking to perform immediate computing work might be better served by an intermediate solution that made more limited use of electronic circuitry. Since the Census

Bureau's interests clearly were grounded by a specific and substantial application, the 1950 decennial census, Stibitz again expressed his doubts about whether an electronic computer was the ideal solution. Stibitz granted that Eckert and Mauchly probably knew enough to build a good electronic computer. However, he roundly criticized them for beginning with the proposed solution rather than starting with a careful study of the Census Bureau's needs.[7]

Stibitz reserved his harshest criticism for the NBS's willingness to issue a procurement contract for what remained a research-and-development effort. If nothing else, his experiences with the NDRC warned him of the dangers of such an arrangement, especially when the underlying components for the system were still under development. Stibitz's principal recommendation to the NBS was therefore to split the procurement process into three phrases consisting of an initial study phase, a component-development phrase, and a final assembly phase. He recommended that parallel contracts be issued at each stage to encourage different technical options while promoting healthy competition among the contractors. Trusting, moreover, that the NBS was in the best position to synthesize the knowledge obtained from the first two phases of the program, Stibitz advised Curtiss to restore the NAML's planned machine-development laboratory back to its original purpose by conducting the final machine assembly itself.[8]

What Stibitz failed to realize, or chose not to accept, was the fact that the Census Bureau's funds transfer hinged on the fact that the NBS could procure a computer through a commercial supplier. The Census Bureau, again, was not authorized to contract for research and development. Yet although this prevented Curtiss from accepting this part of Stibitz's recommendations, it is striking how closely Curtiss followed Stibitz's advice. Curtiss divided the project into multiple phases, and decided to call for competitive bids during each phase of the project. Some of the funds were used to support the component development work of the Ordnance Development Division in an attempt to place private contractors in competition with the technical capabilities of the NBS. Curtiss did collapse the first two phases of the procurement process out of his eagerness to keep to an aggressive schedule. The study phase was therefore allowed to run concurrently with the component development work, and appropriate contracts were let to the Electronic Controls Company and to

Raytheon. Also upholding Stibitz's recommendation that a computer system be designed in relation to a specific application, but compromising somewhat on his emphasis on competition, Curtiss instructed Eckert and Mauchly to approach their design with the Census Bureau in mind, while Raytheon was asked to focus on the NAML's computing needs.[9]

In both accepting and modifying Stibitz's recommendations, Curtiss gave articulation to what it meant for a federal laboratory to serve as a technical contract monitor. In doing so, the NBS assumed more definite responsibilities for technical oversight. The full weight of this burden was something the NBS's mathematicians had yet to realize.

Assembling Computer Component Expertise within the NBS

Even as Curtiss dealt with the implications of Stibitz's report, those within the Ordnance Development Division began delving into their study of computer components. Once formal arrangements were made with the Army Ordnance Department, the dual gates of military and academic secrecy were cast wide open, providing Alexander and his cohort with direct access to the technical work at the Moore School and the IAS.[10]

Alexander's team, now formally dubbed the Electronic Computers Group, quickly emerged as the lead unit for the NBS's contributions to computing components. Other researchers remained wedded to their disciplinary interests, and their enthusiasm for computers waxed and waned with the immediate prospects that their prior work was relevant to digital computers research. For instance, those who worked on a thin film material called "dark traces" showed interest only while their material remained a promising candidate for high-speed memories. In contrast, Alexander and his staff shifted their allegiance from electronics to electronic computing. This was made possible both because of the breadth of interesting projects they could find from a starting point in electronics, as well as the broader perception that electronics engineers were at the heart of the new developments in electronic computing. (In this respect, organizational boundaries served to isolate NBS engineers from the kind of influence Julian Bigelow felt at the IAS.) Alexander's group therefore persisted in their efforts even when their early ideas failed. By early 1947, the ECG had a stable research program that focused on enhancing the reliability of vacuum tubes; evaluating whether the newly available germanium crystal diodes (a non-linear solid-state device that lacked the amplifying characteristics of a vacuum tube) could be used to reduce the costs of an electronic computer; and developing an input-output system for the Moore School and the IAS.[11]

For some of this work, Alexander could build on the NBS's traditional role as a standards organization. For instance, the vacuum-tube-enhancement program grew out of the NBS's established practices in product testing. Alexander's team therefore employed familiar methods and metrics to assess the vacuum tubes' reliability, and they used existing test facilities to analyze the different modes of failure. This work gave them a body of knowledge with which to approach tube manufacturers. But in also drawing on a new wartime practice, namely that of contracting for R&D, Alexander extended the NBS's usual approach to product testing by folding in this new tradition. Specifically, Alexander offered Raytheon a $10,000 contract if they would make a set of changes to their 6SN7s and run these tubes through test production. Although this was hardly enough to pay for the full costs of such an experiment, this contract allowed Alexander, as a representative of the federal government, to nudge a commercial manufacturer into an untried market that it might have been reluctant to enter. Alexander's group would play a similar role during the early 1950s with respect to the development of cathode-ray-tube memories.[12]

The point of all this is to show that just because Alexander and his group were relegated to computer components work, this did not mean that this work was either trivial or inconsequential. Alexander drew on the technical and administrative traditions of a substantial, federal research laboratory to construct a meaningful research program built around electronic components. This was especially evident with the ECG's work on input-output systems, into which the group devoted a substantial part of its early energies. The technical challenge here was that of ensuring that manually keyed data could be accurately and efficiently read into and extracted out of a computer's high-speed memory system. Working along with the IAS and the Moore School staff, Alexander and his team decided that this could best be done by performing data entry using conventional teletypes, where the data would then be transferred from paper tape, to magnetic wire, to the computer's internal memory and back. The NBS was assigned to focus on the first and last part of the procedure—generating the paper tapes and the tape reader that would generate coded electrical signals that could be used by a magnetic wire drive, and the inverse of this for the output—while the Moore School and the IAS focused on what was at least perceived as a more technically challenging task: building a reliable magnetic-wire storage system and its interface with the computer's memory. Magnetic wire posed real technical challenges. But Alexander, for his part, also expanded his work to incorporate an error-detection scheme so that two supposedly identical tapes would be compared automatically to flag data entry errors. Alexander described this work at the January 1947 Harvard symposium, even as the work was tucked in near the end of the program.

This work also presented some unexpected challenges. During an early meeting at the IAS, Alexander was given some basic specifications for the system by von Neumann. Von Neumann assured him that if the NBS designed the system according to these specifications, the system should satisfy the Moore School's requirements. Deferring to von Neumann's authority, Alexander never bothered to run the specifications past the Moore School engineers. It was somewhat late into the project when Alexander received a complaint from the Moore School that the proposed system would not work with their computer. The source of the difficulty was unclear at first, but eventually Alexander and his staff discovered that the basic data encoding scheme—binary for the IAS's computer, binary coded decimal for the Moore School's—meant that his error-detection scheme had to be set up differently for the two machines.[13] This mistake, though not a major technical setback, had broader consequences for Alexander and his staff. First, it was embarrassing. But more to the point, since there was no guarantee that there were no further surprises lurking behind the differences between the machines, Alexander and his staff continued to study the two computers and their detailed differences. In doing so, Alexander's group pried open the black box that had been set up through formal specifications. This meant that the group moved toward an understanding not only of computer components, but of computer systems as a whole.

From Contract Monitor to Procurement Service

Meanwhile, John Curtiss and Edward Cannon (an applied mathematician who had been hired to direct the now misnamed Machine Development Laboratory) were busy dealing with the popularity of their new arrangement with the Census Bureau. As other government agencies learned of the NBS's willingness to oversee the procurement of a digital computer, the MDL began receiving requests to extend this service to other offices within the federal government.[14]

Whatever the merits of Eckert and Mauchly's work for the Census Bureau, Cannon and Curtiss—who remained directly involved—approached these new requests more cautiously. Motivated by Stibitz's recommendation that they pay attention to the application, they instructed their staff to begin by evaluating each request from the point of view of whether they would actually benefit from an electronic computer. Based on these initial assessments, the MDL rejected the requests from the Social Security Administration and the Treasury. Both were judged to be better served by conventional tabulating machinery.[15]

Yet there were other applications—primarily military—for which digital computers did make sense. One of the new procurement contracts the MDL agreed to oversee was for the Air Force's Air Materiel Command. Truman's decision to rapidly demobilize American ground forces in Europe, and the Soviet aggression that ensued, raised broad concerns about military preparedness. Amid Congressional debates, the AMC sought to develop a system that could ensure an adequate parts inventory at its maintenance facilities. So long as the notion of efficiency was tied to military objectives, the traditional cost justifications that were used to rule out the Social Security and Treasury requests did not apply. But the supply problem also presented an interesting technical challenge. Because the AMC divided its record keeping operations between its central administration and major field sites such as its maintenance facility in Rome, New York, there was a need to design a system that could ensure the integrity of the data as distributed across multiple computers. Although this was clearly more of a practical problem and not a problem involving fundamental research in applied mathematics, it nevertheless fell within the scope of what the NAML's mathematicians wished to pursue.[16]

The NAML also opened negotiations with the Office of the Air Comptroller. As against the work of the AMC, the Air Comptroller's Office made far more strategic recommendations about the choice and quantity of the Air Force's arsenal. Air Force generals had always criticized their Army superiors about their technological conservatism. So after the Air Force was established as an independent branch of the Armed Services in 1947, Air

Force officers were eager to have some system that could justify—or at least offer a more objective assessment of the strategic value of their new, expensive weapons systems. Objectivity also had a special place amid the perennial threat of Congressional investigation over corrupt military procurement practices. Here too political, if not strategic interests took precedence over matters of fiscal accountability.

In late 1946, the OAC initiated a project called the Scientific Computation Of Optimal Programs, or Project SCOOP. (The reference to "programs" here was to procurement programs, not computer programs.) The computation of optimal procurement strategies was based on a new area of applied mathematics, advanced by George Dantzig and others, that would become known as linear programming. This work required enormous computational power in order to perform the matrix operations required to solve very large systems of linear equations. This work fell even more squarely within the applied mathematics research agenda of the NAML.[17]

Rounding out the military interest in computers was the more conventional needs of the Army Map Service. The AMS maintained its own coastal and military ordnance surveys separate from the work of the U.S. Geological Survey. All such surveys required extensive use of calculation to rework raw survey data to correct for such factors as the Earth's curvature. Both the scale and kind of data manipulation performed by the AMS fully justified the use of electronic computers.

While the MDL was busy evaluating these requests, ECC and Raytheon carried out their initial design studies, and their reports were due to arrive in September 1947. Curtiss now turned to von Neumann for advice. Because of the growing prominence of digital computers, the National Research Council had established a Committee on High-Speed Digital Computing Machines, with the intent to both assess and promote the use of digital computers in science and engineering. Edward Condon, director of the NBS, was a member; von Neumann was chair. In light of this convenient association, Curtiss (who sometimes attended instead of, or along with Condon) asked whether the committee could review the machine designs that they received as a result of the study contracts. Von Neumann agreed to form a subcommittee to do so, especially since the NRC, by charter, was set up to advise the federal government on matters related to research and development. Curtiss may have had some specific thoughts as to what he hoped to gain from the committee. Nevertheless, in deferring, like Alexander, to von Neumann's general authority, Curtiss simply wrote that he would "appreciate any advice which the Committee cares to offer."[18]

The NRC subcommittee's evaluation, as drafted by von Neumann, fell along the exact same lines as Stibitz's earlier report. Going a step further in its investigation than Stibitz, the subcommittee visited the Census Bureau, and raised the question once more of whether an adequate study was made of the Bureau's requirements. Von Neumann also attacked the NBS's decision to pursue a procurement program, and this in no uncertain terms: ". . . the design and construction of large-scale digital computing machines represents an art which is still in the research and development stage. The sub-committee, therefore, feels very definitely that it would be unwise to attempt at this time the building of duplicate calculating machines. . . ."[19] Acutely aware of the practical challenge of building an electronic computer, von Neumann issued this warning: "It appears that the Bureau of Standards has already incurred certain obligations toward several other government agencies. . . . It is our opinion that it will be difficult to fulfill such obligations to the letter. . . . It seems to us that these risks should not be taken, and that the entire program should be rediscussed between the Bureau of Standards and the government agencies in question, so the solution [to their requirements] may become possible."[20]

It is clear from his other remarks in the subcommittee's report that von Neumann continued to view computers primarily as a scientific instrument and an object of research. Von Neumann was willing to emphasize the importance of the end user, but he continued to conflate computer applications and computer-development research precisely in the way that he had to in order to meet the institutional expectations of the IAS. Difficult scientific problems were therefore supposed to drive new innovations in computer design. From this point of view, the NBS's procurement program ran counter to an effective program for computer-development research, which he continued to see as a technical and national-strategy priority so long as the work remained "in the research and development stage."

Still, it is useful to compare this position against von Neumann's earlier remark that the overall design of computers was "reasonably well advanced." Von Neumann's new recommendations were based as much on an intersubjective evaluation of the NAML's research capabilities, and the work that NBS mathematicians, as opposed to NBS engineers, could accomplish in terms of innovative digital computers design. As with Stibitz's report, von Neumann was making one last attempt to direct the MDL back to the mission he himself helped to define during the original naval advisory panel meetings. Recall that von Neumann had written that he wanted to have a "directing influence." Sitting at the heart of the network of researchers engaged in computer-development work, and a governmental body charged with giving advice, von Neumann could exert some pressure on a group he saw as drifting away from the vision he had put forward.

Von Neumann's recommendation could not be ignored entirely; it had to be presented before the Applied Mathematics Executive Committee, where it would carry some weight. However, as a research administrator, Curtiss had some bureaucratic maneuvers at his disposal with which to mitigate the report's impact. Fortunately, von Neumann had politely sent advance word of the report's major recommendations before its official delivery to the NBS. This gave Curtiss the time he needed to convene a meeting of his executive committee and to orchestrate its response. He suggested at this meeting that von Neumann had misunderstood his request. That the NRC subcommittee had narrowed the scope of inquiry to a specific class of machines—machines with a mercury delay-line memory—rather than offering the broad technical guidance he expected. At the same time, the NRC subcommittee failed to focus on specific differences between the Electronic Controls Company's and Raytheon's design proposals, which meant that their report was useless for advising the NBS on which proposal to fund. In any event, the report's major recommendations, Curtiss suggested, were incompatible with the laboratory's mission. It did not matter that there were inconsistencies between these remarks and Curtiss's original request, which in fairness von Neumann had followed to the letter.[21]

Although there are no surviving records on this score, it is likely that Curtiss also continued to defend the procurement program as a means of mitigating the risks of research and development. This would have resonated with the interests of the NBS's sponsors, who would not have shared von Neumann's interest in conflating research with application, at least more than necessary. Also, because of the time it took the NRC subcommittee to conduct its evaluation, Curtiss had moved ahead in issuing the request for proposals for the machine-development phase. In fact, he already had these proposals in hand. Curtiss therefore had the MDL conduct a hasty, internal evaluation of these proposals, and he advanced his own recommendations at the same executive committee meeting, prior to the arrival of von Neumann's report. While such bureaucratic tactics can produce somewhat cumbersome narratives, they remain a vital part of the historical actors' experience and cultivated practices. Here they provided a means of utilizing organizational structure to limit the influence of a postwar science advisory group.

The MDL had received proposals from Hughes, from Reeves Instrument, and from Engineering Research Associates, as well as from the Eckert-Mauchly Computer Corporation and Raytheon. Curtiss and Cannon considered the first three proposals to be technically deficient. Raytheon's bid, at $400,000, was determined to be too high. This left Eckert and Mauchly as the only valid contractor. By this point, the MDL had commitments from the Air Materiel Command and the Office of Air Comptroller, in addition to the funds transfers from the Census Bureau and the Office of Naval Research (ONR). So at its 22 March 1948 meeting, the Applied Mathematics Executive Committee approved the MDL's decision to issue four machine procurement contracts to the EMCC for their Univac computer.[22]

By separating itself from von Neumann's network, the MDL began to assume even greater responsibility for making its own technical judgments. Not surprisingly, Curtiss and Cannon found themselves turning more and more to Samuel Alexander for advice.

The Ascent of the Electronic Computers Laboratory

As the NRC subcommittee predicted, the Eckert-Mauchly Computer Corporation soon ran into trouble, though not for the reasons anticipated by von Neumann. During a routine security check by a unit of the Navy that was working directly with Eckert and Mauchly, the EMCC was found not to be adequately safeguarding classified information. Coming from a wartime academic environment, where classified documents were treated more casually, Eckert and Mauchly may not have been accustomed to the security procedures military agencies expected of its private contractors. Mauchly's affiliation with the Federation of American Scientists, which certain factions within the military pegged as a communist front organization during the conflict over the civilian control of atomic energy, may have also raised red flags. Whatever the reason, the ONR and the AMC rejected the EMCC as an acceptable contractor. Since the MDL was charged with procuring a computer from any available source, this did not result in the withdrawal of funds from the NBS. However, because their own judgment was that the EMCC was the only viable contractor, this placed people at the MDL in a bind. The success of the MDL's own procurement program depended on Eckert and Mauchly's success.[23]

Fortunately for both, military procurement decisions, including decisions about security, remained decentralized within the military establishment. The Office of Air Comptroller remained willing to accept the EMCC as a contractor. For them, describing their problem in its general form did not constitute any major disclosure of classified information, and their strategic—and political—concerns overrode any residual concerns about security. Likewise, there was nothing especially sensitive about the calculations performed by the Army Map Service. By the end of 1948, the MDL could again boast that it had three procurement contracts in place for the Univac.

Yet, although the NBS may have been content with the new arrangements, it is necessary to look at the situation from Eckert and Mauchly's point of view. Eckert and Mauchly were ready to start their work on the Univac back in May 1946. They had hoped to receive a procurement contract directly from the Census Bureau, but had been forced to accept an arrangement where the NBS deprived them of at least 15 percent of the money the Census Bureau set aside for machine-development work. Stibitz's report caused the NBS to then divide the project into multiple phases. Although they could do some work under the study contract, the new structure imposed constraints on how Eckert and Mauchly could schedule their R&D. Various delays, including those associated with competitive bidding and the six months it took the NRC subcommittee to issue its recommendations, stretched out their work schedule. The security review then added to the delay. All in all, Eckert and Mauchly had to wait over two years before they had a firm commitment to begin machine-development work.[24]

Meanwhile, Eckert and Mauchly had to pay for space, maintain a staff, and absorb other fixed costs. As private firms, the Electronic Controls Company and the Eckert-Mauchly Computer Corporation could and did seek other contracts. Nevertheless, the work for the NBS remained among the firm's best prospects. Certainly, Eckert and Mauchly's difficulties were not unusual for any undercapitalized firm. Still, it is worth noting how the MDL's actions—the very organizational decision making process that it had put in place to mitigate the risks of R&D—introduced the delays that drove Eckert and Mauchly to the brink of bankruptcy.

Wherever the faults lay, Cannon was forced to tighten the NBS's fiscal control over the EMCC. The yearly carryover of unexpended funds triggered auditing mechanisms within the federal government. In justifying the delays to the Bureau of the Budget, Cannon had to issue a revised contract that more precisely specified a scheduled set of tasks. It was one of Alexander's staff members who designed this "Checkpoint Program," in consultation with EMCC engineers. A detailed breakdown of the tasks— the demonstration of the arithmetic unit, the demonstration of the delay-line memory, the delivery of a magnetic tape drive unit—required a detailed understanding of computer systems and a sense for what reasonable technical specifications to assign.[25]

It was during this period that the relative authority of the MDL and the ECG began to shift. In the beginning, the MDL was a highly visible unit of a prestigious new laboratory for applied mathematics. In contrast Alexander's group was but a small team of engineers tucked away in the Electronics Section of the NBS's Ordnance Development Division. But as the ECG began to offer the MDL more technical support, and to pursue other efforts in conjunction with the Moore School and the IAS, it began to garner both the attention and the resources necessary to become a substantial player. This was backed by an organizational change. During a general reorganization of the NBS in 1948, Condon decided to consolidate the NBS's electronics activities into a new Electronics Division, since electronics as a whole was garnering considerable attention (and funding). In the process, Alexander's group was elevated to the status of a "section," and rechristened the Electronic Computers Laboratory. This placed the group on the same organizational footing as the MDL.[26]

In the wake of this reorganization, Alexander found new opportunities to expand his group's responsibilities. The most important one arose in the context of Project SCOOP. The various delays with the Univac, which continued to be exacerbated by the EMCC's fiscal difficulties, created an opportunity for Alexander to put forward his own machine-development proposal. By this point, Alexander and his staff had been studying digital computers and their components for nearly two years. Moreover, in addition to what he knew of the IAS and Moore School projects, Alexander and his staff had privileged access to the proprietary computer designs submitted by the MDL's prospective contractors; insofar as the ECL served as a technical advisor to the MDL, Alexander stood at a crucial node in the emerging network for the circulation of computer-development knowledge. Moreover, the work on the "Checkpoint Program" gave Alexander specific information about the status of Eckert and Mauchly's efforts.

It was not that Alexander believed his staff to be superior to those working on the Univac. Nevertheless, in their interactions with the EMCC, it became apparent that Eckert and Mauchly were struggling to design a commercially serviceable computer, with careful attention paid to such features as automatic error detection and ease of maintenance. The EMCC was also targeting a broad market, so that the Univac had to support a wide range of applications from commercial accounting to scientific computation. This also meant that the EMCC had to devote a good deal of its engineering effort to the development of peripheral equipment needed to support these different applications. The Univac, by necessity, was a complex system. Yet the central arithmetic unit, which gave the computer its power, was enticingly simple.[27]

Alexander had participated in Curtiss's meetings with the OAC almost from the outset. This gave him the opening he needed to propose that his new laboratory build an "interim computer" to satisfy the immediate needs of that organization.

Alexander carefully pitched his proposal in relation to the OAC's interests. He suggested that he could build the interim computer in six to eight months, which would allow the OAC to begin its coding work much earlier than the Univac's then-anticipated delivery date. Although the machine would be somewhat limited in its capacity, it would be large enough to allow the OAC to conduct early experiments in linear programming. Alexander also offered a number of other justifications. The interim computer could be used to test the performance and reliability of new cathode-ray-tube memories, which the OAC was hoping to incorporate into yet another "super-speed" computer. Upon delivery of the Univac, the interim computer could still continue to do useful work, specifically by performing the decimal-to-binary conversions so as to preserve the computational power of the Univac for the more demanding matrix operations. Finally,

Alexander suggested that the OAC could have this computer for just $75,000. All this transpired at a time when serious doubts about Air Force procurement decisions were surfacing, including doubts about the very expensive Convair B-36 strategic bomber. The OAC authorized the work on the interim computer around the end of 1948.[28]

Still, the early work on the Standards Eastern Automatic Computer, the name given to the interim computer, reveals as much about what Alexander's group did not yet know about computer-development work. Among the surviving documents are the minutes from the group's early planning meetings.[29] It is not possible, for reasons of space, to describe the collaborative work process for the SEAC with the same detail as for the ENIAC. Yet it is worth giving some attention to how the work proceeded, if only to provide an additional point of comparison for the work that unfolded at the IAS and at MIT.

These minutes make it clear that the group struggled even with the most rudimentary aspects of computer circuit design. Early work was dedicated to such things as building pulse transformers and understanding the high-speed characteristics of long signal lines—relearning, in other words, what transpired in the course of the ENIAC project. Yet, as with the work on the ENIAC, the SEAC engineers drew upon and extended the body of best electronic design practices in order to cope with a complex and unfamiliar problem. They were quick to separate their initial explorations from a more formal division of labor. They also conducted an early and explicit assessment of the technical interdependencies among their various tasks, and scheduled their work accordingly. As was the case with the ENIAC, one engineer assumed the responsibility for maintaining a set of conservative yet sufficiently permissive design rules that were continuously amended to ensure reliable operation. The historical record of the group's planning meetings was itself the product of a conscious decision. Although these engineers regarded paperwork to be a nuisance, they accepted that the complexity of their undertaking required formal mechanisms of coordination. This included the minutes as well as rigorous documentation of their design decisions. All information was kept in a central file.[30]

But perhaps what was most important to the project's success was the informal channel of communication that ran within the organization. As a small team of engineers, those working on the SEAC were not hindered by bureaucratic structures. So long as various tasks were being conducted in parallel, especially with new and untried technology, it was necessary for all the engineers to remain cognizant of each other's work. A change in the specifications of a certain component could easily require other team members to redo their designs. Amicable work relations made such repetitive design work possible. In this sense, Alexander benefited from both a tightly knit workgroup and their sense that they were racing against bigger groups that were endowed with greater resources. Indeed, when the historical records are considered with an ethnographic eye, it is impossible not to note the youthful enthusiasm of the SEAC engineers. These documents reveal the same kind of technical excitement and group dynamics that Tracy Kidder captured in the 1981 book *The Soul of a New Machine*, which described a successful team of computer engineers working at Data General some 30 years later. But the group working on the SEAC was truly at the vanguard of computer development work.

Still, Alexander and his staff ran into significant difficulties. High-speed vacuum-tube engineering was still an art, not a science. It was September 1949 before the group had sufficient mastery of the basic components to begin building their computer.[31]

Publicity, Context, and Authority

The SEAC became operational in June 1950. The machine was completed in 16 months, not 6, and at a cost of $474,000, not $75,000. There was nevertheless considerable jubilation, since the SEAC was the first real stored-program computer in the United States. Moreover, in the wake of the Soviet detonation of an atomic bomb, the costs of the SEAC faded into the background as multiple uses were found for this machine.[32]

In fact, the SEAC received considerable publicity during its official dedication, largely for the same reasons that the ENIAC had received publicity several years earlier. In 1950, Edward Condon, director of the NBS, was under assault from the House Committee on Un-American Activities (HUAC). Condon's earlier stance on atomic energy, along with the circumstances of his appointment, made him a good target for conservative members of Congress and their military allies. After J. Robert Oppenheimer, Condon was one of the most prominent scientists to receive the HUAC's early scrutiny.[33]

Yet although this investigation was the most publicly visible aspect of the conflict, an important underlying issue was the volume of military funds transfers to the NBS. Condon had indeed succeeded in retaining the wartime practice of funds transfers. The new work in computing was merely representative of other, similar arrangements made in the general area of electronics, ordnance, materials, and other strategic technologies. By 1950, fully two-thirds of the NBS's research budget came from military sources. Whatever the opinion of the senior Army officials and their congressional allies who opposed Condon, many military agencies chose to work with the NBS because of its acknowledged expertise.[34]

Condon would in fact win an important legislative skirmish shortly after the SEAC's dedication. The legislative foundation upon which the NBS executed interagency funds transfers was on somewhat shaky ground, having been established primarily as a cost-cutting measure during the Great Depression. The law simply allowed interagency funds transfers as a way of eliminating needless duplication in federal research facilities and expenditures. Condon therefore sought to maneuver a bill through Congress that explicitly authorized the NBS to conduct research on behalf of other agencies through interagency funds transfers. Public Law 619 was approved by Congress on 22 July 1950. This battle probably cost Condon his job.[35]

The SEAC began operating just as this political battle was taking shape. The secrecy surrounding the work on guided missiles precluded public disclosure of this work—not to mention how an accidental release of such information would play into allegations of Condon's communist sympathies. By comparison, nearly all of the computer-development work took place without any veil of secrecy. The way in which von Neumann and other academics structured the field ensured that postwar computer-development work remained largely open. Still, the SEAC was a valuable project conducted at the request of a military agency. Moreover, from the standpoint of the legislative battle, it was significant that the SEAC was the first generally available stored-program computer. This demonstrated that a civilian, federal research laboratory did in fact possess an expertise that was not matched by U.S. corporate and academic institutions. The SEAC was exactly the kind of artifact that could offer a political defense for military funds transfers. Neither the costs, delays nor the limitations of the "interim" design played a part in these rhetorical constructions.[36]

Finally, the SEAC's prominence was secured through its use by those working on the hydrogen bomb. Though accounts which suggest that the Atomic Energy Commission simply commandeered the SEAC are somewhat exaggerated, nuclear-weapons laboratories did gain significant use of the computer. This was especially true after the AEC contributed the funds to augment the machine's capabilities and to support additional operating shifts. As far as the SEAC's public image, it did not matter that the details pertaining to the AEC's use were never disclosed. This veil of secrecy was itself a mechanism that drew more attention to Alexander and his machine.[37]

The SEAC continued to generate further publicity. Because it was a general-purpose computer, the SEAC attracted a steady stream of visitors from different organizations, all of whom were interested in the use of a digital computer. The NBS itself was set up to facilitate this diffusion, since the NAML's Computation Laboratory provided computational services not only to other government agencies, but to the multiple research divisions inside the NBS. The Computation Laboratory was already using tabulating machines to do so, and it began using the SEAC once it was available. This meant that the NAML's mathematicians were able to introduce the benefits of electronic computing very quickly to fields as disparate as optics, crystallography, and military aviation. This wasn't all. The various computer manufacturers and federal agencies that were associated with the NAML's procurement service all came to see the SEAC in operation.[38]

Officially, the SEAC was transferred to the NAML's Computation Laboratory after the dedication ceremony. However, because of the fear of a prolonged shutdown, the AEC intervened by paying for a three-story building addition so that the SEAC could remain in the building where it was built. Property implied ownership, so that credit continued to accrue with Alexander and his staff. Indeed, each visit gave Alexander an opportunity to sell his group's capabilities. Each visit by a federal agency offered the prospects of a new funds transfer.[39]

The relationship between Cannon's Machine Development Laboratory and Alexander's Electronic Computers Laboratory continued to change. The MDL did not lose all its credibility. Cannon still oversaw a popular computer procurement program, and the SEAC's use contributed to the growth of the Computation Laboratory. Yet it was significant that Alexander and his staff were now able to secure financial support independent of the MDL's procurement program. Because of the separate organizational jurisdictions of the NAML and the NBS's Electronics Division, Alexander could continue to forge new agreements without the approval of the NAML's mathematicians.

Machine Development as a Dead End

While public recognition should have brought additional funds to Alexander, machine-development research proved to be a dead end within the confines of a national lab. The separation between the NAML and the ECL was not delineated quite as neatly as suggested above. Events and opinions inside the NAML did still affect what could be done by Alexander's group.

Much of these constraints emerged through the influence of the Applied Mathematics Executive Committee, whose oversight continued to apply to both groups. A significant disagreement first surfaced over Alexander's decision to undertake CRT memory development work in an attempt to improve the SEAC's performance. When the executive committee reviewed this work in early 1950, they chose to back RCA's work on the Selectron, and recommended that Alexander terminate his work. Alexander accepted this judgment, but expressed his discontent that a committee of mathematicians, and more important, a committee of external advisors had implicit executive authority over his research activities. He pushed for a different arrangement. Drawing on the clout from his work on the SEAC, Alexander convinced Curtiss and other senior NBS administrators that this external body should be reconstituted as the Applied Mathematics *Advisory* Council.[40]

Even then, the influence of an external advisory group was not so easily curtailed. One immediate source of Alexander's difficulties lay in a parallel effort to build a digital computer at the Institute for Numerical Analysis, the NAML's basic research arm that Curtiss decided to locate at the University of California at Los Angeles. Relying on the same rationale that was advanced at the naval advisory panel meetings, INA mathematicians argued that it was even more important for a fundamental applied mathematics laboratory to have access to an electronic computer. When Air Force monies became available for this purpose, Curtiss assigned Harry Huskey, an applied mathematician who was involved with the ENIAC and with an important British computer-development project, to build the Standards Western Automatic Computer. Although Huskey gained considerable knowledge about electronics working with the Moore School engineers, and then from the engineers Alan Turing assembled at the British National Physical Laboratory, he still had serious limitations as an engineer. In particular, he was still relatively inexperienced at managing large-scale engineering projects, and he failed to pay proper attention to such things as reliability engineering. It also did not help that in 1949 there was still no strong infrastructure for electronics engineering in Southern California. When the SWAC was powered up, in February 1951, it failed to operate for more than several seconds at a time.[41]

Unfortunately for Alexander, the SWAC's perpetual difficulties were made painfully obvious during the regular meetings of the NAML's Applied Mathematics Executive Committee. Even after the group was reconstituted as an advisory council, the Air Force representative continued to question the value of pursuing computer-development work inside the NBS. Moreover, so long as the advisory committee remained a body of applied mathematicians neither they nor the senior mathematicians at the NBS had a strong inclination to see computers as an object of research. They were content to view computers merely as a scientific instrument, and to continue the work of developing commercial suppliers for such equipment. Insofar as this council remained the major interface between the NBS and its external sponsors, attitudes formed within the committee influenced subsequent funding decisions. So long as machine-development work was conducted through funds transfers, the external advisors still had a hand on the purse strings.[42]

A broader set of political events accelerated the demise of the NBS's machine-development program. Although the NBS won the political skirmish over the issue of funds transfers in 1950, a deadly blow came in the wake of the "AD-X2 battery additive controversy." A controversial study issued by one of the NBS's traditional product testing laboratories generated the threat of lawsuits and accusations of favoritism and impropriety. Condon's successor, Allen Astin, was accused of neglecting the NBS's basic product testing and standards responsibilities due to its extensive involvement with military research. The new Secretary of Commerce, Sinclair Weeks, who had arrived with the more conservative Eisenhower administration, asked for Astin's resignation. The scientific community intervened on Astin's behalf, forcing a negotiated compromise. In lieu of Astin's resignation, Weeks agreed to convene an investigative committee headed by Mervin Kelly, president of Bell Telephone Laboratories. Soon the specific questions about a private manufacturer's car battery additive became quite inconsequential, as the overall mission of the NBS was brought into question. Moreover, Weeks put a committee of industrialists in the position of advising the NBS on how to reorient its research.[43]

In its final report, the Kelly Committee instructed the NBS to return to its basic mission in standards and testing, and to divest its military development work. In fairness, neither Kelly nor his fellow committee members intended to disrupt the NBS's activities in basic research. They acknowledged that the NBS was a leading scientific organization, and that the NBS should sustain its activities in fundamental research including work of military interest. They simply suggested that all military *development* work be transferred back to military laboratories or to private contractors, while all fundamental military research at the NBS would be supported through direct appropriations from Congress. Yet, although this might have been sound policy in a corporate setting where an organization could be restructured through executive decision, Kelly's recommendations offered no tenable solution for a federal government fundamentally divided about its research policies. Moreover, it was simply unlikely that Congress would double or triple the NBS's budget at a time when politics called for fiscal restraint. Yet a military order, issued in the wake of the Kelly Committee report, implemented policy by requiring the termination of funds transfers to the NBS. Between 1952 and 1955, the NBS lost 60 percent of its budget and 40 percent of its staff, albeit mostly through the divestiture of major military development projects.[44]

The full story of the crisis at the NBS requires more elaboration than can be told here. Research based on military funds transfers did not cease entirely at the NBS, especially in technical domains, such as computing, where the NBS possessed unique capabilities. Even in 1955, a full 94 percent of Alexander's budget came from funds transfers, and most of it from military sources. The NAML, on the other hand, was among the NBS divisions axed during the curtailments and divestitures. Broader disciplinary trends encouraged this action, since by 1953 numerical analysis and applied mathematics generally had found a place within academic mathematics departments. INA became a part of UCLA's math department.

What remained, then, of the NAML's computing services was regrouped in reduced form as the new Applied Mathematics Division of the NBS.[45]

These changes at the NBS were not a complete disaster for Alexander. The Electronic Computers Laboratory rode the crest of the SEAC's success, and computers continued to be very popular. Consequently, computing emerged as one of the areas that Astin was unwilling to let go. In fact, Alexander benefited organizationally, if not financially, from the turmoil. In the course of the breakup of the NAML, all computing activities, except for mathematical research and computing services, were placed under his jurisdiction. This included the NAML's computer procurement service. Moreover, during the reorganization that occurred in 1953–54, Alexander's section was elevated to full divisional status as the NBS's new Data Processing Systems Division.[46]

Still, the broader, commercial ideologies that gave birth to the NAML's procurement program continued to influence computer-development work at the NBS. In October 1951, John Curtiss received a visit from IBM's Applied Science Director, Cuthbert Hurd. Hurd, an applied mathematician, spoke approvingly of the SEAC and its potential for scientific work. But during the meeting, Hurd also suggested that "IBM and other companies engaged in manufacture of machines ought to have an opportunity to review the specifications [for future machines] before it was decided that the Bureau of Standards should engage in the building of the machine." Curtiss concurred. The NBS, after all, was a unit of the Department of Commerce, where such comments had to carry weight.[47]

The most significant reason for the decline in machine-development work at the NBS was the success of the NAML's computer procurement program. Indeed, at no point did the NBS's administrators or its committee of external advisors have to make a firm decision about the NBS's machine-development work. After the release of the Univac I (May 1951) and the IBM 701 (late 1952), most organizations, including federal agencies, turned to commercial manufacturers for their electronic computers. Alexander and his staff continued to work on specialized components. And they had one more chance to build a computer. This was the DYSEAC, built for the Signal Corps. Made up of compact, ultra-reliable, pluggable components, it could be installed in a tractor-trailer. Even this work documents how Alexander and his staff were relegated to specialized niches where commercial firms were unwilling to go. Once commercial suppliers captured the market, the bulk of the funds for machine-development work simply ceased to flow.[48]

Conclusion

In this chapter, following an approach established by Robert Thomas, I focused on the complex relationship between institutional politics, technology, and organizational decision making. While focusing on the NBS, I also attempted to push beyond a view of the firm as the right unit of analysis, in paying greater attention to the different kinds of negotiations that tend to unfold for public institutions. Thus, I gave equal weight to interorganizational relations as to intraorganizational ones. I considered direct forms of authority as exercised within an administrative hierarchy, as well as indirect forms of authority: advisory groups, sponsors, congressional oversight. I also paid close attention to the complex pattern of autonomy and interdependence that emerged at the intersection of organizational boundaries and shared technical interests. Drawing on evidence from a historical time scale, I extended Thomas's observations by focusing not only on situational strategies—strategies that emerge out of an actor's specific institutional and organizational location—but also on the technical and organizational changes that result from a sustained history of interactions.

As Thomas suggested, it is the complex politics of organizational decision making—the kind of bureaucratic maneuver Curtiss employed to mediate the influence of an external advisory board—that give meaning to organizational structures and the relationships they represent. In this respect, this chapter, along with Thomas's original study, follows a line of inquiry opened up by the "new institutionalists" in organizational sociology.[49] These scholars came to view bureaucratic organizations as dynamic entities that emerge from the detailed interactions that define the work of an organization (including research organizations), rather than seeing these routine interactions as emerging out of formal articulations of organizational structure.

This occurred at various levels in the chapter. Edward Condon's tenuous relationship with the military, and a specific history of military funds transfers, allowed him to chart a distinctive course for the NBS after World War II. John Curtiss's disciplinary interests and his interactions with outside groups gave articulation to the responsibilities of the NAML's Machine Development Laboratory. Samuel Alexander, meanwhile, carved out an expansive role for himself through an opportunistic strategy of matching institutional contexts to his group's growing technical expertise. Alexander's expanding role at the NBS demonstrates most clearly how research organizations emerged from a continuous history of local negotiations and decisions about organizational roles and responsibilities.

If this chapter adds to the observations offered by the new institutionalists, it does so by suggesting how a rather eclectic mix of factors contributes to the politics of organizational decision making within technical organizations. This follows from the "ecological" outlook that governs this book; it is also consistent with notions of the "seamless web" of sociotechnical practice, and "heterogeneous engineering" as defined by John Law. In each case, the emphasis is on how an elaborate system of social and technical meanings come to bear upon, and come to be constituted in turn through everyday actions.

Consider, for instance, the early decision made by Curtiss regarding the role of the Machine Development Laboratory. This was a situation where the *applied mathematicians* at the NBS, faced with a *proposal* from an *external civilian* agency, chose to reorient a machine-development laboratory into a federal *procurement program* based on *funds transfers*. He and his *superiors* decided that *applied mathematicians* at a *federal research facility* could adequately *monitor* the activities of *engineers* working in *private enterprise*. The use of italics here is somewhat arbitrary. Yet each of these terms required articulation, or more properly, a re-articulation during the course of the history just described. All this was at stake in building digital computers and the research organizations necessary to carry out this work.

Indeed, in looking at the broader sequence of events, it is clear that when the NBS's applied mathematicians considered launching a commercial procurement program, they simultaneously invoked concerns about free enterprise, disciplinary interests, military priorities, institution building, fundamental research and the nation's economic recovery. There is ample evidence in the historical records to show that all these concerns were brought to bear. Moreover, invoking a decision—here to favor commercial procurement—meant reinforcing certain associations at the expense of others. The cumulative effect of such decisions was to give specific definition to the NBS's role as a federal research laboratory in the realm of digital computer research, as well as the kind of computer that emerged from this work.

But if history has its many turns, they are not without direction. Free market ideologies, expressed not in the abstract but through very specific circumstances and actions—the NBS director's expressed preference for private contracts; a history of the NBS's activities in promoting commerce; federal bureaucrats inclined to think in terms of procurement rather than research and development; and even the market-like mechanism of inter-agency funds transfers—all ensured that Curtiss and his staff experienced recurring pressures to define a line of research that remained compatible with industrial interests. Military strategies and policies also defined and simultaneously set limits on what the NBS could and could not do.

This account, then, is ultimately about sociotechnical closure, precisely as defined by Pinch and Bijker (1987/1984). Still, the point of taking a social historical approach is to take seriously the options and points of flexibility that remain open to historical actors. Clearly, neither Curtiss nor Alexander was simply constrained by their respective contexts, and they used this flexibility to diverge from von Neumann's dominant vision for computer-development research. This raises the specific question of whether Curtiss could have done better by following von Neumann and Stibitz's advice, namely by abandoning his goal of helping other federal agencies to procure a digital computer. Counterfactual histories of any sort have their dangers, but it seems quite possible that the NAML's applied mathematicians could have sustained some authority in the broader realm of computer design had they done so. But, once Curtiss succumbed to his disciplinary inclinations—and perhaps as well to the financial opportunities presented by a computer procurement program—contractual obligations, recalcitrant artifacts, and a host of other concerns served to define the trajectory of his machine-development laboratory.

But the very same circumstances ensured that Alexander and his staff did not suffer the fate of the invisible technician. Alexander and his staff were able to reconstitute their knowledge as technical authority in the context of the entrenched goals of Curtiss's program. As a result, he and his engineers slowly usurped many of the responsibilities first assigned to the mathematicians. What was one group's constraint served as another group's opportunity, pointing to some of the complex ways in which contingency manifests itself in research.

Moreover, if broader political events and the disciplinary developments in applied mathematics served to bury Curtiss's ambitions, this was not true for Alexander—in spite of the fate of machine-development work at the NBS. Through his various successes, Alexander built up his own network, garnered new resources and expertise, and assembled an organization that continued to find new opportunities in digital computers research. As the head of the NBS's Data Processing Systems Division, Alexander was able to take his group in new directions by constructing ever new rationales. This also contributed to the further diversity of digital computers research. Ironically, a good deal of this work would build on the study of computer applications begun under the NAML's computer procurement program. As I have described elsewhere, the NBS went on to play a significant role in the systems analysis of electronic data processing within the federal bureaucracy.[50] But I now turn my attention to the relationship between research and rhetoric in the context of another computer-development project: Jay Forrester's work on Project Whirlwind at MIT.

Figure 5.1
A section of the Whirlwind I computer. At far left: Jay Forrester and Norman Taylor.
Photograph used with permission of MITRE Corporation. © The MITRE Corporation.
All rights reserved.

5   Research and Rhetoric: Jay Forrester and Federal Sponsorship of Academic Research

This chapter revisits Jay Forrester's work on Project Whirlwind at the Massachusetts Institute of Technology. At the end of World War II, Forrester redirected the resources of a major wartime research contract to assemble the largest postwar computer-development project in the United States. As compared to the dozen or so engineers at the Institute for Advanced Study and the National Bureau of Standards, Forrester employed more than 200 engineers and technicians. However, the computational demands of his application—flight simulation—drove Forrester in technical directions that lay beyond his group's immediate capabilities. Thus, whereas engineers slowly gained authority over the mathematicians at the NBS, the opposite became true for MIT. As Forrester ran into successive technical difficulties, schedule delays, and cost overruns, the mathematicians at the Office of the Naval Research began to assert greater control over his work. They nearly terminated the project in 1949. The Soviet detonation of an atomic bomb that year, followed shortly afterwards by the Korean War, breathed new life into the project, and Project Whirlwind was brought to successful completion as part of a new, continental air defense system. Nevertheless, machine-development work also proved to be a dead end for Forrester. He left the field in 1956 to pursue a career in industrial management.

While the general outlines of this story are well known, owing especially to Kent Redmond and Thomas Smith's 1980 book *Project Whirlwind*, I wish to revisit this story for two reasons. The first is again a matter of historiography. While I consider most existing accounts of Forrester, beginning with Redmond and Smith's, to be solid academic works, it is interesting how so many historical studies evaluate Forrester's project through retrospective assessments of its "accomplishments"—Forrester's contributions to real-time computing, the development of core memory, Forrester's visionary spirit when others, especially in the Navy, were supposedly more conservative.[1] These serve as important parables for discussions about the value of research and development; it should be evident by this point that this is a common feature of the digital computer's historiography. But if our goal is to really understand the process of innovation, then this too stands as a violation of the "symmetry postulate" in constructivist analyses. What remains necessary is a social historical approach that pays close attention to local circumstance and the different actors' points of view.

This leads me to a second, related point, namely that the early history of Project Whirlwind provides a unique window into the emerging relationship between MIT and the Office of Naval Research. It was no accident that Project Whirlwind occurred at MIT, or that it was funded by the ONR. These were two of the most important players in fundamental scientific and engineering research during the early postwar period. The very fact of the sustained tension generated by Forrester's project provided an occasion for these two institutions to mutually articulate[2] what an appropriate relationship between the leading U.S. postwar research university and its federal sponsor should be. This was true irrespective of the specific fate of Project Whirlwind. Project Whirlwind is to the relationship between MIT and the ONR, what the history of computing is to the emerging infrastructure for Cold War research. Redmond and Smith offer some tentative speculations on this score, but this remains mostly a missed opportunity.

What I specifically examine in this chapter are the detailed rhetorical strategies by which Forrester tried to uphold the validity of his project in the eyes of the ONR and his own institution, MIT. Forrester's ability to sustain his project amidst considerable adversity has attracted the attention of Forrester's admirers and his critics alike. To understand how he could have done so, it is necessary to take a closer look at the institutional contexts, the technical developments, and the rhetoric through which Forrester gave specific form to his project and to the relationship between MIT and the ONR.

I hasten to add that I do not invoke the notion of rhetoric in any pejorative sense. As described by Bruno Latour (1987), rhetoric is part of the mundane practice of science and engineering. While this is not the place to provide an extensive review of Latour's ideas, it may be noted that the main rhetorical mode employed by scientists and engineers is that of evidentiary accumulation, where qualifying statements, or "modalities," are used to both support and challenge a primary assertion. Layers of interpretation, such as commentary regarding a researcher's qualification, may be embedded within these modalities. Both an increase and decrease in the modalities may contribute to the hardness or softness of a scientific fact, or in a case like Forrester's, the perceived validity of a particular approach to computer-development research. Latour's notion of translation is also relevant here. Since one of Forrester's burdens was to uphold the validity of his research program in the eyes of cognizant MIT and ONR officials, he could do so only through the successful translation of their interests through his technical accomplishments and rhetorical constructs. The success and ultimate failure of Forrester's effort to understand and translate the interests of the ONR and those of MIT would define the extent and limits of Project Whirlwind. Finally, Latour makes it evident that rhetoric in science and engineering does not function alone, but always in relation to what he calls "machinery," namely technological artifacts *and* social networks of the sort captured by the phrase "political machine."[3]

If I add to Latour's observations, it is by focusing on rhetoric itself as a form of situated practice. In Latour's more abstract analyses, researchers appear free to alter their arguments and research agendas. In contrast, attention to historical circumstance helps to reveal how rhetorical constructions themselves become anchored within the network of artifacts and social relations forged during the course of research. Like players caught in a difficult game of chess, each argument made by Forrester and his cohort would entail a series of technical choices, organizational arrangements and financial commitments that would prove difficult to reverse. Each decision would set the stage for subsequent actions. While I have already presented this general argument in the context of John Curtiss's struggles at the NBS, the significance of rhetorical practice will appear in sharper relief with Forrester's project. In fact, what will be most striking in this chapter is the extent to which Forrester was able, by skill or by circumstance, to mobilize rhetoric to maneuver through and give articulation to the institutional objectives of his sponsor and host institution.

In many respects, the following analysis once again treads upon the ground covered by Andrew Pickering in his studies of scientific practice. In addition to what I have said so far about practice and contingency (see chapter 2), Pickering explicitly sees his work as an extension of Latour's analyses. Moreover, Project Whirlwind is one of four case studies he uses in *The Mangle of Practice* (1995). But whereas Pickering, in drawing on Noble's (1984) prior account of Forrester, focuses on his subsequent involvement with numerically controlled machine tools, I focus on the initial exchanges between MIT and the ONR. This is more relevant to a study of emergent Cold War research institutions.

Pickering also tends to miss important historical cues about institutional contexts and overemphasizes, as a result, the contingency and indeterminacy of research. By contextualizing Forrester's actions, it will become apparent how flexibility, for Forrester, came to depend increasingly on developing a body of rhetoric that could justify a specific approach to digital computers research. This attention to Forrester's rhetoric is necessary, in any event, if we are to challenge historical accounts that are themselves based on the rhetorical closures Forrester himself helped to bring about. By the end of this chapter, it should be apparent how Forrester's rhetoric shaped not only the relationship between MIT and the ONR but also the historical accounts of Project Whirlwind.

The Origins of Project Whirlwind

Project Whirlwind emerged from MIT's wartime Servomechanisms Laboratory, when this lab's interest in control-systems engineering merged with the Navy's work on flight simulators and with wartime developments in electronic computing. The Servomechanisms Lab was an important component of the National Defense Research Committee's work on fire-control systems. As described by David Mindell, the laboratory's director, Gordon Brown, received the resources to create the laboratory in exchange for allowing a very important paper on control-systems theory to be placed under military classification. The relationship between Brown and Warren Weaver, who was then still head of the NDRC's Section D-2, was quite strained. Brown tended to delve into engineering problems associated with the production and deployment of military systems instead of keeping to laboratory-based research. But if this created tensions with Weaver, his pragmatic approach and willingness to cooperate with military patrons earned a fairer reputation among his colleagues and those within the military services. Thus, when the Special Devices Division of the Navy's Bureau of Aeronautics approached other MIT faculty members to talk about a new "universal" flight simulator, Brown was invited into their conversations.[4]

Captain Luis de Florez, the director of the Special Devices Division, approached MIT out of his own wartime aspirations. De Florez, an MIT graduate, amassed considerable reputation building flight simulators for the Navy's flight training schools. Because of the pace with which new aircraft were being put into active service, there emerged an incentive to develop a universal simulator capable of mimicking the behavior of any aircraft. De Florez also recognized that a universal simulator could play an important role in the design of new aircraft. Even by World War II, aeronautical engineering remained an art. Among other things, aircraft design engineers continued, justifiably, to distrust the analytical models offered by aerodynamicists. Experimental traditions remained integral to the practice of aircraft design, so that experimental data about an aircraft's stability, including impressions collected from skilled test pilots, remained integral to the design process. A universal simulator could accelerate this process by offering a simulation tool, one that retained the ability of test pilots to offer their judgments literally through the seat of their pants.[5]

De Florez turned to his alma mater because of the technical challenges associated with a universal simulator. The SDD's operational trainers, built under contract by Bell Labs and Western Electric, used individually machined cams to emulate flight equations. In contrast, a universal flight simulator required a more versatile computing device. MIT's differential analyzer, along with new analog electronic devices developed by Bell Laboratories, offered immediate prospects. The proposed simulator also raised difficult questions about whether a mechanical system could have sufficiently well-characterized responses to produce meaningful judgments about flight stability. This second issue made it easy for Brown to interject his knowledge, since it sat squarely within the tradition of servomechanisms research. As a consequence, Brown was given the initial study contract for the Aircraft Stability Control Analyzer (ASCA) in November 1944.[6]

Brown assigned Forrester to carry out this study. Forrester entered MIT's master's program in Electrical Engineering in 1939, and he had just completed a thesis under Brown. Forrester was intimately involved with Brown's burgeoning wartime contracts. Although this delayed his thesis work, by late 1944 he was also serving as one of the laboratory's assistant directors.[7]

Brown had clearly inherited much of Vannevar Bush's thoughts about research and graduate education. Brown therefore gave Forrester considerable latitude in how he conducted his work. As was common for

Servomechanisms Lab projects at the time, Forrester began with a fresh, analytical approach to the problem. He began by studying existing trainers, by examining their hydraulic components, and by exploring different options for computing aerodynamic equations as required by the ASCA analyzer. This last inquiry included a careful study of the physical and mathematical properties of various computing devices ranging from analog electronic devices, to mechanical integrators, to hydraulic transmission devices that could be made to serve a similar purpose. It was in piecing together the various aspects of the system that Forrester recognized how complex the ASCA analyzer was. Especially after further conversations with the aeronautics faculty raised the number of flight equations from 33 to 47, Forrester grew skeptical about the aeronautical faculty's earlier assessment about feasibility. At this point, Forrester took the notion of a study contract seriously. Drawing on the professional practices of a consulting engineer, Forrester sought to advise the Navy not only on design options, but their practicability. Forrester suggested to Brown that he should ask for an extension of the study period.[8]

Brown brushed aside Forrester's reservations and pushed for the full development contract, owing in part to certain changes within the Navy. In May 1945, Rear Admiral Harold Bowen came to head the new Office of Research and Inventions (ORI), which set out to centralize naval R&D. Bowen had his own reason for creating such an office—breaking the Army's nuclear monopoly—but this translated into a need to win over the nation's civilian scientists. Somewhat manipulatively, Bowen placed considerable emphasis on providing unrestricted support for fundamental research. De Florez and Brown both interpreted this to favor a project like the ASCA analyzer, precisely because of its risks and rewards. The Navy issued a letter of intent on 30 June 1945 for an $875,000 R&D contract. Forrester continued to head this project, joined now by a handful of engineers who transferred in from other wartime projects that were drawing to a close.[9]

After the letter of intent was signed, de Florez hired Perry Crawford, another MIT graduate, to serve as his technical advisor and thus the project's de facto program manager. De Florez, saw his own success as based on a record of technical achievement, and so he was more concerned about establishing technical communications than of ensuring accountability. He allowed Crawford and Forrester to develop a strong, collegial relationship, building on the fact that they were former classmates. It was Crawford who introduced Forrester to electronic, numerical computation. Crawford worked on MIT's Rapid Arithmetic Machine. He was also familiar with Claude Shannon's 1937 master's thesis, and he had operated MIT's differential analyzer in conjunction with the NDRC's work on fire-control systems. Piecing the relevant ideas together, his 1942 master's thesis described a fire-control computer designed using numerical electronic circuitry. Crawford relayed this knowledge to Forrester during a conversation shortly before joining the SDD.[10]

Still, digital electronics remained a remote possibility for Forrester until the October 1945 computing conference at MIT. Both he and Crawford were members of the audience. The conference disclosed not only the wartime work on the ENIAC to them, but the more innovative EDVAC design. The sudden change in Forrester's research program attests to the fact that he and his group were swept up by the enthusiasm for digital computers. Especially amid the group's growing doubts about the adequacy of analog devices, this new approach must have seemed like a godsend.[11]

Forrester directed his staff to investigate the new approaches, which one of the staff members noted "are highly thought of in mathematical circles at present." This allusion to the higher authority of mathematicians was no accident. Forrester and Crawford were clearly swayed by the prestige accorded to the mathematicians gathered at MIT. Yet in evaluating digital electronics Forrester and his staff drew as well on the local technical traditions of the Servomechanisms Lab. In particular, the wartime technical work on fire-control systems had led to the formal articulation of terms such as precision, accuracy, sensitivity, and noise. These concepts were essential to the proper characterization of servomechanisms, and were an integral part of the emerging field of control-systems theory. The concepts were portable to other domains, and Forrester's staff did so to evaluate the wind tunnel data they hoped to use for the ASCA analyzer. Having just done so, they were well prepared to transpose the concepts back into the domain of control-systems engineering in assessing the relative merits of analog and digital computation.[12]

Forrester drew up a memo summarizing his and his staff's initial findings. One of their concerns was that the differential analyzer would freeze at low rates of rotation. By their estimates, the mechanical components of a differential analyzer had to have 60 times their present sensitivity to accurately evaluate the pitching oscillation of an aircraft. Analog electronics offered a possible work around, and it rivaled the speed of its numerical counterpart. However, it seemed unlikely that the signal to noise ratio of analog electronics could be raised to 20,000 : 1 as needed to guarantee accurate results. In contrast, it was evident that high-speed numerical computation could deliver arbitrary degrees of accuracy and sensitivity simply by making the interval of calculation small enough.[13]

Other, more subjective considerations also figured in the abstract calculus by which Forrester established the validity of the new technical approach. For instance, Forrester was quick to note that a digital computer offered the promise of radical simplicity. Whereas an analog simulator would have to have a separate computing element to mimic each of the flight equations, resulting in a "large, interconnected, simultaneously operating network," a digital computer made it possible to spread out the calculations across time, so long as the computer was fast enough to keep up with the overall simulation. A stored-program computer would also allow a universal simulator to be reconfigured quickly and automatically. In fact, it allowed for instantaneous changes, as required during a transition from normal flight to a landing or a stall. This too meant that they would need less equipment. Drawing on the engineering design practice of assessing tradeoffs, Forrester judged that the simpler design of a digital solution would lead to lower construction costs, but that it might involve higher development costs. He also suggested that the development time would be longer, although he was quick to add that the construction time should be shorter. The foremost disadvantage of a digital approach was that although the circuits appeared "ready for development," the ASCA application still required pushing digital techniques well beyond anything then in existence. Still, Forrester judged that the digital approach was attractive enough that, "a careful study must be made."[14]

As is evident from the tenor of their evaluations, Forrester and his staff were intrigued with digital electronics at a much more fundamental level. For control-systems engineers plagued with concerns about noise and insensitivity, the digital abstraction, with its promise of arbitrary accuracy and precision, presented a special allure. They therefore noted enthusiastically that digital electronics presented "no calibration problem," and how "any voltage of 50% (or over) of the value it is supposed to have" would register. Working through Crawford, Forrester secured the SDD's authorization to carry out a more intensive study of digital circuitry. Between January and March of 1946, Forrester used the resources offered by the ASCA contract to conduct the experiments necessary to establish that digital electronics could be made ready for use in the ASCA analyzer.[15]

Meanwhile, continued conversations between Forrester, Crawford, and senior administrators at the SDD and the ORI ensured that Forrester's evolving plans would continue to find support within the Navy. Postwar negotiations over science funding continued to provide a context favoring advanced, risky research projects. Bowen continued to push to centralize naval research, now under the proposed Office of Naval Research. Some academic administrators remained suspicious of Bowen's motives, and remained especially skeptical about whether the Navy could, or would, fund fundamental research on a sustainable basis. Therefore, in early 1946, it was not a foregone conclusion that Bowen could find good academic programs willing to accept his money. His ability sustain a relationship with a laboratory with the Servomechanisms Lab's reputation was therefore itself something of a coup.[16]

But if the circumstances with the Navy were highly auspicious, the situation inside MIT was more complex. MIT contributed disproportionately to the NDRC's wartime programs. The total amount of NDRC contracts awarded to MIT was approximately $116 million, or 21.7 percent of the NDRC's total expenditures. It should be noted, on the other hand, that 90 percent of this amount went to radar systems development at MIT's Radiation Laboratory.

MIT president Karl Compton studiously followed Vannevar Bush's directive of 11 January 1945, which asked all academic institutions to prepare for the termination of NDRC contracts at the end of hostilities. Toward the end of the war, Compton also instructed his faculty to retain or accept federal contracts only to the extent to which they were consistent with MIT's peacetime mission. However, both the scale and success of MIT's wartime research programs led Compton to see military research as part of the rightful activities of an academic institution, and more so than the presidents of other leading research universities. Indeed, Compton saw such research as part of MIT's obligation to public service, as long as it did not interfere with MIT's principal mission in education and the advancement of knowledge. In discussing MIT's postwar research program, Compton wrote in his 1944 annual report: "The temper of the times justifies the expectation that this type of contribution by M. I. T. to the national welfare will continue to be substantial."[17]

In any event, the weight of Compton's concerns, as shared by academic administrators elsewhere, lay with the corruption of the sciences, not engineering. Although the scientists' participation in many wartime engineering projects complicated matters, the conversation quickly shifted from whether military research should continue on academic campuses to that of ensuring that sponsored research remained driven by academic agendas.[18] This was nowhere more apparent than in the postwar conversion of MIT's Radiation Laboratory into the Research Laboratory for Electronics. Compton's initial plan was to terminate nearly all of the Rad Lab's activities by December 1945. However, John Slater, chair of MIT's Physics Department, intervened by brokering an agreement with the OSRD and the armed services. By March 1946, the Research Laboratory for Electronics had the support of the new Joint Services Electronics Program, and an annual budget of $600,000 provided by the Army and the Navy. This sum was offered as a kind of "block grant," which left the RLE free to set its own research agenda. Most of the senior physicists associated with the Radiation Laboratory returned to their former institutions, or went on to accept academic appointments elsewhere. However, Slater and the laboratory's new director, Julius Stratton, created a special program that allowed the Rad Lab's junior staff members to remain at MIT as an advanced degree candidate while earning a near-professional salary. All this came with Compton's approval. It also set multiple administrative precedents for Forrester's project.[19]

There is no direct evidence that detailed conversations about Forrester's project took place in Compton or Bowen's office. Nevertheless, Forrester, and more important, MIT's director of sponsored research, Nathaniel Sage, were acutely aware of the evolving institutional policies. As the director of MIT's Division of Industrial Cooperation (DIC), Sage helped to administer all of MIT's wartime research contracts. Drawing on this experience, Sage could help Forrester craft a proposal that played to MIT's and the ORI's interests in fundamental research, even as it supported Forrester's newfound interest in digital computers. Like von Neumann, Forrester was able to use the changing institutional context to push his research in a direction he himself wished to go.[20]

Sage was also careful to draw up a contract that attempted to limit the institutional obligations of, and risks to, MIT. His solution, not dissimilar to the one advanced by George Stibitz, was to divide the project into multiple phases. Phase I required MIT to pursue the "research, development and construction necessary to demonstrate digital techniques of the type required for the final computer" to be used in the ASCA analyzer. The initial work was defined almost entirely around components research, though it also allowed for the mathematical studies that would be needed to make digital computers applicable to military problems. Forrester did also agree to build a "small digital computer" for realistic evaluation of the new components. Phase II required MIT merely to design a digital computer capable of supporting the ASCA application. The actual construction of the computer and analyzer, and their experimental operation for aircraft design problems, were left to phases III and IV of the contract. The ORI formally accepted a slightly amended version of this proposal in April 1946, committing itself to $1.19 million for the first two phases of the work.[21]

Knowledge and Project Organization

Having received a contract he felt he could execute, Forrester set out to assemble the staff and research organization necessary to perform the work. The size of his contract prompted Forrester to think rather grandly. He turned to the multi-divisional structure of the Rad Lab, and perhaps the NDRC itself for an organizational model. He therefore created separate "divisions" to deal with the aerodynamics, mathematics, electronics, mechanical systems, high-speed storage and other aspects of the ASCA analyzer. Uncomfortable with Forrester's lofty designation, his staff would use the words "division," "section," and "group" rather interchangeably in their reports and memoranda. To ensure proper communication, Forrester scheduled weekly planning meetings and required all engineers to submit biweekly progress reports. Faced with blatant noncompliance, Forrester soon agreed to monthly summaries submitted by each division head.[22]

Forrester's project, now designated Project Whirlwind, continued to bear the imprint of MIT's tradition in graduate education. Another important legacy left behind by Vannevar Bush was the idea that talented engineers ought to be able to pursue the interdisciplinary work necessary to apply their knowledge to different areas of application. Forrester inherited this outlook through the Servomechanisms Lab. Forrester therefore proceeded to look for exceptional young engineers of "doctor's degree caliber," rather than recruiting accomplished figures in the field. The project's scale precluded exclusive use of graduate students, but it is significant that Forrester hired none of the engineers who worked on the ENIAC. Nor did he begin by hiring applied mathematicians and aeronautical engineers with substantial reputations.[23]

Originally, Forrester envisioned that the technical coordination across his divisions would unfold through a linear process. The aerodynamics group would pass their specifications on to the mathematics group, which would define in turn a set of computational requirements for the block diagram group. The block diagram group would then produce the technical specifications required by the various electronic components groups. But the work could not be so linear. Both the absolute and relative performance of electronic components influenced the block diagram design. Meanwhile, different machine designs presented mathematicians and aeronautical engineers with different possibilities for setting up their equations. Yet the work was plagued with uncertainty. Those working on the electronics were barely prepared to make the transition from 100 kilohertz to 1 megahertz, even though they were shooting for much higher speeds. Meanwhile, Forrester's mathematicians were only beginning to acquire a rudimentary knowledge of numerical analysis. Robert Everett, as the head of the block diagram group, was forced to concede that he could not offer a final design for the ASCA analyzer until their knowledge was advanced on almost every front.[24]

How and in what order this knowledge unfolded is a critical part of Project Whirlwind's history. For the moment, the process began with yet another revision of the ASCA flight equations by the aerodynamics group. This was in April 1946. Everett was a "consultant" to these conversations, and this allowed him to see—even though he lacked an extensive mathematical background—that the latest list, which contained more than 90 equations, required a fresh review of machine requirements. Everett proceeded to describe a series of options, from the "completely serial or sequential method" described by "von Neuman (sic)" in the EDVAC report, to a machine that contained multiple computing elements as would be the case with "analogy type" machines.[25]

It was some time after this preliminary evaluation that Everett came across a copy of the Institute for Advanced Study's latest block diagrams. These diagrams revealed von Neumann's latest scheme for transmitting all the digits of a number in parallel, while continuing to handle operations sequentially. Everett was swayed by the relative efficiency of this design, and by von Neumann's reputation. He began to favor this "parallel" design, even as he admitted that further evaluation was necessary to establish its validity for the ASCA analyzer. These evaluations, conducted during the summer and early fall of 1946, allowed Everett to align Project Whirlwind to von Neumann's network. Such evaluations were part and parcel to the growing consensus that transformed von Neumann's design into a recognized "architecture."[26]

Yet a decision to build a "parallel" machine entailed a series of commitments, as von Neumann himself had already surmised. The parallel transmission of digits strongly favored the use of cathode-ray-tube (CRT) memories, which, unlike a delay line, could give direct access to all the stored data. Forrester therefore created a special group—a "special problems" group within the Electronics Division—to evaluate this possibility. Working from one of Everett's tentative specifications, the group reached the conclusion that only an internal tube development effort could produce a storage tube with the stated design goal of a 6-microsecond access time. (Forrester and his group referred to their memory system as *electrostatic storage tubes* as opposed to CRT memories because they used a technique where electrostatic charges were stored on a separate dielectric surface embedded inside the cathode-ray tube. This allowed the use of a separate set of circuits to help retain the charge, making it possible to read the contents of the memory using a system of secondary electron emissions that improved access time.) The limited parallelism of the IAS computer offered efficiency and elegance, but only at the cost of this new research.[27]

Forrester would subsequently explain away the first eight to ten months of the project as the time it took to study electronic components and alternative machine designs. Yet the view from the ground was that of considerable frustration; they could not even settle on a basic machine design. To arrive at this important decision, Everett and his counterparts had to devise specific tools to transgress the boundaries erected by their own organization. The process began in April during Everett's initial reassessment, when he went ahead and produced a set of preliminary design specifications "for the purpose of discussion." This allowed the various electronics groups to spend several months experimenting with the necessary components. It was at this point that Everett worked alongside the mathematicians in producing the first, realistic estimate of the analyzer's performance. Working together, they composed an imaginary instruction set with which to code up a portion of the ASCA equations. Conjoined with the performance targets now offered by the electronics groups, Everett found that a bitwise-serial machine with a 100-microseconds addition time would solve only six solutions per second. A serial machine with a dedicated interpolator could raise this speed to 20 solutions per second. But only a parallel machine with a high-speed multiplier could achieve a rate of 30 solutions per second. These were still estimates, but Everett could now toss the problem back to the aeronautical engineers to have them specify the rate of solution they needed for an adequate simulation.[28]

It was November before Forrester and his staff turned their attention to the small digital computer they were required to build during Phase I of their contract. Faced with a choice between missing the June 1947 deadline and producing a limited, "experimental" machine, Forrester and his staff opted for the latter. They did decide to implement the parallel design favored by Everett. Otherwise, they accepted that this would be but a very small computer that would simply "work." They agreed that the machine would have but an 11-bit word length, only basic arithmetic circuitry, and require something on the order of 600 vacuum tubes. To prevent any misunderstanding, Forrester labeled the machine a "pre-prototype," hoping to dispel any illusion that this machine was designed to handle actual ASCA equations.[29]

However, it was difficult for Forrester's staff to keep to this limited machine. Because of the project's overall emphasis on research, the divisions came up with very different ideas as to what constituted a good experiment. Those working on electronic circuits wished to test their components at full operating speeds. Those working on the high-speed multiplier asked for a few more circuits that would enable them to put their components to test. The mathematics and aerodynamics groups insisted that the pre-prototype perform "actual computations." Meanwhile, those concerned about construction and the pressing deadline pushed for the use of standard circuit racks to facilitate assembly and access. Some of the changes had an exponential effect. The request to extend the machine's word length to 15 bits, for instance, which made it possible to test the electrostatic storage tubes, required every part of the machine to be scaled up accordingly. This was clearly a machine designed by committee.[30]

The most important issue was what constituted a valid "experiment" for Everett's block diagram group. While both he and his staff were trained as electrical engineers, they had already begun to see their expertise as lying with computer designs. This gave them little incentive to limit the complexity of the pre-prototype. Everett added a fully functional A-register, B-register and five flip-flop registers. He also added 32 words of manual "dead-switch registers" with which to emulate the behavior of the high-speed memory. Most significantly, Everett gave the pre-prototype a substantial programming capability. "Actual computation" proved to be a slippery phrase in the hands of the block diagram group. Because Everett was put in charge of the pre-prototype's overall design, his group's interests dominated the group's first major experiment.[31]

As the head of construction, Harris Fahnestock warned his colleagues about impending deadlines. In order to begin construction in April 1947, it was necessary to order all the parts by 1 January. But in order to know what parts to order, it was necessary to have detailed circuit drawings, which required a final block diagram. Everett and his staff worked hard over the Christmas holiday to deliver this diagram on 27 December. Yet, as was noted by one of the staff members who studied Everett's report, the latest design called for over fifty drawings, presumably of circuit diagrams. Since few, if any of these diagrams were produced to date, the immediate implication was the futility of trying to meet the parts deadline (1 January). In fact, one of the lead engineers was already asking whether they could even have their machine fully specified by the construction deadline (1 April).[32]

The actual complexity of the pre-prototype became apparent only during the course of working out the details. Even as late as early December, Forrester's staff had estimated that the final ASCA analyzer would have about 3,000 vacuum tubes. But with all the proposed extensions, the pre-prototype was now set to have more tubes than this recent estimate for the final machine. This was not simply a matter of added cost. If the primary purpose of Phase I was to ascertain the feasibility of a digital solution, then the implied changes in the overall scale of the ASCA analyzer should have generated some concern.[33] However, this was an issue that remained unspoken in the group's recorded deliberations. Perhaps Forrester decided that there was, as yet, no compelling reason to doubt the overall validity of the work. New technical options continued to present themselves. Moreover, it helped that at this point Forrester was being encouraged to increase rather than decrease his rate of expenditures. Forrester authorized Everett and Fahnestock to proceed with the pre-prototype, regardless of the implications this had for their project and schedule. Forrester kept Crawford informed of the changes, and offered suitable rationales. Neither Crawford nor his superiors intervened. This was precisely the kind of discretion the SDD was willing to grant its investigators in the new context of postwar research.[34]

Research and Accountability

Forrester was able to inaugurate his project with very little critical oversight, but this changed during the following year. Wartime research generated new bodies of practice for contract administration and oversight, and these procedures became more formalized as a result of the many conversations about postwar science policy. During 1947, Forrester experienced no less than four occasions during which he was asked to justify his work. Each inquiry brought further scrutiny. Each generated further articulations about the goals of Project Whirlwind, and the expectations of his sponsor.

The first was apparently a routine evaluation carried out by the Naval Research Advisory Committee. When the ONR was established, in August 1946, Congress gave statutory authority to a committee of civilian scientists and engineers charged with advising the ONR. Warren Weaver was asked to chair the committee. Weaver, who had long been involved with computing efforts at MIT, decided to pay a visit to Project Whirlwind. Indeed, Weaver must have arrived quite curious about the project; it was displacing MIT's other computer-development project, which he had funded through the Rockefeller Foundation.[35]

With his seasoned experience as a research administrator, Weaver quickly discerned that Forrester's enthusiasm for digital computers was causing him to stray from his original mission. As both Stibitz and von Neumann had done for the NBS, Weaver doubted whether Forrester had conducted an adequate study of the ASCA analyzer's mathematical requirements. He also questioned whether Forrester was making the right technical choices, asking whether the project was "failing to be good biscuits by trying to be a cake."[36] Weaver noted, on the other hand, that both Forrester and Crawford seemed quite competent, and that the staff at MIT was hard at work. Weaver also left Forrester with what was, for Forrester, the somewhat ambiguous comment that computer development was fast becoming a field more appropriate to a school of applied science and engineering, rather than one devoted to theoretical research.[37]

At the same time, Weaver was careful in his subsequent conduct. Sitting at the heart of the postwar negotiations over science policy, Weaver was fully aware of the limited authority of an advisory group. Instead of taking a confrontational approach, as Stibitz and von Neumann had done, Weaver chose to work through his own network. He reported his observations to Mina Rees, who by then was the head of the ONR's Mathematics Branch. Although he was careful to note that no judgment should be made as a result of any single visit, he clearly alerted Rees to a project that required closer scrutiny.[38]

Forrester, having been made a bit anxious by Weaver, sent Nathaniel Sage, the DIC's director, a memo describing the visit. For his part, Forrester asserted that Weaver had arrived somewhat confused about whether the ASCA analyzer and the digital computer were one or two projects. He explained how he thought he convinced Weaver that the two had to be one in the same.[39] Still, the visit made Forrester uneasy enough that he thought to articulate a formal rationale for his project, which he wrote down in a separate memo titled "Electronic Digital Computer Development at MIT."

Though it might seem like an oxymoron to say that a general-purpose computer could be built for a specific purpose, this was exactly the line of reasoning Forrester followed. Weaver's visit reminded Forrester of the applied mathematicians' interest in digital computers. Not coincidentally, this was exactly when the ONR was helping the NBS to set up the National Applied Mathematics Laboratories. In drawing certain inferences from comments Weaver made during his visit, Forrester described how his computer was well suited for current topics in mathematical research. His computer had a design speed of 5 megahertz. It would have a 40-bit word length, a 50-microsecond multiplication time, and 16,000 words of high-speed memory. All this would be valuable for the solution of difficult partial differential equations and other difficult mathematical problems. What was not noted in the memo was the fact that these specifications referred to their tentative thoughts about the final ASCA analyzer, and not their present plans for the pre-prototype. This was a detail that could be left obscure in the memorandum.[40]

Forrester agreed that this kind of computing speed was necessary only where there were highly iterative calculations. However, he pointed out that such iterative calculations also occurred in "certain engineering applications where the computer will control mechanical or electrical devices operating in real time." As early as January 1946, Forrester had recognized that "dynamic-systems" would be one type of application for digital computers. Nevertheless, it was the pressure generated by Weaver's visit that brought him to give more precise articulation to real-time control systems as a distinct area of application that stood alongside scientific computation in its need for computing speed.[41]

Forrester also wrote about the strong qualifications of his group, placing Project Whirlwind directly within the lineage of established research traditions at MIT. Forrester therefore began by casting his work as a "natural outcome" of Vannevar Bush's work and the long history of work on mathematical instruments at MIT. But by also citing MIT's work on electrical transients, Forrester drew more generally on the credibility of MIT's Electrical Engineering Department, as well as its mathematical approach to engineering analysis. Meanwhile, his reference to control-systems engineering placed his project firmly within the Servomechanisms Lab tradition, where he was on surer ground. These were all important rhetorical "modalities" for asserting the validity of his project.[42]

It may seem paradoxical that, although Weaver represented the ONR, the overall effect of his visit was to draw Forrester away from the ASCA analyzer and toward the broader applications for digital computers. Yet this was one of the explicit purposes of the ONR's scientific advisory committee. It was set up in part to prevent undue military influence over the course of academic research.

The second review was carried out by the SDD, which became the Special Devices Center when the organization was folded into the ONR at the time of the ONR's founding. Since every project received its allocations on an annual basis, negotiations over funding usually began well in advance of the start of a fiscal year (1 July). This conversation would have been routine had it not been for the anticipated delays. Forrester initiated the conversation with his first semi-annual progress report, submitted in January 1947.

Forthright about the changes in his schedule, Forrester asked to extend Phase I of the project out to the spring of 1948. He offered a carefully constructed rationale. He explained how the revised performance estimates for the ASCA analyzer made it necessary to drive the circuits at higher than expected speeds. This also meant that the project had to go after basic advances in the components, including its main switching circuits and pulse transformers. Their evaluation of the ASCA requirements also revealed the need for parallel digit transmission, which added to the preprototype's complexity. Since reliability was a major concern, Forrester informed the SDC that he also needed enough time to develop new error checking circuits suitable for the ASCA analyzer. Most of all, Forrester was careful to cast the delay as a considered technical judgment, rather than the product of technical frustrations and an accidental machine design.[43]

In April 1947, Forrester also had to amend his request, both with a request to extend construction out to June 1948, and to tap into the $528,000 designated for Phase II of the project. He wrote that since the machine they were referring to as the pre-prototype was now more nearly like the computer that would be used for the ASCA analyzer, funds earmarked for its design were reasonably spent on the construction of this first machine. Incidentally, Forrester never built the pre-prototype. Or, rather, it was the pre-prototype that simply became the Whirlwind I. Forrester instructed his staff to stop referring to their machine as a pre-prototype.[44]

The SDC's current director, who had replaced de Florez some time back, granted Forrester's request and accepted these changes as part of the inherent risks associated with fundamental research.[45] Still, someone inside the SDC, most likely Crawford, was concerned enough with the delays to ask Forrester to exercise an option stated in his contract. This was a standard provision that allowed MIT to employ a subcontractor. This clause dated back to the Rad Lab days, when it was important for MIT not to be saddled with mundane engineering responsibilities; reread in the postwar context, subcontracting also provided a way to ensure the academic integrity of the work on campus. The SDC nevertheless interpreted the clause as a way of speeding up Project Whirlwind.[46]

Forrester was not yet ready to hand over production drawings. But being eager to please his sponsor, Forrester approached the Boston Division of Sylvania Electric Products, a former subcontractor to the Radiation Laboratory. As a private firm, Sylvania had a 137 percent overhead. They estimated that the work would cost about $320,000. Forrester could not shoehorn this sum even into his Phase II allocations, so he had no choice but to ask the SDC for more funds. This generated the third round of review,[47] which consisted of two separate inquiries from the SDC's director. The first inquiry was about the cost. However, in formulating his query using customary language, the director simply asked Forrester to justify his use and choice of a subcontractor. This allowed Forrester to treat the inquiry as a matter of fact, stating that the request had originated with the SDC. Forrester offered no further evidence on this score. Regarding his choice of contractor, Forrester presented an impressive list of Sylvania's qualifications, owing to its wartime electronics work. What was left unspoken was the question of whether the SDC should spend an additional $320,000 on the ASCA analyzer.[48] In the second inquiry, Forrester was asked to justify the costs of his electrostatic storage tube work. By mid 1947, a full quarter of the project's staff members were tied up with this one effort, and the work seemed to have no end in sight. Here Forrester chose to submit a highly technical reply. He described how the current tubes were providing signals on the order of 0.1 volts, with a write time of 50 microseconds, a read time of 1–2 microseconds, and a storage density of five digits per inch. He then described the additional work that was needed to bring the tubes up to ASCA specifications. Although he did not have sufficient data to offer a firm schedule, he presented a self-imposed deadline of November 1948 for having a full complement of tubes, since he expected the Whirlwind I to be operating by that date. If the SDC's director circulated Forrester's response among any engineers knowledgeable about such work, none came forward with a critique strong enough to stop the work. Toward the end of September, the ONR signed a letter of intent approving the supplemental request.[49]

The added costs and the new delays prompted an independent inquiry by Mina Rees. By fall of 1947, Rees had come to believe that the "consensus of visitors to the project is that there is too much talk and not enough machine."[50] Technically, Project Whirlwind lay outside of Rees's jurisdiction. The Mathematics Branch was part of the ONR's Physical Sciences Division, whereas the SDC remained an independent unit that reported to the Chief of Naval Research. However, as noted by Redmond and Smith, this was already something of a demotion for the SDC, since it went from being a *division* under the ORI to a *center* under the ONR. In following the various turns in postwar science policy, the ONR came to emphasize work that was conducted along disciplinary lines as opposed to the project-oriented approach of the wartime NDRC organization. Under these circumstances, the naval interest in fundamental research, which had been working in Forrester's favor, now worked to his disadvantage. The projects orientation of the SDC, as reinforced by the center's desire to extend its reputation as a naval simulation and experimental facility, came to be seen as an oddity in the ONR's organization. The SDC's work was tolerated, but vulnerable. This gave Rees room to maneuver, and she gained special permission to review the mathematical aspects of Project Whirlwind.[51]

Unfortunately for Forrester, with the ASCA analyzer the mathematics was inseparable from the machine. So long as a computer remained tagged to a specific application, the mathematical adequacy of a project lay with issues of the machine design—just as von Neumann had argued. Rees therefore began to inquire into the engineering aspects of Project Whirlwind. As she began to understand the Whirlwind I's design, she grew troubled by the fact that the work seemed far more expensive and extensive than von Neumann's work at the IAS. Why, Rees asked, should the Whirlwind I cost more than the original?[52]

The issue came to a head in October 1947 when Rees pressed the SDC and Forrester to defend the mathematical adequacy of their work and to explain why Project Whirlwind cost so much more than the work at the IAS. Crawford intervened by agreeing to an independent review. Rees and Crawford settled on Francis Murray, a Columbia University applied mathematician who had worked directly with von Neumann and was therefore acceptable to all parties.[53]

Forrester had no choice but to treat this inquiry seriously. He focused his energies on the comparison with the IAS. He first turned to Everett for a good answer, but to no avail. Insofar as Everett based his work on the IAS design, he could provide no significant technical distinction between the two machines. There was some difference in speed, but not so much that it could explain away the cost differential. Desperate for a better answer, Forrester turned instead to the comment Weaver left him with at the end of his visit. Discarding his earlier effort to cast Project Whirlwind as a fundamental research effort, Forrester now cast his work as a practical, engineering endeavor quite different from the work at the IAS—an endeavor highly suited to an accomplished engineering institution like MIT.[54]

During Murray's visit, Forrester took the pain to describe his staff's engineering efforts in detail. He then followed Murray's visit with a 22-volume technical and administrative manual describing the work of Project Whirlwind. Forrester was careful not to deny the validity of von Neumann's work. Yet the known difficulties associated with von Neumann's project only bolstered his case.

In fact, it was precisely the expansive scale of Project Whirlwind that impressed Murray the most. Murray endorsed Project Whirlwind: "It should be clear from [this report] that engineering development is not the real concern of the Electronic Computer at IAS and this is quite proper. However, it is also true that engineering development is absolutely necessary for the development of electronic computers and delaying engineering development, say by postponing Whirlwind, will delay the use of digital computers in the type of problem with which we are concerned."[55]

Murray was eager to use computers. This was a result of the year he had spent as a visiting member of the IAS. Thus, although Murray conducted his investigation with the ASCA application in mind, the report's tenor belied his own interests. The 'we' in the above recommendation referred as much to the community of applied mathematicians interested in using digital computers, as it did the SDC's interest in flight simulation. Murray did agree that the mathematical analysis conducted to date was inadequate. However, he decided that such mathematical work would not impact the current engineering effort, since it would only affect later decisions about the final machine configuration. Once Rees was handed this positive evaluation, there was little she could do to derail Forrester's project.[56]

Technical Difficulties and Transparency

Forrester may have won this rhetorical skirmish, but his project kept heading into deeper waters. The root of the matter was that Rees was right. Forrester had committed himself to engineering development work without an adequate mathematical analysis of the ASCA requirements. Mathematics and engineering were not fully integrated in his project. Moreover, a simulator, especially one that could reproduce the "feel" of an aircraft's control stick through its real-time response, presented many physical subtleties. As Forrester's mathematicians came to realize the full scope of the problem's complexity, it became necessary to increase the power of the machine. But insofar as Everett committed the project to the limited parallelism of the IAS design, this placed tremendous demands on the computer's components and subsystems. This is not to say that there was a better technical solution. Nonetheless, these choices forced Forrester's staff to undertake fundamental components research. The various electronics groups were left trying to push their circuits to 5 megahertz and beyond, as opposed to the 1-megahertz design speed of the IAS computer. Even more daunting was the work on the electrostatic storage tubes.

To provide some sense of the group's technical difficulties, when the various electronics groups began designing their circuits to a specification of 5 megahertz, they had yet to design the test equipment they needed to check such circuitry. By February 1947, Fahnestock, as head of construction, could report that they had some working models, but that all their circuits required substantial improvements. The project suffered an additional setback when they began interconnecting their various components, only to discover that their flip-flops (bi-stable circuits that retain a binary state) could not properly drive an array of gates (the computer's logic circuits). They had to add additional circuits (buffer amplifiers and cathode followers), which made it necessary to add additional power supply voltages. Their new power supply had to deliver twelve closely regulated DC voltages to a machine that occupied an entire room, which was itself a "very serious" engineering challenge. Then in August the staff discovered an intermittent failure—a "black-out effect"—that required several months to isolate. Even after authorizing Sylvania to begin construction, MIT had to insist on significant design changes, including the complete replacement of the machine's most common vacuum tube. Other major changes occurred as late as August 1948. Despite the appeal of digital circuits, where "any voltage of 50% (or over)" supposedly registered, the digital abstraction turned out to be quite difficult to sustain.[57]

Similar difficulties plagued the work on electrostatic storage tubes. Here the principle challenge had to do more with issues of fabrication and process control. Storing a small electrical charge on a dielectric surface reliably and repeatedly presented a formidable physical challenge, one that plagued similar efforts elsewhere. This left the group chasing fundamental studies in dielectric materials, material deposition and anodizing techniques, and a series of design studies that aimed to reduce the amount of electrical shorts and vacuum leakages. Forrester's staffing policies did not help matters, for at first he hired only one engineer who had any significant experience working on similar devices. And while much of their investigations could have been carried out sequentially and more cost effectively, their self-imposed deadline forced them to explore many alternatives in parallel. By the end of 1948, the storage tube work was claiming 33 percent of Forrester's current expenditures.[58]

Above all else, the subcontract with Sylvania compounded matters. This was partly due to the familiar difficulties of technology transfer. But other difficulties resulted from Forrester's decision to push his experimental circuits into production. In his original negotiations with Sylvania, Forrester insisted that Sylvania allow for midcourse changes in the design, since Project Whirlwind was a fundamental research project. At the same time, because the subcontract was supposed to accelerate their work, Forrester arranged for Sylvania to produce the mechanical drawings and physical layouts (drawings of the actual location of the components) based on circuit diagrams submitted by MIT. The decision to use standard circuit racks instead of a denser packaging strategy also complicated matters, since it made the Whirlwind I a very large machine. The Whirlwind I required 72 full-height circuit racks, whereas the IAS computer was housed in a single frame.[59]

This arrangement with Sylvania was supposed to free Forrester's staff to focus on their research and on general design problems. However, each of the mechanical drawings and physical layouts had to be examined to ensure that they introduced no undesirable electrical effects. Once the construction of the Whirlwind I began in earnest, Forrester's laboratory became a veritable production shop for checking and rechecking all of

Sylvania's drawings. By the summer of 1948, Forrester had 70 professional staff members and 105 "non-staff" employees, most of whom were technicians. Over and above this figure were the several dozen engineers and draftsmen employed at Sylvania.[60]

Not surprisingly, new tensions erupted over funding. In early 1948, the SDC, being concerned about Project Whirlwind's accelerating expenditures, approached Forrester with the expectation that he would adjust his fiscal 1949 spending downwards. There was, moreover, the underlying issue that Project Whirlwind had a fixed contract, and was never intended to be an ongoing research program that could claim annual appropriations. But Forrester worked with Nathaniel Sage in developing a budget projection that was based specifically on a linear and sustained extrapolation of current expenditures. As Forrester's expenditures rose from $30,000 to over $100,000 a month, and with Sylvania submitting an estimated cost overrun of $287,000, Sage put Project Whirlwind's fiscal 1949 budget at around $1.8 million.[61] This was approximately 10 percent of the ONR's entire 1949 allocation for new university contracts, and almost the entire sum the ONR hoped to spend on mathematics research. As a result of the scale of this request, both Crawford and the SDC were called to task. Concerned from the outset that Project Whirlwind might require a great deal of fundamental research "which had no physical equipment as a product," Forrester had already instituted a rigorous program of technical reporting that ensured that the Navy would see their productive efforts. But it was becoming all too evident that despite having spent over a million dollars, all Forrester could show for his efforts was a pile of 480 engineering memos, 150 technical reports, and still no working computer. Crawford was transferred.[62]

The Extent and the Limits of Rhetoric

The future of Project Whirlwind now hung on Forrester's ability to reestablish the validity of his project in the eyes of the ONR. The ONR's outgoing chief, Admiral Paul Lee, issued a notice in June 1948 indicating that he would limit Forrester's fiscal 1949 expenditures to $900,000. As Forrester's work came to fall under the direct scrutiny of the ONR, he set out once again to rethink his project's identity. A new "staff indoctrination program," along with presentations at academic conferences, provided him with an opportunity to develop his ideas before a sympathetic audience. At the February 1949 meeting of the American Institute of Electrical Engineers, for instance, Forrester presented his latest ideas about "systems engineering." His staff had begun to install the components manufactured by Sylvania, only to discover that the system had various unanticipated characteristics. These technical difficulties added another year to the project's schedule, and budget. Although the term "systems engineering" was already beginning to have a somewhat different connotation within the field of electrical engineering, Forrester used the term to describe how systems integration was a separate and technically valid phase of computer-development research.[63]

Forrester's most important rhetorical tool was his "Limited Distribution" ("L-series") memoranda and reports. This series emerged somewhat fortuitously from an earlier inquiry by the SDC's director. Sometime after Weaver's visit, and as the cost of his project began to rise, Forrester and Crawford set out to cultivate a broad interest in digital computers within the SDC. After some conversations with the SDC staff, the then current SDC director asked Forrester to write a report on how digital computers could be used for anti-submarine warfare. Forrester and Everett delivered two reports describing an elaborate command and control system for tactical naval warfare. This system would collect target information from multiple vessels and aircraft carrying sonar and radar, and issue immediate recommendations regarding the deployment of depth charges and other weapons systems needed to meet an impending threat. Given that these documents dealt with delicate matters of military strategy, Forrester created the L-series memos to limit their circulation. A year later, Forrester revived the memos to discuss the sensitive issue of his project's future.[64]

What required immediate attention was a letter sent by the new Chief of Naval Research, Admiral T. A. Solberg, to MIT's president, Karl Compton. Solberg reaffirmed his predecessor's decision to limit Project Whirlwind's budget, and asserted, moreover, that he wished to "reexamine both the technical and financial scope of the project" through a direct meeting between the ONR and MIT. Compton asked Forrester to explain the situation. As noted by Redmond and Smith, this generated a "flurry of activity."[65]

Forrester organized a "planning committee" made up of himself, Everett, Fahnestock and two others. Their task, quite explicitly, was to find a sufficient justification for their project. In an initial meeting with Compton, Forrester gave him a preliminary account of how important digital computers were for the military establishment. Even though the Whirlwind I was still under construction, the project had attracted many visitors, each of whom introduced Forrester to different applications from military logistics to air traffic control. These remarks caught Compton's attention. Compton proceeded to instruct Forrester, via Nathaniel Sage, to prepare a report on the military applications of digital computers.[66]

In the first report, Limited Distribution Report L-3, Forrester and his committee set out to literally estimate the strategic importance of digital computers to the military establishment. They began by comparing their efforts to the wartime work on radar, and argued that computers had a similar and perhaps even greater potential for broad military application. They cited applications such as "air traffic control, integrated fire-control and combat information centers, interception networks, scientific and engineering research, guided missile offense and defense, and data processing in logistics." Supplemented with the claim that digital computers were technically more complex than radar, the committee estimated that the total national program in digital computers research would cost $2 billion, as expended over a fifteen-year period. By then, the first production equipment would enter military operation, with full deployment costing $1.97 billion more.[67]

Report L-4 then offered three alternatives for *expanding* Project Whirlwind's expenditures to an annual level of $1.8 million, $3.8 million and $5.8 million, respectively. At $3.8 million, Forrester promised to add a "substantial operating force for efficient solution of scientific and engineering problems." At $5.8 million, there would be an additional group "devoted to the application of digital computers to fields of control and military uses."

The two reports operated as a pair. By demonstrating the digital computer's applicability to a wide range of military applications, Forrester hoped to justify the level of expenditures associated with Project Whirlwind.[68] Both reports also employed a new modeling technique. Forrester and his committee fused the DIC's techniques for generating budget forecasts with the Servomechanism Laboratory's own traditions in mathematical modeling to construct a more elaborate model for plotting project expenditures. (See figure 5.2.) They began with the claim that "the

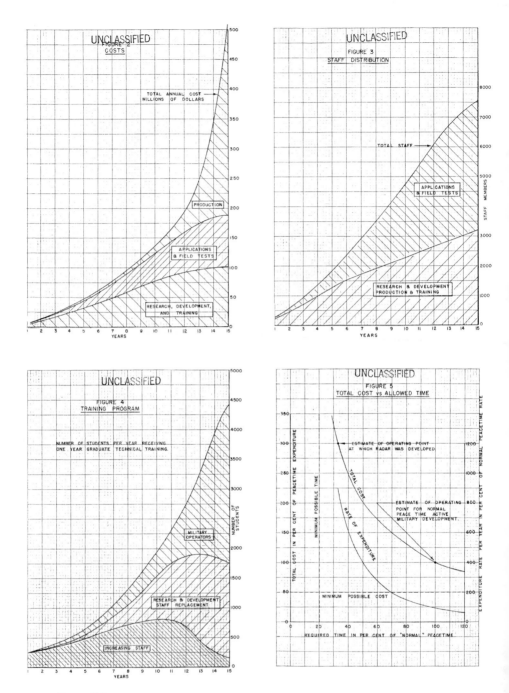

Figure 5.2
Forecast costs and scale of national program for digital computers research in the military establishment. From Jay Forrester et al., Report L-3, AC 290, Whirlwind Computer Collection, Box 15/4, Archives Center, National Museum of American History, Behring Center, Smithsonian Institution.

cost of making this equipment available to the Military Establishment has been much underestimated by linear extrapolation of past laboratory programs." They then took their own budgetary data to be a matter of fact, and reasoned that an exponential curve offered the best fit. The estimated cost of $2 billion was more or less a direct extrapolation of Project Whirlwind expenditures on a fifteen year exponential curve. On the other hand, the model was not simply an assessment of total costs. Using once again their own experience as a point of reference, the committee plotted out the staffing levels and training requirements of a national program. The latter estimate included separate calculations for staff increases, staff replacement and the training of military personnel. Given their familiarity with the costs of graduate education, and their difficulty hiring qualified staff, the group also folded this information into their analysis of total expenditures.[69]

The limits of a quantitative approach to objectivity are most apparent in the cases where they fail. Compton was pleased enough with Forrester's reports to encourage their distribution in Washington. He also allowed Forrester, Sage, and a sympathetic member of MIT's electrical engineering visiting committee to serve as his delegates in a meeting with Solberg.[70] This occurred on 22 September 1948. However, the meeting revealed severe weaknesses in Forrester's rhetoric. Nothing may have been said to directly dispute Forrester's claims. But the ONR simply expressed its discontent with Forrester's fiscal irresponsibility, and its growing skepticism about the technical merits and rate of progress of Project Whirlwind when compared against other computer-development projects. Sage and Forrester stood their ground and pushed Solberg to accept a compromise. By the end of the meeting, they had a signed agreement in which Solberg offered to provide $900,000 for the first 9 months of the fiscal year (through 1 April 1949), with the implication that additional support would be forthcoming if Forrester proved able to bring his rate of expenditures down to $100,000 a month.[71]

At this point, Forrester suffered not only austerity measures but also further scrutiny from MIT officials. Memos L-5 and L-6 were designed to reestablish the validity of Project Whirlwind in the eyes of MIT. Memo L-6 revisited the earlier comparison between Whirlwind I and the IAS machine, which had resurfaced during the ONR meeting. Here, Forrester and Everett placed in writing the opinions they had conveyed to Murray.[72]

Memo L-5 was a more reflective document. Written by Forrester, it was a document that defended Project Whirlwind as a legitimate undertaking of the Servomechanisms Lab. The memo was titled "Project Whirlwind: Principles Governing Servomechanisms Laboratory Research and Development." Forrester accepted that the main purpose of the Servomechanisms Laboratory was to undertake projects that "combine engineering research and development with systems considerations." In plain language, this was an admission that the laboratory preferred to build new systems out of existing components instead of undertaking fundamental components research. Forrester also conceded that in embarking on a project that was closely tied to graduate education—as much as a third of his technical staff were masters or doctoral students—he incurred an obligation for the fiscal stability of his project. Forrester granted that the Servomechanisms Lab preferred to undertake projects with a demonstrable application. But here, Forrester defended his actions, asserting that all the work to date supported the requirements of the ASCA analyzer.[73]

Forrester did concede that a flight simulator, in itself, might not justify the scale of his current expenditures. However, rather than regarding his work to have been misdirected, Forrester took the work to be valid and considered his failure to lie instead with not finding a suitable application—and sponsor—for Project Whirlwind. Put differently, Forrester was completely taken by digital computers. This was nowhere more apparent than the first principle he offered for Project Whirlwind research, which he placed alongside the principles of the Servomechanism Lab. This was the principle of Computer Importance: "We believe digital computers to be of great future importance in control systems for both military and civilian application." This was an axiom, not a question to be addressed through engineering inquiry.[74]

There were more subtle constructions to his argument. The principle of "Urgency" suggested how it was appropriate for Project Whirlwind to undertake fundamental components research. The principle of "Speed" adapted the argument he made after Weaver's visit, by claiming that control-systems applications would make use of all available computing speed. Meanwhile, Forrester rewrote his project's history in describing the principle of "Reliability." Instead of admitting that that size of the machine resulted from an early decision having to do with the testability and ease of assembly of the pre-prototype, it was now a feature designed to ensure ultra-reliable operation.[75]

Taken as a whole, Forrester used these memos to reconstitute a coherent mission for Project Whirlwind. Fundamentally, Project Whirlwind was now a research program in computer applications, especially in the realm of real-time control systems. However, in order to pursue this research, it was necessary to have a fast and reliable computer. This required large initial expenditures on the physical aspects of computer engineering work. It also required fundamental components research of the kind "the Servomechanisms Laboratory has always undertaken . . . to guarantee success of the work." With this re-articulation, Forrester could justify, at least to himself, all of Project Whirlwind's past and present activities.[76]

As the technical difficulties dragged on for another year, Forrester set out to construct a budget more mindful of the ONR's priorities. Although he wished to expand his study of computer applications, he had to accept that "the present loss of confidence, by groups outside of MIT" made it necessary to focus on completing the Whirlwind I. Rees had told him to do so. Forrester continued to note that caving to budgetary pressures might "cost heavily in future preparations." He turned once again to his new modeling technique. Memo L-10 described how a 1950 budget of $1.2 million would sustain a "balanced organization" that would complete Whirlwind I by January 1950. A budget of $1.075 million would extend construction an extra month, resulting in a savings of only $40,000. If the budget was reduced to $975,000 or less, this would necessitate "personnel loss and reorganization" at the peak of the project. This would generate major inefficiencies and increase the final cost of the project.[77]

The going was not easy. In February 1949, technical oversight for Project Whirlwind was transferred to the Mathematics Branch. The ONR not only limited Forrester's fiscal 1950 budget to $750,000, but withheld these funds until he could demonstrate that his project could overcome its technical difficulties. As of 1 July 1949, Forrester had only $40,000 in unexpended funds, and a hundred staff members on his payroll. Desperate, his staff set out to assemble enough of the Whirlwind I to collect meaningful data on the source of the machine failures. On 17 August they ran the first test program on the machine. Although this did not bring Whirlwind I into useful operation, it was enough to convince the ONR to release its money.[78]

Closure and Rhetorical Practice

As Forrester's relationship with the Navy soured, the Air Force stepped in to help. Driven by the need to deploy a continental air defense system, which became a national priority in September 1949 when the Soviet Union tested an atomic bomb, the Air Force had the deep pockets Forrester needed to bring Project Whirlwind to a successful end.

Yet from the standpoint of a more formal definition of "closure," as found in the science-studies literature, it is necessary to separate out these two stories into one of failure, and one of success.[79] Especially when viewed from the standpoint of rhetoric, Project Whirlwind did in fact reach closure with the Navy. In the eyes of the Navy, there was no longer a viable program for computer-development research. By the same token, Forrester's rhetorical strategies, when taken along with the artifacts and research organization he assembled, provided an important foundation for his negotiations with the Air Force. Forrester's rhetoric, forged over the course of his difficult negotiations with the ONR, survived as a body of practice that opened a path to a very different endpoint.

In terms of the actual historical events, Forrester's relationship with the Navy ended as a result of the investigation carried out by the Research and Development Board. I have already described the RDB as an organization created to help the Department of Defense maintain a sound and unified national research program. (See chapter 3.) What remains to be discussed here is the extent to which Project Whirlwind instigated the review by the RDB's Ad Hoc Panel on Electronic Digital Computers.

Even Vannevar Bush referred to the situation at MIT when he created in early 1948 the RDB's Committee on Basic Physical Sciences (CBPS).[80] But it was Karl Compton, once again, who intervened directly in the committee's affairs. Compton had agreed to serve as the interim chair of the RDB even before the 22 September meeting at the ONR. Apparently unaware of Compton's pending appointment, Admiral Solberg suggested to him that Project Whirlwind be reviewed in the context of an overall evaluation of digital computers research conducted by the RDB. After the 22 September meeting, Compton implied that he would make sure this was done. Whether or not this meant that Compton had lost confidence in Forrester, Compton apparently saw the RDB as a just forum with which to bring the relationship between Forrester and the ONR to some kind of resolution. This would avoid any unnecessary confrontation between MIT and the ONR. Compton conveyed his wishes through MIT's Dean of Engineering, T. K. Sherwood, who held a seat on the CBPS. This in turn produced a private meeting that included CBPS Chairman William Houston and ONR Chief Scientist Alan Waterman. Compton, citing a conflict of interest, recused himself from any further decisions. Nevertheless, he had interjected a local issue into the RDB's machinery.[81]

In considering how harsh the ad hoc panel was about Project Whirlwind, there is irony in the fact that the panel drew heavily on the concepts and language laid out by Forrester. The ad hoc panel's notion of a "prototype" built on a category Forrester (and von Neumann) helped to construct. One of the panel's recommendations, to increase computer components research, also derived from Forrester's emphasis on electrostatic storage tubes and fundamental components research. Nyquist, who worked on control-systems engineering at Bell Labs, took note of Forrester's claims about systems engineering and system integration, and this led him to bring his own knowledge of systems engineering to bear on the panel's report. On a less promising note, the notion of end use emerged from the debate over whether Forrester designed his computer with a specific appli-cation in mind. Meanwhile the multiple investigations and criticisms of Project Whirlwind, and Forrester's rejoinders to them, provided rich fodder for the panel's investigative approach.[82]

The harshest criticisms of the ad hoc panel's report were in fact reserved for Project Whirlwind. The ad hoc panel described the Whirlwind I as a "very large machine with but five decimal digits capacity, [and] extremely limited memory capacity." Although the computing circuits were said to be 85 percent complete, there was limited progress on crucial components of the machine. The electrostatic memory system employed a "very expen-sive construction" and was far from producing acceptable results. Also, there were as yet only indefinite plans for the input-output equipment. Despite Forrester's claims about systems engineering, the panel found his work to be the worst from a systems point of view. On the criterion of "end use," during the fiscal crisis of mid 1948, Forrester and the SDC had decided to terminate all work on the cockpit and other parts of the ASCA analyzer so that he could focus on completing the Whirlwind I. This, along with the project's transfer from the SDC to the Mathematics Branch, led the panel to conclude that the project was not being pursued with any well-defined end use.[83]

Most important, the panel found that Forrester had spent $3 million out of an estimated $5 million spent to date on all digital computer development paid for by the federal government. If there was one project whose termination could affect a significant savings, it was Project Whirlwind. The panel found that "the scale of effort on this project is out of all proportion to the effort being expended on other projects having better specified objectives." The panel recommended that Forrester's project be terminated, if no appropriate end use could be found.[84]

Forrester turned once again to his Limited Distribution memoranda to defend his project. Once more, he produced two complementary documents. L-17 was written to undermine the credibility of the ad hoc committee and its report. L-16 responded to the specific criticisms against Project Whirlwind. Forrester defended the notion that broad military objectives, as opposed to a contractually specified application could justify digital computers research. He also attacked the report where it was most vulnerable, namely its endorsement of the Harvard Mark III as a suitable prototype.

Forrester also continued to refine his notion of real-time computing, describing how it and scientific computing were "based on very different considerations." There was less need for precision, which justified the Whirlwind I's word length. On the other hand, real-time applications imposed tremendous demands on fast access to main memory, which made it necessary to undertake the work on electrostatic storage tubes. Citing what he saw to be a contradiction in the panel's report, Forrester suggested how the panel condemned his work on CRT memories even when it called for more components research. He also complained bitterly that "Whirlwind ideas have fared well in the hands of the Panel but not the actual project or the Whirlwind I computer." Many of the arguments Forrester constructed in an attempt to save his project were all too easily refashioned into critiques.[85]

In spite of his practiced rhetoric, this time Forrester's words served only to make his own cage. Forced to assemble his arguments in one place, the various omissions and inconsistencies in his arguments became all too transparent. Moreover, Forrester had written a 57-page rejoinder to a 55-page report—of which only three and a half pages were devoted to Project

Whirlwind. The overall impression conveyed by his new memos was that of an overly ambitious project burdened with too many technical challenges. Forrester was no longer able to occlude the difficulties associated with his project through his use of rhetoric.

Although the ad hoc panel clearly exceeded the RDB's statutory authority in recommending the central administration of computer-development research, this did not mean that the report had no effect on specific projects. (Indeed, this was true of the RDB at large.) The ONR now had the evidence it needed to bring the project to a halt. On 6 and 7 March 1950, ONR representatives met with MIT officials. This time Forrester was excluded from the first meeting. During this session, the decision was made to curtail the ONR's support to $330,000 for the 1951 fiscal year. Moreover, MIT's new provost, Julius Stratton, put forward a suggestion that made it easier for the ONR to redirect all future support to a study of computer applications. From fiscal 1952 onwards, the ONR did in fact specify that its support be directed to a study of scientific and engineering computing rather than Forrester's work in computer development.[86]

At this point I return to the historiographic issue I raised in the opening passage of this chapter. Andrew Pickering, in *The Mangle of Practice* (1995), also examines the many unforeseen turns in Forrester's work, specifically during the developments that followed the introduction of numerically controlled machine tools derived from Project Whirlwind at General Electric's aircraft engine manufacturing facility in Lynn, Massachusetts. At one level, this chapter simply takes this story out to the beginning in describing how contingencies drove Forrester's project from its outset.[87]

Yet I believe Pickering misreads Noble's account of what happened at General Electric in a way that deserves attention. Pickering's focus on technological practice and "material agency"—the importance of paying attention to how physical objects affect historical outcomes—serves as an important foundation for this chapter, and indeed my book as a whole. Nevertheless, in his desire to pay close attention to artifacts and how they co-determine historical outcomes, Pickering tends to downplay the importance of social and historical contexts. Specifically, Pickering criticizes Noble for referring to a managerial ideology of control as an important determinant of the events at General Electric, where management ultimately deprived workers of an experimental program for operating numerically controlled machine tools through a system of worker self-management. His discontent with Noble is that this reference to managerial ideology amounts to an unwarranted claim that social institutions determine historical outcomes.[88]

Yet, from the standpoint of a social or labor historian, it is Pickering who naively misreads Noble's reference to social institutions. In considering the time and place where the events occurred—a postwar struggle for workplace control in an old-line East Coast firm and an industry with a strong tradition of labor organization—there is little doubt that managerial ideologies of control, and worker strategies of resistance, were well encoded into the practices of those at the GE Lynn facility. Noble's account is rife with such evidence. Moreover, Noble saw that these practices were being reworked in the context of competing managerial ideologies, one traditional and one more progressive. The threat of automation also brought many issues to the fore. It makes sense then that social institutions, as instantiated through local practices, contributed as much to the historical outcome as anything having to do with the technical difficulties encountered with early numerically controlled machines. Historians may be somewhat careless in their use of language in attributing causality to social institutions, but Pickering's criticism also amounts to a misreading of the disciplinary conventions of the historian's craft.[89]

The same general motif regarding how institutions structure historical outcomes applies to my interpretation of Project Whirlwind's history, but as slightly transposed. What I see in Noble's account is the suggestion that when institutional relationships are substantially defined, there tends to be little room to maneuver. Experiments that test entrenched social relations remain vulnerable, and can be quickly brought to an end. In contrast, Forrester found very little in the way of socially specified constraints during the early stages of his project. He manipulated institutional relationships as easily as he manipulated technical choices and organizational structures. By the same token, as the group became committed to particular technical approaches and organizational solutions, and bound itself to institutional commitments that became increasingly enforceable, the choices open to Forrester came to lie increasingly with a body of rhetoric that could justify his approach to digital computers research.

Still, the flexibility with which Forrester could cast his rhetoric suggests the extent to which the *institutional arrangements* for Cold War research were not yet well defined. There was little in the way of social conventions that dictated what Forrester, Crawford, Sage, or any of the senior officials at the ONR and at MIT could or had to do. At the same time, elevating Project Whirlwind's technical difficulties to a level of rhetoric ensured that these conversations would not be just about how to build a digital computer, but about defining the proper relationship between MIT and the ONR. History should not be too harsh on Forrester. He was a young engineer, with limited administrative experience, placed in charge of a major research project precisely because historical circumstances called for the rapid expansion and articulation of this larger relationship.[90]

Redmond and Smith reserve their judgment as to whether Project Whirlwind served as a "test case" in defining the relationship between MIT and the ONR. But as I have already suggested, I believe their question is miscast. What mattered as much, if not more, was how the crisis brought specific ideas and concerns to the attention of MIT and senior ONR administrators. The actions taken in response to Project Whirlwind would become a matter of institutional practice, and policy, at MIT. In the wake of the controversy over Project Whirlwind, MIT's administration, beginning with Compton's successor, James Killian, would push for more long term funding commitments from the ONR and other agencies. They would express an aversion to large, mission oriented research projects. The administration would also discourage its faculty from entering into subcontracts, and insist instead that military agencies issue these contracts directly to the manufacturers enlisted to support or complete the work begun by MIT. A good deal of MIT's personnel ˙policies, including its commitment to its graduate students and "non-staff" employees, were articulated as a result of the considerable hiring and layoffs associated with Project Whirlwind.[91]

These were not trivial matters. Much of the administrative negotiations over MIT's subsequent work on the SAGE[92] air defense system—Killian's hesitations, the decision to set up Lincoln Laboratories as an independent research facility, and even MIT's desire to avoid the ultimate systems engineering responsibility for the project—all appear in a different light when viewed against the backdrop of the multiple crises surrounding Project Whirlwind. This is not to say that MIT's attitude toward postwar research administration was shaped by Project Whirlwind alone. MIT's wartime experience with the Radiation Laboratory, and the negotiations over the postwar Research Laboratory for Electronics contributed as much if not more to MIT's postwar institutional policies. Other articulations of policy occurred because of Charles Draper's Instrumentation Laboratory, as described by Donald MacKenzie (1990) and Stuart Leslie (1993). Still, it is significant that at the time of Forrester's fiscal crisis in 1948, Project Whirlwind was the single largest research project on campus. Moreover, no other project butted up against institutional concerns with such alacrity. The fact that Compton chaired the Research and Development Board and Killian served as Special Assistant for Science and Technology in the Eisenhower administration also suggests how local events may have contributed to larger conversations about national research policy.

So long as the bulk of the ONR's records remain closed to general historical study, it is more difficult to get a clear picture of how the ONR's encounter with Project Whirlwind shaped its policies and administrative practices. However, it is possible to draw some inferences by considering Project Whirlwind in the context of Harvey Sapolsky's (1990) commissioned history of the ONR. The ONR's Contract Research Program was the principal source of funding for basic research at U.S. universities during the first decade after World War II. As such, the ONR encountered similar administrative problems with many of its other dealings with academic institutions. In this sense, Project Whirlwind's effect on the ONR may have been less extensive than it was for MIT. But what is striking about Redmond and Smith's account of Project Whirlwind, as well as my own, is the extent to which the image of the ONR that it portrays stands in contrast to Sapolsky's, namely that of a hands-off agency that sought to court university researchers with the promise of academic freedom. Far from being a contradiction, this only reinforces Redmond and Smith's original interpretation, reiterated in their more recent study (2000), that the mission oriented activities of the SDC were an anomaly within the ONR. Sapolsky makes it clear that this only became truer as the ONR came to define itself increasingly as an agency dedicated to supporting basic research in the physical sciences.

On the other hand, Project Whirlwind may well have encouraged senior ONR administrators to shift more rapidly to this position. Mission-oriented research was always subject to the risk that a difficulty would crop up that would require the kind of intervention demonstrated with Project Whirlwind. If the ONR's goal was to develop sustained relations with the universities, as suggested by Sapolsky, then this was clearly a liability. Not surprisingly, although the ONR's enabling legislation gave the office the authority to conduct and coordinate applied research in the Navy, this was the kind of work that the ONR grew increasingly willing to leave to the Navy's principal materiel bureaus.[93] The difficulties the ONR experienced with Project Whirlwind must have contributed to the articulation of this problem.

For instance, at the March 1950 meeting during which the ONR decided to seriously curtail its support, the ONR accepted that only one-half of the Whirlwind I's computer time would be dedicated to general scientific and engineering use, with the other half being devoted to military applications supported by the Air Force and other military agencies. Notes from this meeting make it clear that this was seen neither as an improper appropriation of a Navy project by the Air Force, nor as a way to eventually kill the project, but a welcome transfer of the project to mission oriented military entities. The ONR's chief scientist, Alan Waterman, who would come to serve as the National Science Foundation's first director, explicitly noted that this would be a satisfactory arrangement.[94] Beyond such speculation, it is difficult to ascertain the precise effects of Project Whirlwind without a close study of the ONR's administrative records. Nevertheless, it would appear that the ONR's encounter with Project Whirlwind played a part in supporting the course it charted for itself in the context of postwar research.

What saved Forrester was the new work on the SAGE air defense system. Rees labeled Forrester "lucky" for this turn of events. Certainly the timing of the Air Force's new interest in Project Whirlwind was fortuitous. Yet in the course of his struggles with the Navy, Forrester had assembled an Applications Study Group to demonstrate that his computer could have a valid "end use." He was actively searching for new sponsors. Indeed, he had already begun conversations with another group within the Air Force on using the Whirlwind computer to study military air traffic control. He also had in hand the two reports that described a tactical command and control system for anti-submarine warfare. Not surprisingly, in the wake of the public panic over Soviet atomic capability, Forrester was able to offer his L-1 and L-2 memos to Chairman George Valley of the Air Force's new Air Defense System Engineering Committee. Valley was a physicist and a professor at MIT. By then, Forrester also had a digital computer that was working well enough so that Valley could be given an impressive tour.[95]

But equally important to Forrester's subsequent success was his practiced rhetoric. Forrester could offer Valley a smooth explanation of real-time computing. He could speak about how such a computer had to be engineered for ease of manufacturing and reliability. A high-speed memory system was necessary for integrating combat and weapons systems data in real time. Forrester also spoke to Valley about the value of preserving a sizable research organization with a well-trained staff that could serve as the nucleus of a major new strategic program. Although there was a need for further studies, the decision was eventually made to place Project Whirlwind at the heart of a new project for continental air defense. Once this was arranged, Forrester had little difficulty getting the additional $2 million he needed to complete the Whirlwind I.

Forrester's rhetorical practices outlasted Project Whirlwind. Echoes of his arguments can be discerned in even the most recent historical accounts, as when historians tout Forrester's contributions to real-time computing. Most important, Forrester's battle with the ONR gave him a novel forecasting tool. In speaking of "balanced organizations," and in using mathematical equations to model such things as staff levels and expenditures, Forrester transposed certain concepts from the field of control-systems engineering into the study of social organizations. These were tools, born of a crisis, which Forrester brought to his subsequent studies in system dynamics. Forrester's interest in industrial management may in fact have stemmed from the very difficulty he experienced in trying to devise a quantitative method convincing enough to shape matters of national policy. In this field as well, Forrester would amass a substantial reputation as a practiced rhetorician.[96]

Figure 6.1
Cuthbert Hurd (second from left) and others at IBM's laboratory in Poughkeepsie,
New York. From Box 6/23, Cuthbert Hurd Papers, Charles Babbage Institute for the
History of Information Processing, University of Minnesota Libraries, Minneapolis.

6   Institutions and Border Crossings: Cuthbert Hurd and His Applied Science Field Men

If the Cold War created many opportunities for universities in the United States, it also created new opportunities for businesses there. Backed by military commitments to a more technologically advanced arsenal, federal research expenditures continued to rise steadily after the initial contraction at the end of World War II. Expenditures rose to wartime levels by 1950 before redoubling itself with the Korean War. Yet if the period between 1946 and 1953 represented the origins of the "military-industrial complex" as identified by Eisenhower in his famous presidential farewell address, the extensive institutional allegiance between business and government, and between science and industry, constituted a new direction for many firms. This was especially true for firms that were outside of the traditional sphere of the U.S. defense industry.

Owing to historical concerns that arose from the radicalism of the 1960s, our historiography has tended to focus on the question of how the Cold War affected U.S. research universities. The previous chapters have tended to follow this line of inquiry. However, there remains a need for comparable studies of the equally important and extensive transformation of U.S. firms. In this chapter, I turn therefore to a very different institutional ecology, or rather a new intersection of institutional ecologies forged during the effort to draw more U.S. corporations into Cold War research.[1] Specifically, both this chapter and the next approach this issue through case studies involving the International Business Machines (IBM) Corporation. This chapter focuses more or less inwards on IBM and the Applied Science Department the firm established in 1949. Chapter 7 focuses on the IBM users' group, Share.

IBM's Applied Science Representatives and their director, Cuthbert Hurd, helped facilitate IBM's entry into the postwar technical computing market.

They also secured IBM's commitment to digital computers. In following the early activities of the Applied Science Department, I abandon, for the moment, any direct interest in what may have been typical about IBM and its place within postwar changes in U.S. industry. I do so in order to elucidate, in some detail, the process that was necessary for this one firm to span the gap between its corporate traditions and the tradition of the emergent postwar scientific community. Nonetheless, by the end of this narrative, it should be clear that the kind of work performed by IBM's Applied Science Representatives was an effort all firms had to undertake to enter the expanding military and scientific sectors of the Cold War economy.[2]

There was nothing radically new about the work Cuthbert Hurd and his Applied Science Representatives did to cultivate a market for technical computing. The existing literature on the history of sales and salesmanship demonstrates how sales representatives, in their various guises, learned to foster product loyalties, develop inter-firm relations, and educate consumers about new products and technologies. Quite often, they also served as important conduits for customer concerns and requirements, which successful firms learned to integrate into their corporate strategies and product-development plans. Such work began during the late nineteenth century, and continued with the growth of the modern corporation. This was the work performed by Hurd and his Applied Science Representatives. More broadly, Olivier Zunz, in his subtle analysis of the sales agents of the McCormick Harvesting Machinery Company, describes how these agents brought western farms into the fold of an increasingly industrialized U.S. economy. As suggested by Zunz, they took part in "making America corporate." Though on a somewhat different scale, this was the work IBM's Applied Science Department did in managing yet another important transition for American corporations.[3]

Still, the Cold War era presented unfamiliar opportunities and challenges, many of them having to do with the postwar authority of scientists. After the wartime development of the atomic bomb, radar, and penicillin, not to mention the computer itself, scientists entered the postwar years amid new public perceptions about the efficacy of their esoteric knowledge. This had many implications beyond new corporate commitments to fundamental research. It affected, for instance, IBM's sales relationships. While these effects do not give way to any simple reductive explanation, among their more interesting features were the customers' ability to push IBM in new technical directions, IBM's perceived need to hire specialists conversant in the new language of science, and the permeation, arguably, of academic norms into the corporate practices of IBM. For instance, IBM decided not to commodify certain kinds of knowledge, and learned instead to encourage the free and open circulation of ideas among its customer installations. This was clearly a reflection of the increased dominance of academic institutions in this area of advanced technology. IBM, in any event, could not treat the postwar scientific community as yet another customer. Although IBM, in the end, would find a way to assert its corporate interests, they would do so only after the firm absorbed the indelible imprint of a broader scientific culture into its corporate identity.

Background

It is worth recapitulating several aspects of IBM's early history. IBM was established in 1924 as an outgrowth of a company created by Herman Hollerith in 1896. It remained a mid-size firm throughout the 1920s and the 1930s, specializing in the manufacture and lease-service arrangement of tabulating machines. Unlike bigger firms like General Electric and AT&T, IBM had no modern industrial research facility. IBM and its competitors retained traditional inventions and engineering departments that tied research quite closely to product development. Throughout the interwar years this remained an asset rather than a liability. Best understood within the context of the literature on the professionalization of management, an important element in the success of IBM and its rivals was an ability to capture new niches for corporate accounting and accountability through product innovation.[4]

IBM's success was also attributable to the strength of its sales culture. It was John Patterson, the president of one of IBM's competitors, National Cash Register, who pioneered such modern sales practices as assigned territories and statistical quotas, along with sales schools, a "Hundred Point Club," and the associated motivational rhetoric designed to cultivate a sales staff's sense of loyalty. Patterson's efforts to rationalize his sales force were part of a much broader movement among all national manufacturers to control the presentation of their products. But whereas mass advertising came to dominate consumer goods, sales representatives continued to play a vital role in the producer goods sector where complex negotiations continued to occur with respect to pricing and customer requirements. The increased professionalism of a sales force could also play into a firm's ability to collect information pertinent to product development. IBM, along with other firms in the industry, adopted NCR's practices lock, stock, and barrel. IBM President Thomas J. Watson specifically inherited Patterson's sales practices during his early tenure as an NCR executive.[5]

During the 1930s, office equipment manufacturers, including IBM, sought to augment their customer relations through a "system selling" strategy. The effort to sell a "system," as opposed to a product, originated with the electrical industries, when the Edison companies sought to sell an integrated system for electrical lighting based on the strength of a single patent. While a technical rationale often underlay all system-selling strategies, a separate if not dominant motive was that of raising the customer's cost of switching to a system produced by a competing manufacturer.[6] System selling never lost this connotation. Nevertheless, the approach did take on new meaning as customers for office machinery struggled to integrate expensive equipment into their increasingly complex accounting and paperwork operations. To support such efforts, IBM first established a Methods Research Department in 1931, followed by a Systems Service Department several years later. The latter group, which was staffed entirely by women, played a part quite analogous to that which would be played by Hurd and his Applied Science Representatives. IBM's systems service women, who were formally assigned to "assist" IBM salesmen, became specialists in the use of IBM machinery. They helped customers select and install new equipment, circulated useful knowledge among IBM customers, and helped bolster IBM's image for service.[7]

Despite this organizational precedent, IBM's entry into Cold War markets did not emerge directly from this path. Considering the role that computing and applied mathematics played in technical developments during World War II, IBM executives proceeded to cultivate a direct relationship with the postwar scientific community. As described in chapter 3, this led Watson to create a new Department of Pure Science in collaboration with Columbia University. However, Watson's primary concern, once again, was not that of pursuing an undeveloped market, but of absorbing wartime computing and electronics developments into the firm's core business and product line. As a consequence, IBM tried to treat its technical computing installations as regular customers. But in the months after World War II, the IBM methods promulgated by Wallace Eckert proceeded to spread rapidly across many fields of science and engineering, and especially across the defense-related industries. Applied mathematicians associated with the nuclear weapons program and former members of the Applied Mathematics Panel contributed at least as much to this diffusion as Wallace Eckert's group.[8]

Thus, even as many laboratories struggled to build the first stored-program computers, researchers in fields as diverse as nuclear engineering, aeronautics, geology, and optics turned to existing IBM machinery to move on with their work. IBM's first electronic units, the IBM 603 multiplier introduced in 1946, followed soon after by the IBM 604 electronic calculator and the somewhat more elaborate Card Programmed Electronic Calculator (generally known as the CPC), were designed partly in response to the perceived needs of these new scientific customers. Nevertheless, IBM's normal sales representatives continued to find it difficult to decipher the peculiar requests emanating from scientific and engineering installations.[9]

It was IBM's commitment to the CPC, introduced in late 1949, that elicited Watson's decision to establish an Applied Science Department to help deal with scientific customers. Watson clearly understood that marketing, as opposed to research, required a separate organization. But as another indication of the customer driven nature of IBM's entry into scientific computing, Cuthbert Hurd was already at IBM. A research mathematician and a former dean of Allegheny College, Hurd set up the high-speed computing and statistics section at Oak Ridge National Laboratory. At Oak Ridge, Hurd turned to IBM equipment to solve statistical problems associated with nuclear materials processing. Keen on introducing industrial statistics to other firms, Hurd decided of his own volition to join IBM.[10]

The first several months that Hurd spent at IBM were quite unstructured. Seeing himself more as an academic than a business person, and a scientist rather than someone committed to technical sales, Hurd continued to spend much of his time attending to his professional correspondence and scientific affairs. He was given an office at IBM's corporate headquarters in Manhattan. He was asked to field any inquiry about scientific computing he received from IBM's sales agents. Nevertheless, he spent as much time organizing a conference on a new statistical method for the physical sciences known as the Monte Carlo method, a task he had agreed to do before joining IBM. But then Hurd proceeded to improvise on this academic tradition by organizing a pair of seminars describing the CPC. It was this initiative that earned the attention of IBM's top executive.[11]

With Hurd's new appointment as the director of the Applied Science Department, IBM now had a mathematician in their midst who was prepared to help sell the CPC. Shortly after the appointment, a colleague from the Rand Corporation wrote: "I am pleased to hear that IBM corporation has finally decided to set up such a group as yours. May I ask how you plan to make contact with the customer?"[12] Hurd recognized that he needed a more permanent arrangement than an occasional seminar and whatever polite references to IBM equipment he could include in his professional correspondence. Fortunately, several of the district sales offices had already hired individuals with technical backgrounds to deal primarily with scientific customers, and the sales manager of IBM's Electric Accounting Machines Division had already told these men to report to Hurd on all matters related to the CPC. When clients from other districts began asking about the CPC, Hurd offered one of "his" men for the job.[13] Realizing that more of these men could be stationed in the field to act proactively, Hurd replied: "You asked an interesting question concerning how the group which we are now establishing plans to make contact with the customer. We have not arranged all details, but tentatively we plan to have men assigned on what would be a regional basis in the United States. We must, of course, obtain the men."[14]

It was Hurd who specifically bypassed the systems service women. If this decision was based on real differences between technical computing and the accounting methods that were the women's expertise, it was also consistent with broader trends in postwar labor history. Although women had made substantial inroads into accounting and mathematics during the 1920s and again during World War II, the immediate postwar years saw the aggressive retrenchment of a gender-based segmentation of labor. This segmentation was easiest to implement when establishing a new category of work. Moreover, in considering the isolated work of promoting IBM equipment in the field, Hurd specifically sought out young men, preferably with an advanced degree in science or mathematics. Although Hurd did hire a number of women who were involved with wartime computing work as part of the Applied Science Staff that worked with him in New York, all the Applied Science Representatives, whom Hurd referred to as his "field men," were in fact men. These men had to reacquire much of the knowledge that the systems service women had presumably already learned about supplementing IBM's sales organization.[15]

The Applied Science Field Men

Hurd obtained his men and began casting them out into their disparate territories beginning in late 1949. IBM's district sales offices provided a valuable infrastructure for placing his "special representatives." Moreover, since IBM equipment was installed just about everywhere, IBM salesmen were often able to provide a necessary introduction. To avoid possible tensions, Hurd placed his men on strict salary. They in fact never closed a sale, but simply forwarded a prospective client to the usual salesman assigned to a particular territory. This made the field men a popular asset, as the regular salesmen came to see the field men as a valuable aid in making an expensive sale.[16]

Still, Hurd could offer no straightforward plan with which to "make contact with the customer." Every field man was given some training on the CPC. Beyond this, Hurd could offer his men little guidance. Each of the men came to understand that they had to act like a salesman in some ways. But they also knew they had to carve out a different identity, and find a specific justification for their work that was grounded in their technical expertise. Regardless of their somewhat similar backgrounds—most had an undergraduate degree in science or mathematics, although some also had a PhD—each of the field men had to forge their own strategy. After all, each of the sales districts presented a distinct market and conducted its business in particular ways. Each of Hurd's men had to assess his own market, learn to work with the local sales culture, and cultivate an appropriate demand for technical computing. To be effective, they had to rework their own identity, much as Hurd had already begun to do. They all did so, with varied success. The challenges of this work, and the access each field man gained, are best seen through an account of individual representatives.[17]

The Professional
Donald Pendery stepped onto the arid terrain of Los Angeles, leaving behind the damp weather of Seattle. Southern California was the heart of the postwar military aviation industry, and much of the region's postwar

boom was due to the output of this one industry alone. Pendery could see many new buildings toting such names as Hughes, Lockheed, and North American Aviation from his first glance across the airport's tarmac.

Pendery was one of Hurd's first representatives. He first worked for Hurd as part of what was then an undifferentiated Applied Science staff, although Pendery spent much of his time at trade shows and with distant clients. One of these early journeys brought him to Seattle, where he had spent several weeks installing Boeing's first CPC. He then collaborated with Boeing's computing staff in drawing up their first CPC programs. Next, Pendery was sent to Los Angeles to deal with a minor crisis because of this aviation experience. He soon found himself reassigned to Southern California on a permanent basis. In April 1950, Pendery took up the work of introducing the CPC to the region's aviation firms.[18]

The West Coast aviation industry had already made an aggressive foray into technical computing. After the United States rapidly reduced the size of its ground forces, strategic air power emerged as an important substitute.[19] The Air Force, which again was established as an independent branch of the armed services in 1947, undertook a major initiative in aviation research. Up until then, analog computation and relatively simple numerical calculations dominated aviation engineering work. But with the new emphasis on advanced weapons systems, which included new systems such as guided missiles and jet aircraft, many engineers began to seek increased precision through digital computing techniques.[20]

In fact, the CPC was an invention that emerged from Southern California's aviation industry. Research mathematicians at Los Alamos, or more precisely their staff, could afford to spend considerable time setting up complex problems on minimally modified IBM machinery. However, a typical aviation-engineering calculation was considerably simpler than those undertaken at Los Alamos, taking hours rather than weeks to complete. Moreover, given the relatively smaller computing staff of an aviation firm, the frequent reprogramming, which still required manual rewiring, became a significant concern. It was therefore neither the staff at IBM nor Los Alamos, but two staff members in Northrop Aircraft's central job-shop for engineering calculations, who in early 1948 discovered a way to skirt the difficulty through a novel arrangement of IBM machinery. Bill Woodbury and Greg Toben, both of whom worked in the "IBM Department" at Northrop, found a way to extend some wires across two machines to produce something closer to a real computer. A set of cards loaded into one machine sent a sequence of signals to the other thereby controlling the latter's operation. Soon dubbed the "poor man's ENIAC," this was the machine that IBM decided to reengineer and market as the CPC.[21]

Pendery approached the Southern California market with some apprehension, since his customers had been using IBM equipment now for several years. They knew more about technical computing than he did. Fortunately, the competitiveness of the industry, along with the fact that one of the industry's own firms developed the CPC, ensured that most aviation firms were eager to learn whatever they could about the new machine. With some assistance from the local salesmen, Pendery paid a visit to each firm to give an overview of the CPC. Then, when the firms acquired their first CPC, Pendery revisited each site to help install the machine. Pendery also brokered arrangements with local universities to set up short-courses on CPC programming techniques. He helped the firms recruit suitable candidates for technical computing work.[22]

Just like any sales representative, Pendery's initial contact with a client always required a bit of "personality work." He too had to rely on the art of social improvisation, establishing a quick rapport by appealing to common interests and a projected sense of familiarity. However, Pendery's mathematical knowledge, and the fact that he was not there to make a sale, gave him an entry the IBM salesmen could not enjoy. This was backed by the technical intricacies of the machines that were his charge, along with real differences in how the relevant knowledge was distributed between the customers and the manufacturer. Compared to the usual salesmen, Pendery could base his rapport less on image and persuasion as against a sustained technical exchange.[23]

Take, for one example, the particular circumstances surrounding Pendery's knowledge about CPC programming techniques. The CPCs, again, were still programmed in part through elaborate wiring configurations. Efficient setups for floating-point calculations required clever wiring. The users valued Pendery's visits, in part, because of his growing expertise in these matters. On the other hand, Pendery did not produce this knowledge by himself. More often than not, he picked up a new idea from the users as he traveled from site to site. While some firms were at first hesitant to share their knowledge, Pendery convinced them that they had more to gain from an open exchange rather than through their exclusion. By establishing himself at the fulcrum of a local technical exchange, Pendery made himself indispensable.[24]

Such access provided Pendery with a way to glean other valuable information for IBM. Occasionally, Pendery gathered information about the technical design or even the pricing data of a competitor's product. Equally important, Pendery could learn about a customer's applications. He sought out information about a firm's pending military contracts, which allowed him to estimate the demand that would be placed on their computing installation's machinery. Nor did Pendery have to coerce his clients to obtain such information. If there was coercion in this relationship, it was often the customers who tried to convey their knowledge through Pendery in order to persuade IBM to alter its products or to accelerate delivery.[25]

Pendery, who took all this information quite seriously, produced a weekly report of his visits. Moreover, he remained attuned to what Hurd seemed to need back at "World Headquarters." In early May of 1951, Hurd became concerned that he did not understand the precise relationship between customer applications and the performance of the CPC. He sent a memo to his men asking them to describe each CPC installation. Hurd also specified the information he wanted. While all the field men responded to this request, Pendery took the extra step of rearranging his subsequent "call reports" to address similar concerns. Hurd recognized Pendery for his professionalism. Perhaps more important to Hurd and IBM, Pendery revealed the nature of Southern California's emerging technical computing market. As described in greater detail in the next chapter, IBM's sales to Southern California would become the mainstay of the firm's computer sales during the 1950s.[26]

The Ambassador

The Los Alamos Scientific Laboratory stood above a precipice whose walls and cliff dwellings were home to the Anasazi Indians some 600 years ago. The shade of the canyon walls, and the small stream that ran beneath it, provided a fertile valley that offered some protection for life. In contrast, the new structures built at Los Alamos paid little heed to the region's history or geography. Its chambers dealt with the most abstract artifice of nature, and then the horrific reality of the hydrogen bomb. The Anasazis left this site long ago.

Lloyd Hubbard was a man doubly displaced. He was the first representative assigned to District 9, which covered all of the Southwest, including Southern California. But once Pendery established himself in Los Angeles, Hubbard was left with all the other installations that were scattered across the district. Customer inquiries took Hubbard to the Bureau of Reclamation in Colorado, the California Research Corporation (in Richmond), and various weapons laboratories in Texas and New Mexico.[27]

Despite the urgency and scale of the computing work at Los Alamos, Hurd first assigned no one to cover that laboratory. He had his own contacts there. IBM engineers were also working directly with Los Alamos scientists because of a directive from Thomas Watson. On those occasions when Hubbard did visit Los Alamos, he had to commute from Albuquerque because of security restrictions.[28]

But it soon became apparent that even a sophisticated customer like Los Alamos could benefit from the regular visits of a special representative. Hubbard made his first extended visit there in February 1951. This coincided with a period of intense calculation, and he was put straight to work. Hubbard assisted with programming and hardware installation. By May, he was providing a course on CPC programming techniques to thirteen of the laboratory's mathematicians. By the latter part of the summer, it was clear that Hubbard was spending most of his time providing technical assistance to this one client, perhaps with some ambiguity as to whether he really still worked for IBM.[29]

As his affiliation intensified, Hubbard became one of the most important channels of communication between Los Alamos and IBM. Hubbard established working relationships with Carson Mark, the head of Los Alamos's powerful theoretical division and with Preston Hammer, the head of the division's computing branch. Both began to rely on Hubbard to convey their requests, ranging from machine modifications, to changes in IBM's delivery schedule, to improved machine maintenance. IBM did honor the special priority of Los Alamos's work, but it could not support every request. As such, Hubbard was placed in the difficult position of having to evaluate the significance of a particular request in spite of the delicate issue of nuclear secrecy. Hurd commended Hubbard for the "excellent relation" he established with the laboratory's senior scientists. Eventually, a special arrangement was made for him to live at Los Alamos. Hubbard carved out a place for himself in this esoteric community.[30]

But just as Hubbard began to settle into his new routine, an incident erupted in the laboratory's computing branch. Hammer, an erudite and demanding research mathematician, alienated his staff during the latest period of intense work. In the breach that followed, Hammer was asked to head a separate section for applied mathematics research. Hammer's lead programmer, Richard Stark, also transferred out of personal loyalty. It was Stark who then exacerbated matters by submitting a machine modification request directly to IBM. The local IBM sales representative wrote to Hurd, warning him that "Mr. Stark is entirely in error in writing such a letter." The Purchasing Department at Los Alamos, perpetually upset about the irregular arrangements that always seemed to incur unexpected financial obligations, was "looking for just this type of material to hang Stark on." Hubbard lamented the subsequent turn of events. He wrote privately that, "Since Dr. Stark has done almost all of the development of programming on the 604, his loss would be very keenly felt." But Hubbard chose not to intervene in local affairs. IBM's principal business would continue to be with Los Alamos's computing branch. He shifted his loyalties, if not his sympathy, to the new head of the computing branch. [31]

The Novice

Not all of Hurd's field men were as successful as Hubbard and Pendery. A case in point was the Boston area representative, Daniel Mason. After World War II, Boston emerged as a leading center for military electronics work based largely on the wartime radar development work at MIT's Radiation Laboratory. Hurd hired Mason straight out of MIT's master's program in mathematics.[32]

Having spent some time in the area, Mason already knew of the various firms in the region that were beginning to make a name. Unfortunately for Mason, familiarity was a detriment. Bound by the weight of academic culture, Mason remained confused about the nature of his work. Unsure about whether his job was about mathematics or marketing, the very effort to resolve his identity produced the hesitations that kept him from charting a successful career. During the period that he served as one of Hurd's field men, Mason acquired neither the deep technical expertise nor the savvy posture of the other representatives. He erected his own barrier between himself and his clients.

Hesitation is in fact a serious bane for any salesman. Mason's difficulties in this respect are most evident in his interactions with MIT. MIT maintained an Office of Statistical Services, which had become the technical computing service for MIT researchers after the eclipse of its Center of Analysis. The office's director, Frank Verzuh, considered himself a sophisticated user. He had contributed to technical developments in the use of the pre-electronic IBM 602 calculating punch, specifically for generating mathematical tables and solutions to simultaneous linear equations. He was widely recognized for this work. In 1951, Verzuh acquired an IBM electronic calculator, the IBM 604, and invited Mason to assist him with its operation. After a somewhat frustrating visit, Mason wrote: "Mr. Verzuh is causing me considerable anxiety. Although his 604 has been installed only four weeks, already he has procedures requiring big, long wires between the 604 and 521 control panels. I have a feeling he is somewhat perturbed because I do not answer questions such as 'When a negative balance selector is picked up, how many program cycles does it take before the selector is actually transferred . . . ?'"[33] It is not that Pendery and Hubbard would necessarily have known the answer to such a question. However, both would more likely have answered, with confidence, that they would get an answer from someone at IBM. But it took Mason three months to revisit MIT. Upon his return, Mason discovered that the IBM 604 MIT acquired was not entirely useful because Verzuh had failed to order the appropriate peripherals. Mason proposed to analyze MIT's operations "as thoroughly as possible" to determine the best machine configuration. Hurd caught this misstep. In addition to the delay, a bad recommendation could incur obligations. In the margin of Mason's memorandum, Hurd scribbled "Show Verzuh list of CPC-604 attachments!"[34]

Mason differed from the other representatives less in kind than in degree. He too visited various customers. He analyzed their requirements, helped install equipment, and relayed an occasional request back to IBM. Most clients welcomed the information Mason brought with his visits, and sometimes this facilitated a sale.

Yet Mason's work was sloppy. In contrast to Pendery's reports, Mason wrote in a rambling prose that occasionally included personal matters. He wrote of his efforts, for example, to join the MIT graduate mathematics club. Mason found it difficult to break from the paternalism of academic institutions that had come to dominate the region's technical work. As such, he remained somewhat disjointed during his first year or two at IBM.[35]

But instead of reprimanding Mason, Hurd remained sympathetic toward his young employee. Hurd allowed Mason to stay on in Boston until he became more acculturated to the work. Then he drew Mason into the corporate fold as part of the Applied Science Staff, where he went on to have a successful career. Indeed, part of Mason's early difficulties stemmed from the fact that Hurd had kept the job so vague. Intentionally or not, this allowed Hurd to develop a highly diverse staff in which one member's talents complemented another's. George Petrie, an applied mathematician with a PhD, used his credentials to gain access to military officials in Washington. Truman Hunter, the Philadelphia representative, became one of IBM's experts on training and machine installation. John Chancellor, who worked out of Chicago, absorbed the norms of the local commercial culture that valued accounting over technical expertise. This was not a detriment, since he helped to find an efficient way for a single IBM facility to handle problems in technical computing and commercial data processing.[36]

Yet this diversity was both functional and contingent. As demonstrated by Mason's troubles, it relied on an individual's ability to discover a place within a local environment. Those who made the adjustment were able to be of considerable service to IBM. They facilitated an exchange in knowledge that ensured broad use of IBM machinery. They established themselves as trusted experts, giving IBM the advantages of a soft and hard sell: Customers turned to the field men for honest appraisals of IBM machinery, while a salesman followed to close the sale. Put in more general terms, the access each field man gained allowed him to fulfill a larger role as a technical intermediary. Insofar as Mason was a young college graduate, his initial confusion was understandable. Nevertheless, the field men, who often served as Hurd's recruiters, came to recognize young men who were "not suited to sales work."[37]

The Knowledge of a Network

Individually, each of the field men served as an important intermediary between IBM and its scientific customers. But regardless of their individual success or failure, collectively the field men provided Hurd with an information network. By collecting the field men's disparate observations, Hurd could begin to construct a larger picture of the technical computing market. He himself could then serve as an intermediary by contributing to larger decisions within IBM. His most important contributions pertained to IBM's decisions about product development. IBM could not build a machine to suit every customer, any more than it could hope to service all requests from Los Alamos. It was only in matching the customers' demands to IBM's engineering capabilities that IBM could devise an appropriate product line. Hurd was not alone in making these decisions. Nevertheless, the efficacy of Hurd's network, and the consequences this had for IBM, is best seen through an account of specific technical developments.

The CPC Model II

IBM's scientific users extensively influenced the firm's hardware. A good case in point was the CPC. After all, the original prototype for this machine was designed by one of IBM's customers. The decision to commercialize the CPC reflected not only IBM's willingness to adopt their customer's innovations, but an acknowledgement that technical computing was a distinct market that required its own machinery.

This emerging market was first surveyed by IBM's Future Demands Department. This was a group different from Hurd's that had a sporadic life in the history of IBM. While the Future Demands Department's authority was often circumscribed, it was revived in 1947 to also help evaluate IBM's postwar markets and new technologies. Steve Dunwell, an engineer deeply involved with IBM's contributions to wartime cryptographic work, led the effort to survey the extent of the interest in the Northrop prototype. Based on what he found, Dunwell drew up the initial specifications for the CPC.[38]

Dunwell and the engineering staff at IBM then proceeded to reengineer Northrop's machine. While IBM consulted with Woodbury and Toben, the two Northrop "engineers," neither was made a part of the design team. No doubt this was done to respect corporate boundaries of propriety. Dunwell also had to make sure that the CPC was designed with sufficient attention to manufacturability, maintenance, and ease of operation. Following a preliminary design study, Dunwell and his engineers decided to replace

Northrop's modified IBM 405 with an IBM 417. They preferred this solution because they could use the IBM 417's built-in data channels in place of the loose cables that interconnected the two units of the Northrop prototype.[39]

Meanwhile, Woodbury and Toben proceeded to explore the different uses of their machine. The various control mechanisms and functionality of its two units afforded considerable latitude in finding new setups. Woodbury and Toben's specific coding techniques are a little too technical to repeat in their full detail. But basically the first unit of the CPC prototype, an IBM 405, was itself a substantial accounting machine that contained its own tabulators, "selectors" and other special circuitry. Northrop had also asked for and received several modifications that specifically enhanced the programmability of this unit. By using complex control sequences, the combined machine could be made to outperform the two units operating independently.[40]

However, these experiments ensured that Woodbury and Toben's knowledge diverged from that of the IBM engineers. Woodbury and Toben did protest IBM's decision to replace the 405 with the 417. But separated by geographic distance, and by the fact that they had not been formally trained as electromechanical engineers, Woodbury and Toben did not insist on reviewing a copy of the engineering diagrams. Given additional assurances from Dunwell that the new machine would perform as well as the prototype, the two chose to trust IBM. But since Northrop devised its coding techniques to suit its machine, any change that did not preserve the machine's esoteric features risked disabling certain capabilities. Meanwhile, Dunwell and the IBM engineers had no experience running a computing facility. Nor did they make extensive use of calculation in their own engineering work. They never understood how Northrop was using the prototype.[41]

IBM delivered the first production model CPC to Northrop in November 1949. Woodbury and Toben immediately discovered the machine's limitations. This was the crisis that Donald Pendery had been sent to investigate. Pendery used his mathematical skills to assess the situation. He analyzed the mathematical requirements for each of the main categories of computational problems handled at Northrop. In each case, he was able to demonstrate how the changes that IBM made had catastrophic consequences. Even without specific knowledge of Monte Carlo simulations or the more conventional calculations Northrop used in its structural mechanics work, Pendery could quantify the overall changes in computation time. Once framed in the common language of mathematics, Hurd could see the seriousness of the situation. Armed with Pendery's report, Hurd convinced IBM to develop the Model II CPC.[42]

In mediating the interactions between customer and manufacturer, Hurd and his men also redefined the roles of the respective parties. This time the IBM engineers chose to work directly with Woodbury and Toben, asking them to supply the new machine's specifications. However, this did not afford a quick resolution of their difficulties. As experienced users, Woodbury and Toben had their own idiosyncrasies. Since they ran a service facility, Woodbury and Toben kept finding new applications that required further extensions to the machine. They could not arrive at the final specifications. Complicating matters, IBM sent engineers, sales representatives and Hurd himself on separate visits to Northrop. Each returned with a different set of specifications, causing conundrum within IBM. In other words, an organizational response to crisis generated its own delays. In the end, this delay proved more costly to Woodbury and Toben, who eventually chose to leave Northrop. Meanwhile, the IBM Model II CPC sold well under its new design.[43]

The IBM 701

It is not at all surprising that IBM made a commitment to add an electronic computer to its product line in 1951. One of the Northrop managers had written to Hurd about a computer built by the Eckert-Mauchly Computer Corporation (the Binac), even as it was being uncrated there in 1949. Hurd and his field men subsequently followed other computers through their various stages of construction. IBM's scientific customers, for their part, could not understand why IBM would not enter the market. To them, IBM had already established itself as the leader in the field, and the CPC could not meet their future needs. But IBM had limited engineering resources. And even in 1952, IBM's electronic equipment sales accounted for only $9 million out of a total of $334 million in revenues. Hurd was still the head of a small operation.[44]

Two events in 1950 changed Hurd's prospects. One was the Korean War. Watson committed his firm to the war effort as he had done during World War II. This opened the possibility for new products that served the narrower aims of national defense. The other was Remington Rand's acquisition of the Eckert-Mauchly Computer Corporation. Remington Rand was

IBM's traditional, if junior rival. When the firm promised to put the full weight of its marketing organization behind Eckert and Mauchly's nearly completed Univac I, this created a real competitor in the field.[45]

Other accounts are correct to point out that both Thomas J. Watson and Thomas J. Watson Jr. (then IBM's Executive Vice President) played major parts in IBM's transition to electronic computing. IBM was still in the midst of its transition to a modern corporation, and its dynastic traditions continued to vest considerable authority in its top executives. The enthusiasm for computers was highly infectious, and both of the Watsons chose to guide the firm in this direction. The senior Watson, however, continued to be cautious about permitting an interest in electronics to interfere with IBM's principal business in office machinery. It was Tom Watson Jr. who played the larger part in accelerating IBM's entry into the electronic computers business.[46]

All the same, Hurd was well positioned to manipulate the will of IBM's top executives, precisely because of their interest in computers. He was, after all, their resident expert. Thus, when Tom Watson Jr. instructed James Birkenstock, the head of IBM's new Military Products Division, to survey the need for new equipment, Birkenstock turned to Hurd for his advice. By this time, Hurd had gathered enough information to compile a list of some sixty computing installations engaged in defense related work. Hurd and Birkenstock proceeded to visit more than twenty of them. At each site they could see the significance of the defense work being done on IBM machinery. After these visits, Hurd was given the authorization to survey the interest for a new "Defense Calculator." This was in February 1951.[47]

Hurd's associations with the scientific community, and the postwar computer-development community in particular, also had an influence on the Defense Calculator's design. Hurd followed the consensus that was emerging around general-purpose computers built according to one of von Neumann's two designs. Yet there were also conflicting reports of important differences even among general-purpose machines—such as the difference between the Whirlwind I and the Institute for Advanced Study computer. This led Hurd to recommend, initially, that IBM plan on designing four different machines, each designed to serve a specific military application. But as he and Birkenstock toured the various facilities, they were swayed by the prevailing opinion that favored the latest IAS design, which they kept seeing during their visits. The Defense Calculator became one version of the IAS machine.[48]

At this stage the electronic computer clearly remained a technology driven by customer-based innovation. Unlike the persuasive efforts that are associated with most consumer goods, information and persuasion tended to flow the other way. By aligning themselves with the new postwar scientific and military engineering communities, Hurd and his field men came to act more like the customer's agents rather than being the agents of IBM. For the moment, Hurd's own academic inclinations helped to alter the course of IBM's products to serve the needs of the scientific and engineering community.

The field men continued to assist Hurd in his efforts. There was still internal concern about, if not outright opposition to the Defense Calculator. It helped, therefore, that Hurd could take an entourage of IBM engineers and executives out to various customer installations. By 1951 several of the field men had already been stationed in the field for a year. They could lead visitors to installations most keen on IBM's new Defense Calculator. The field men also primed the users, suggesting that they might want to voice their interests. These visits served the dual purpose of internal and external marketing.[49]

The fears of those who opposed the Defense Calculator—soon the IBM 701—were well justified. Other projects languished because of the attention paid to this one machine. In part, the shift in resources was consistent with IBM's decision to reorient its business in the interests of national defense. Nevertheless, the enthusiasm for electronic computers remained infectious, especially among IBM's engineering staff. This made it more difficult for IBM to keep its focus on its larger business in commercial data processing.[50]

Data Reduction

General-purpose computers embody an inherent paradox. In standardizing the central devices that perform the calculations, it becomes all the more necessary to program and configure a machine to suit a particular purpose. Data reduction, or the automated analysis of experimental data, posed a special challenge to IBM, not because IBM lacked the expertise with which to build the necessary equipment, but because it had to learn to see this application.

Various kinds of data can be collected automatically as an experiment unfolds and transcribed onto a recording medium such as paper or magnetic tape. While it is possible to analyze this data using both analog and manual computing techniques, the same pressures that were reshaping aviation engineering and other disciplines as a whole brought high-speed digital computing to their experimental facilities.

Since data reduction was commonly performed in scientific research, Hurd might have been expected to recognize this market. However, Hurd, along with many of his mathematical colleagues, had come to regard the electronic computer as a simulation tool. Here, the computer itself generated the numerical data that modeled a complex world. In contrast, data reduction sat squarely within a more traditional experimental approach where nature itself produced data for analysis. The mathematics of the latter tended to be simpler. For Hurd, data reduction did not require the intense computational power of a modern electronic computer.

IBM's customers were not so easily misled. As new wind tunnels and missile test ranges came online during 1951, researchers began producing massive volumes of data that required automated analysis. The field men relayed this information to Hurd. And while Hurd might have discounted individual reports, the field men collectively served as a wired alarm. It was not only the repetition of a message. In fact, the field reports generated by the applied science representatives varied quite widely in terms of their language and content. Yet in accepting a work routine in which Hurd scanned and underlined whatever he found interesting about each field report, he ensured that his own mind would begin to reconstruct data reduction in a coherent picture. "THINK," a long-time motto for IBM, could be found on each and every piece of IBM stationery.[51]

To illustrate how Hurd could assemble such knowledge, the field men first reported how several customer installations were beginning to build or obtain a transcription device that could transfer analog magnetic data onto IBM punched cards. They also reported when the first device of this sort was working at one of the installations. Hurd began to ask for more detail as the volume of these reports increased. The field men adjusted their observations and inquiries, and obtained information such as the rate at which the magnetic tape was being read (e.g., 1 inch per second) as well as the method and exact reduction in data transfer rates employed during the transcription process (e.g. a two-stage reduction with a combined ratio of 625 : 1). Through such details, Hurd could see that the volume of data these facilities hoped to analyze did in fact require the resources of high-speed electronic equipment.[52]

However, IBM took a relatively conservative approach in its effort to support this application. While it decided to offer magnetic tape input for the IBM 701, it required its customers to transcribe data onto IBM punched cards when employing the CPC. This decision came to haunt IBM when the Computer Research Corporation began to offer the CRC 102, and then the larger CRC 107 in 1952. Both were intermediately priced computers that specifically targeted the data reduction market. The modest processing requirements that made data reduction so uninteresting to Hurd also allowed computers to be built of a different design. Specifically, the CRC cast aside a high-performance memory system in favor of a cheaper magnetic drum that was still suitable for many data reduction problems. Hurd had all the information he needed to come to a similar conclusion. But this meant synthesizing difficult details about possible machine designs and the scattered descriptions of an unfamiliar market. The field reports, in any case, were contradictory. They addressed every subject under the sun. In contrast, CRC engineers were already familiar with the problem of data reduction and were quick to synthesize the requisite knowledge. They took a chance on the market and the opportunity it represented.[53]

IBM salesmen and Hurd's own field men pestered Hurd with calls for a competitive product announcement. But Hurd declined to pursue such a recommendation, realizing in part that an early product announcement would jeopardize existing IBM 701 orders. Even if competition caused some erosion, ambiguity would defer action on the part of many customers. Hurd replied[54]: "We cannot however talk about specific new machines at this time in accordance with our policy—which I believe has generally been the correct policy."[55]

Hurd learned to use his field men not only as a means of responding to the customer, but of incorporating the business interests of IBM. Both Hurd and his field men continued to serve as a valuable intermediary. But now Hurd's wisdom included the well-established marketing and product-development strategies of IBM. Hurd was acquiring the cast of a seasoned executive. In doing so, both he and IBM grew more distant from the academic roots out of which technical computing emerged.

Conclusion

In December 1951, in a memo to his staff, Hurd summarized the group's accomplishments as follows: "This is a good time to consider where we are in Applied Science. . . . Progress has been made: more than 150 CPC's installed or on order, 25 Defense calculators ordered, a successful Technical Computing Bureau, the computing word spread around through lectures and visits, and best of all a staff of young men and women who are intelligent, alert, and active and who enjoy the support and backing of IBM."[56]

Hurd's success was not only in installing new machinery, but in installing a social mechanism that conveyed valuable information back and forth between IBM and its scientific and engineering customers. This entailed not only marketing intelligence, or knowledge about how to use computers, but technical information quite valuable to IBM's product-development strategies.

Much of the work was based on doing what sales representatives had always done, but as augmented to accommodate the language and stature of the postwar scientific community. But there was another side to this relationship of mutual dependency. IBM executives were left wondering whether they had come to pay too much attention to the needs of the scientific community, and whether Hurd and his men had "gone native" in their missionary zeal.

In early 1952, T. Vincent Learson, the powerful sales manager of IBM's Electric Accounting Machines Division, complained about one of Hurd's men who was attending a session at IBM's sales training school: ". . . I talked to Swanson of the Chicago office at the recent Sales School. He objects to being a peddler. He believes that University of Illinois and another Illinois research company are good prospects for a CPC, but he does not want to accept them as a quota assignment and accomplish the result. . . . I am sure in the Chicago case we could get an order for two machines if the man were assigned on a commission and quota basis in such accounts."[57] Learson added that he could not afford to have forty-odd men spending a majority of their time making courtesy calls and wiring up customer IBM 604s and CPCs. Viewing the situation quite strictly in terms of labor costs, Learson pushed Hurd to "put this operation on a profitable basis," and suggested, specifically, that Hurd ought to transfer some of the work back to IBM's "Systems Service girls."[58]

Hurd was able to resist such a move. Even for IBM, participating in new Cold War markets meant conceding to a pluralistic institutional context that contained unresolved tensions between corporate, military, and academic interests. Some of these tensions, as demonstrated in this chapter, were quite functional. Hurd and his men built up an extensive trade in knowledge, and they did so by drawing on what Hurd explicitly saw to be the open norms of academic exchange. Admittedly, the move to harbor a technical exchange one step removed from a system of monetary exchange was already a feature of the earlier system selling strategy. What Hurd and his men did was to appropriate and extend this older tradition at IBM.

But instead of pursuing a linear, teleological narrative, this account considers how a scientific and academic culture provided a separate if not entirely independent point of origin for what became a practice that was very important to the future of IBM. Hurd's field men eventually became IBM's first "systems engineers." In this new position, these men—and soon women—continued to help customers install and program their early computing machinery. More broadly, the early computer programs produced by IBM remained both free and open. The early exchange networks created by Hurd's applied science staff would evolve into the more formalized exchanges that would occur within the IBM users' group, Share. (See chapter 7.) Both practices contributed to IBM's subsequent policy of "bundling" its computer hardware, software and other services. IBM's strategy of selling a "solution," as opposed to a "product," was important to the firm's postwar success.[59]

Moreover, not all of the new patterns of interaction were so functional. In 1955, IBM initiated an effort to build a powerful scientific computer known as Stretch. Influenced by the demands of Los Alamos's scientists, IBM engineers set out to achieve performance goals that were beyond their immediate capabilities. When the project failed to meet many expectations, Tom Watson Jr., who had succeeded his father as president and then CEO, felt that the failure had an adverse effect on the firm's image, and hence its business in commercial data processing. Although the precise details and historical significance of this episode have yet to be explored, Hurd was a party to the technical decisions. He escaped the worst of the wrath expressed by IBM's top executives. Nevertheless, Hurd never made it into the innermost circle at IBM. He left the firm in 1962. Moreover, the actions taken by Tom Watson Jr. and other IBM executives were more significant than their effects on any individual's career. It represented an effort to curb what these executives saw to be the excessive influence of their own scientists.[60]

Still, the tension between IBM's marketing and engineering organizations did not disappear. The continued influence of science on IBM, and on U.S. businesses at large, has been largely unexplored or understated in the postwar business literature in what constitutes a subtle historical disengagement with the distinct Cold War technical culture that became embedded in U.S. corporations. Thus, although there has been considerable debate over the efficacy of basic research, few studies have considered how postwar scientific authority affected other important aspects of corporate conduct. The focus here has been on the influence scientists had on IBM's sales culture and its product-development strategies. What occurred here availed itself at the specific intersection of the history of Cold War science, the history of sales, and the history of IBM's corporate organization. Studies that examine other institutional and historiographic junctures would undoubtedly shed greater light on the history of Cold War industrial research in the United States.

Figure 7.1
John Lowe, Manager of Engineering Tabulating, conducting an early demonstration
of technical computing at Douglas Aircraft. From Box 4/14, Cuthbert Hurd Papers,
Charles Babbage Institute for the History of Information Processing, University of
Minnesota Libraries, Minneapolis.

7 Voluntarism and Occupational Identity: The IBM Users' Group, Share

In November 1956, Paul Armer stood before an audience of data processing managers to expound on the virtues of cooperation. Armer was one of the founders of Share, a computer users' group made up of IBM's major scientific and engineering customers. He had been asked by his commercial data processing counterparts to speak about Share's successes. Share was indeed a successful organization, if a somewhat curious one. The idea to create the first nationwide computer users' group began with a group of computing center directors intent on improving the operating efficiency of their facilities. They envisioned an organization that would exchange computer programs. But they hoped to do a whole lot more. Among their concerns was the shortage of skilled programmers, their high labor costs, and the need to have some technical standards. Their concern, above all, was the inefficiency of the fact that every firm that purchased an IBM computer had to write practically all of their programs themselves. This was a situation, Armer and his cohort reasoned, which resulted in a massive and needless duplication of effort.[1]

Share was established in 1955, when scientific and engineering installations still made up the majority of IBM's biggest computer customers. The group was created three years after the release of the IBM 701, and a year or two before the general distribution of Fortran. At the time many groups both within and outside of the universities were beginning to come up with new programming techniques. Share's principal contributions lay in helping to develop early operating procedures, operating systems and a body of knowledge that would come to be known as systems programming. Equally important, Share gave a group of early computer programmers an occupational identity. But Share had to appeal to volunteerism to justify what was, in effect, a collaboration among many firms belonging to the same competitive industries. More broadly, the series of choices Share's founders had to make reveal the complex, opportunistic entanglements between esoteric knowledge, institutional loyalties and professionalization that came to unfold within this sector of the computing field.[2]

This chapter provides an opportunity to revisit traditional narratives of occupation formation and professionalization. Historians and sociologists have made substantial inroads in overturning early models of professionalization. Yet the basic realization that the professions emerge through historical and situation specific strategies can only be the beginning of a robust treatment of the subject. Indeed, despite the widespread acceptance of occupational change as a pervasive feature of the current postindustrial economy, scant attention has been paid to historic changes in the process of occupational formation since World War II. The literature on occupational formation still applies mostly to an older industrial era. They continue to deal primarily with the familiar processes of routinization and mechanization, or else the process of successive specialization that occurs within an already established occupational or professional hierarchy. If our goal is to understand the massive proliferation of technical specializations after World War II, then studies that pay attention to other, less scripted strategies for creating new occupations during the Cold War era remain necessary.[3] In this respect, the unusual mission and makeup of Share points us in the right direction. The institutional dispersion of knowledge, competing institutional priorities, and the diverse set of antecedent organizational models and practices from which a nascent group of computer professionals set out to build an occupational identity point to the inescapable remix of esoteric knowledge and institutionalized practice characteristic of the Cold War years. Equally important to the rise of new technical occupations was a dynamic labor market that encouraged many young men to seek new technical careers in search of upward mobility. This study pays close attention to Share's organizing strategies in order to reveal the complex set of opportunities, constraints and especially contingency as associated with occupational formation in this historic context.

Share's founders did not set out to establish a professional society; they sought to establish a technical collaboration that would reduce their programming costs. But circumstances put the organization in an ambiguous position of being, simultaneously, a corporate collaboration and a voluntary society. Share clearly established itself as an important intermediary between IBM and its end users, not only displacing, but building on the work done by Cuthbert Hurd and his applied science field men.[4] Yet the continued corporate commitments and loyalties of the individual representative to Share would also limit the scope of the organization's activities. In the end, this would wind up reproducing the same, limited path to professionalization as experienced by U.S. engineers.[5] But regardless of this familiar outcome, my focus will be on the era's vast sense of uncharted possibilities for reaching this circumscribed end.[6]

This chapter also pays close attention to *how* esoteric knowledge serves to define new occupational identities. Those who study the professions have come to see esoteric knowledge as crucial to dynamic changes in the professions. Yet many accounts, including Andrew Abbott's 1988 book *The System of Professions*, continue to reduce knowledge to a contested object. They fail to regard knowledge itself as a socially constituted entity. This once again creates an opportunity to incorporate various advances in constructivist scholarship, here for the study of occupations and professions.

In examining Share, I pay particular attention to the fact that, as was the case for so many new postwar areas of expertise, computer programmers drew their recruits from a wide variety of established occupations. One of the major roles of an organization like Share was to resolve competing claims of competence by establishing a coherent set of occupational identities. Here, esoteric knowledge provided an important organizing principle around which to construct essential distinctions. Share could and did mediate this process by advancing certain kinds of knowledge, forging tacit alliances between certain groups, and by helping to determine which groups could then go on to claim professional competence. Significantly, the distinctions that Share helped to establish between operators and programmers, programmers and end users, and systems and applications programmers would continue to define the larger field of computing even after the group ceased to be the principal instrument for the professionalization of computer programmers.[7]

A Tradition of Collaboration

Share clearly grew out of the early technical computing activities in Southern California's aviation industry. Even before World War II, extensive computing was already a part of aviation engineering work, especially in the structural analysis of new aircraft designs. Young mathematics graduates were hired as "stress analysts," and worked mostly as human computers operating desktop calculating machinery. From the standpoint of these workers and their supervisors, Wallace Eckert's "IBM methods" were indeed a welcome development. Convair was the first aviation firm to use IBM equipment for stress analysis.[8]

As I suggested in the previous chapter, the United States' emphasis on strategic air power, and the competitive structure of military contracts created an incentive for aviation firms to demonstrate advanced research and engineering design capabilities. Computers clearly played to this interest. But the competitive spirit of the aviation industry was also matched by a good deal of cooperation. Organizations such as IBM, and individual representatives like Donald Pendery helped to foster this cooperative attitude in computing. There were also other, regional organizations that actively cultivated an open exchange of knowledge about computing. Most notable were the Rand Corporation and the Institute for Numerical Analysis (which again was set up at UCLA), both of which took up the challenge of introducing new forms of applied mathematics and computing that they felt were relevant to the region's defense industries.[9]

Still, it is important to note that California's aviation industry had its own tradition of cooperation. During the interwar years, military procurement strategies, especially that of multiple sourcing, stretched the resources of many contractors to their limits. This prompted aviation firms to diffuse competition in areas they deemed to be peripheral to their core business. Cooperation gained further impetus during the war when the Aircraft War Production Council organized a collaborative effort in production engineering. While it was understood from the outset that the AWPC would be disbanded at the end of hostilities, the effort left a deep imprint on the participating organizations.[10]

Thus, when the new computing installations in Southern California began experiencing common difficulties, they were quick to turn to various precedents for cooperation. For instance, when Bill Woodbury and Greg Toben helped invent the Card Programmed Electronic Calculator ("CPC"), Northrop allowed its two employees to help IBM introduce the CPC to other aviation firms. Woodbury and Toben met with their counterparts and even offered a general course in CPC programming. IBM's training sessions continued to offer a place where the region's computing specialists could gather. Then, in 1952, the managers from several of Southern California's computing installations decided to create a group called the Digital Computer Association. Although the DCA never pursued any collaborative projects, its meetings served to articulate the common challenges faced by those in the region's computing installations. While differences in the internal organization of these facilities persisted, by the early 1950s there was clearly a group of people who were willing to wed their careers to the new technology.[11]

California's computing center directors launched their first major collaborative project shortly after IBM delivered the first IBM 701s to the region, in 1953. The principal motive for this collaboration was the limits of the programming system, or "assembler," written by Hurd's Applied Science Staff. At the time, users of Remington Rand's Univac I, along with the various academic one-of-a-kind machines, had already begun to standardize general coding techniques. It was common, then, to draw up computer programs using a set of mnemonic codes and then to translate these programs into the binary notation recognized by the computer. Since this was quite an arduous task, assemblers, which were programs that automatically translated mnemonic codes into binary, came into general use. Hurd's staff wrote the assembler that was shipped with the IBM 701. But few users found it to be satisfactory. Some installations proceeded to write their own assemblers and began coding their programs using their proprietary systems. But recognizing the disadvantages inherent to such separate efforts, several Southern California installations set out to design a more versatile coding system, which they hoped would meet their common needs. In the end they failed. This coding system, PACT-I, was released in 1956 after substantial delays and never received much use.[12]

When IBM announced its plans for the IBM 704 in mid 1955, Paul Armer of Rand Corporation, Lee Amaya of Lockheed Aircraft, and Jack Strong and Frank Wagner of North American Aviation decided to launch a separate initiative.[13] These senior computing managers, all of whom continued to participate in PACT, felt that the advance notice they received for the 704 created a new window of opportunity. Pragmatic, as opposed to professional interests remained foremost on their minds. All four men faced the same problem, namely that their staff would have to recode their entire program libraries because the IBM 704 was not compatible with the IBM

Table 7.1
Share's founding members. Membership was by computing installation, not by firm or research organization. Large multidivisional firms were therefore likely to have several installations, especially as computers became more common. Representatives of Douglas Aircraft (El Segundo, California) attended the first meeting; however, Douglas is not considered a founding member, since it had not ordered an IBM 704.

Boeing Airplane Company (Seattle, Washington)

California Research Corporation (La Habra, California)

Curtiss-Wright Corporation (Clifton, New Jersey)

General Electric Company (Cincinnati, Ohio)

General Motors Research (Detroit, Michigan)

Hughes Aircraft Company (Culver City, California)

IBM, Computing Bureau (Poughkeepsie, New York)

IBM, Scientific Computing Center (New York, New York)

Lockheed Aircraft Corporation (Burbank, California)

Lockheed Aircraft Corporation, Missile Systems Division (Van Nuys, California)

Lockheed Aircraft Corporation, Georgia Division (Marietta, Georgia)

Los Alamos Scientific Laboratory (Los Alamos, New Mexico)

North American Aviation (Los Angeles, California)

National Security Agency (Silver Spring, Maryland)

Radiation Laboratory (Livermore, California)

Rand Corporation (Santa Monica, California)

United Aircraft Corporation (East Hartford, Connecticut)

---

701. They began to consider how this task could be distributed "evenly over the field of 704 users." There was an added technical challenge in that the 701 had become a veritable Tower of Babel ever since the different installations began using different, proprietary assemblers. In August 1955, with the dual objective of designing a common assembler and promoting a computer program exchange, representatives of 18 computing installations met at the Rand Corporation to establish the computer user's group, Share. Representatives from IBM were also invited to this meeting, supposedly because IBM would also have its own 704 facilities. After Armer declined the position, Strong became Share's chairman.[14]

Share represented the culmination, not beginning, of a long-standing effort at cooperation. Yet in establishing an organization that was more clearly situated in between IBM and its customers, Share promised to garner the resources it needed to achieve a more substantial collaboration. Share represented a viable organization, a going concern, around which computing specialists could assemble their expertise. The extent of this collaboration would depend on how well Share could build upon, and indeed articulate the common interests of the participating firms and individuals.[15]

The Gatherings

Share's early meetings were rambunctious affairs. Most firms sent two or more representatives, meaning that there were some forty or fifty men talking wildly about their technical affairs.[16] Nominally, Share was set up to represent the member installations, not individuals.[17] Each IBM 704 installation was allowed just one voting representative, who was usually the head of the installation or the senior-most member of the programming staff. But insofar as technical discussions dominated the early proceedings, concerns about status often gave way to open participation by delegates.

Indeed, there was a great deal of excitement in the air. The participants' first concern was the common assembler, but a quick succession of other ideas emerged during their conversations. The delegates realized, for example, that they would have to have a common punch card format to exchange programs and data. A common printer configuration was needed to enable programs written at one installation to be printed at another. Similarly, to ensure that each of Share's computer programs would run at every facility, the group specified a minimum configuration for the member installations' machines. Such technical matters were established as "Share Standards," with each installation being asked to work within the limits they had set.[18]

These standards were a point of possible contention. Establishing a minimum configuration could compel smaller installations to purchase equipment for which they had little use. Moreover, since every site already had an established work routine, new standards could force a staff to abandon familiar work practices. Even something as simple as changing an assembler's mnemonic codes could require programmers to alter well-accustomed habits. Coordinating work across different firms also required some sensitivity to the proprietary interests of each firm.[19]

Yet the most notable thing about Share, and a point of pride among its founders, was the strong sense of collaboration that emerged among the representatives of competing and traditionally rivalrous firms. Technical disputes arose on occasion, but on the whole what was most striking was the smoothness of the early proceedings. Common goals and technical interests overtook corporate loyalties.

This collegial atmosphere drew on broader traditions of cooperation, broader than that which could be traced back to PACT, the DCA, or the Aircraft War Production Council. Engineering standards work, as well as the common practices of professional and academic societies provided important organizational models. Only a fraction of Share's representatives were active members of a professional society. Nevertheless, the general conduct of professional society meetings and academic conferences were both familiar and easy to replicate. Share circulated papers, arranged programs, and issued printed proceedings. The working committees, voting procedures, and other protocols associated with technical standards work also became a routine part of Share. All this reflected the representatives' desire to try out different organizational models in an attempt to forge a productive collaboration. All the same, the broader significance of these habits would all come into play as the group began to discover their professional aspirations.

After specifying a standard assembler, Share directed its energies toward producing a basic suite of computer programs, including the mathematical subroutines, input-output utilities, and higher-order "abstractions" like matrix inversion and numerical integration that all facilities were expected to need. In fact, once the group agreed on the set of mnemonic codes for their assembler, every installation was free to begin coding other programs even as one installation took up the task of coding up the assembler. Not all facilities were prepared to begin work, since the IBM 704s were on a staggered delivery schedule. After weighing their mutual interests and capabilities, North American, General Electric, and several others (including IBM) agreed to accept Share's first coding "assignments."[20]

Some tension developed during this initial coding effort. The presentation and circulation of program write-ups provided an important occasion for feedback. Yet although every site acknowledged that this criticism was valuable, it was sometimes difficult to distinguish inefficient code from code adapted to the needs of a particular facility. Opinions also differed about the obligation each installation had to maintain the code they wrote. Share's leaders began to codify their cooperative sentiments by fixing the rules of their exchange. They asked the originator of each program to assume responsibility for making and distributing prompt corrections. A statement titled "Share Membership and What It Entails," adopted in February 1956 at Share's fourth general meeting, noted: "The principal obligation of a member is to have a cooperative spirit. It is expected that each member approach each discussion with an open mind, and, having respect for the competence of other members, be willing to accept the opinions of others more frequently than he insists on his own."[21]

The rule about corrections, along with others, was incorporated into Share's new Reference Manual. This manual described Share's Standards, its rules of membership, and its procedures for distributing computer programs and for mail balloting. The document dealt as much with the conduct of the organization as it did with its technical work. More broadly, through common standards, programming efforts, and norms, Share representatives were beginning to articulate a shared identity.[22]

Delays in IBM's delivery schedule also helped dissipate some of the early tension. Then, as the first IBM 704s entered operation, Share representatives discovered a whole new host of common concerns about things such as machine configuration, equipment failures, and magnetic tape reliability. Share organized special surveys and panels on these and other topics. By collecting operating statistics, Share gave the member installations a way to isolate their problems and to hold IBM accountable for its engineering work.[23]

More significant differences surfaced in mid 1956, toward the end of Share's first year. The new focus on machine operation generated its own discontent, as some of the "programming types" began complaining that this discussion sacrificed their interests for those of their "chiefs." But others asked whether the general meetings were an appropriate place to discuss detailed matters of programming. So long as Share offered some of the best programs and distributed them for free, all new IBM 704 installations became members of Share. However, with 52 installations represented at the August 1956 meeting in Denver, many of the new representatives lacked the expertise, background or inclination to enter ongoing discussions about programming. New representatives also raised issues considered settled by the veterans. Legal concerns, an election

Table 7.2
General meetings of Share, 1955–1957.

		Installations represented	Location
Share I	22 August 1955	18	Los Angeles
Share II	12 September 1955	16	Philadelphia
Share III	10 November 1955	22	Boston
Share IV	6 February 1956	27	San Francisco
Share V	9 May 1956	37	Chicago
Share VI	22 August 1956	52	Denver
Share VII	13 December 1956	68	New York
Share VIII	24 April 1957	67	Dallas
Share IX	1 October 1957	76	San Diego

dispute and disagreements over how to train new personnel bogged down the proceedings. These difficulties struck at the heart of Share's self image as a productive organization. As one representative succinctly put it, "one hundred and fifty people in one room produce a hell of a lot of gas."[24]

The Southern California representatives decided they had no choice but to seize control of the organization. Representatives from eleven installations met privately in El Segundo, California, to hold a "postmortem" of the recent meeting in Denver. Based on this discussion, these installations demanded that Share institute a new committee structure that made a formal distinction between permanent and ad hoc committees. Simultaneous committee meetings, along with technical panels on well-focused themes, became a regular part of the now semiannual gatherings. The general sessions were to be conducted under parliamentary procedures, with a formal agenda that began with a report from each committee chair. A special committee was established to work out new rules for election and governance. More novel was the new "indoctrination committee." Perhaps a choice of language that emerged from the McCarthy era, this committee's charge, in no small part, was to prevent idle inquiry on the floor of the general session.[25]

This was all about bureaucratization and rationalization. Initially, common interests and an urgent need for collaboration served to overcome individual differences. But as Share and its agenda grew, significant tensions appeared. Far from being destructive, these tensions forced a set of organizational changes that allowed the group to remain an effective intermediary between IBM and its customers.

In making this transition Share continued to draw on familiar organizational models. Its general meetings came to resemble even more the conferences held by established professional societies. And although these changes meant giving up some of Share's early flexibility, Share's new committee structure allowed focused groups to carry on with the work. Programmers now met separately to discuss new programs and programming techniques. Computing center directors met to discuss "sensitive issues, policy and SHARE business." Meanwhile, the mathematics committee, led by John Greenstadt of IBM, assembled its own recruits in an attempt to improve Share's mathematical subroutine library. Those affiliated with Share reiterated their commitment to a productive collaboration even as they began to recognize their own subspecialties.[26]

Defining a Productive Effort

A Measure of Efficiency

The emphasis Share placed on sustaining a productive effort raises the question of what, exactly, motivated this interest in productivity. The answer lies in the details of Share's origins. Corporate tabulating facilities, which were intricately involved with the earliest experiments in technical computing, were closely tied to an earlier reform movement in systematic management. Those who supervised these facilities were themselves strong advocates for careful measures of their own efficiency. The stress analysts who first began collaborating with their "IBM departments" were quick to adopt this measure, especially since it resonated with their careful planning of their mathematical labors.

When a demand for new technical computing facilities first emerged, corporate executives turned to these tabulating department supervisors and analysts for advice. Having been molded by the methods experts' attention to labor costs, both were quick to emphasize the laborsaving potential of computing machinery. Justifications for computer equipment gradually became divorced from direct measures of mathematical labor, but similar arguments persisted. This is not to say that those in charge of new computing facilities would have otherwise been unconcerned about operating efficiency. But the older methods of cost accounting employed by tabulating departments set the standard by which to gauge the performance of new computing facilities.[27]

In other words, new electronic computer installations continued to emphasize production planning and procedural efficiency. From a technical standpoint, this was quite beneficial, for it spurred many new developments. Initially, during 1953 and 1954, many installations used their digital computers like tabulating machines, loading programs one at a time. But in considering the cost of leaving a machine idle, computing centers began to routinize their operations. Moreover the cost of writing an entirely new program for every application brought them to develop standard program libraries. These two developments were directly related. Many of the early standardized programs, including the assembler, simply encoded manual work practices. Even mathematical subroutines were based on hand computing algorithms employed by human computers. Moreover, the early "utilities," such as the programs used to input and output data, were quite often based on standard procedures machine operators devised for the effective use of the facility. The stored-program computer was exactly the kind of technology that could automate routinized procedures by encoding them directly into the machinery. It was this kind of programming work, except for the task of writing mathematical subroutines, which came to be known as systems programming.[28]

Directors of computing centers convinced their own managers that a substantial internal coding effort was needed to make effective use of the equipment. IBM originally planned to provide its users with a basic suite of programs. But IBM made little use of computers in its own engineering work, and most IBM programmers had little or no experience with the routine computing work of other engineering industries. Members of Hurd's Applied Science Staff who worked at IBM's Technical Computing Bureau in New York did have broader experience with computer applications, but the wide-ranging and sporadic nature of the problems they took on did not inspire the same concern for efficiency. Indeed, the initial set of IBM 701 programs written by the Applied Science Staff reveals a different set of priorities when compared against those written by programmers from the aviation firms. Much has been written about IBM's lack of interest in programming; experienced computing installations were equally uninterested in entrusting their coding efforts to IBM.[29]

Each computing center director proceeded to hire a sizable staff, although the specific arrangement varied. North American Aviation installed its first IBM 701 computer within an established administrative computing unit and set up a separate group of mathematicians and analysts inside its engineering department. More often, a new technical computing unit was organized in a firm's engineering department or in central research, where an integrated staff of programmers, operators, and analysts provided a range of computing and programming services. Regardless of the specific arrangement, individuals within each facility began to specialize in different types of programs—utilities, assemblers, subroutines, and applications.[30]

However, computing center directors found it difficult to maintain a successful coding effort. All of them underestimated the costs of programming, especially since the programmers themselves were just beginning to recognize the costs of debugging code. Staffing difficulties exacerbated matters. Even before Sputnik, the shortage of "technical manpower" was considered a national crisis, and this shortage was especially acute in Southern California. And since there were no formal training programs in computer programming yet, these directors had no choice but to hire young men and women from very different backgrounds. Within a year or two, most had twenty or thirty programmers on their staff. The cost of programming began to exceed that of machine rental. Such circumstances tested the credibility of the computing center directors' claim that they were running an efficient operation.[31] These directors pursued a number of strategies for maintaining strong in-house programming capabilities. Some weighed the advantages of more traditional approaches to controlling workers, such as measuring the amount of time a programmer spent on specific tasks. But the wide-open labor market and the esoteric knowledge required for computer programming made it difficult to impose such discipline. While most groups set up in-house training programs, the labor market again made it difficult to retain their newly trained staff. In this context, programmers had little difficulty asserting their autonomy. In fact, directors of computing centers were quite dependent on their programmers. Having promised their superiors that they would deliver an efficient facility, these directors came to rely on, and thus identify with their own programming staff.[32]

A versatile programming language such as PACT-I offered the promise of a technological fix. To be precise, PACT-I was a "compiler," a program that allowed programmers to write relatively complex procedures and mathematical expressions using a simpler, "high-level" language, which the computer would then automatically translate into executable code. Those involved in PACT hoped that a compiler would not ease only their coding burden but also their training requirements.[33] But efficient compilers were hard to design. With PACT-I still unavailable at the time of the IBM 704 announcement in 1955, Share was the technically conservative alternative. It offered the possibility of distributing the coding effort over the entire field of 704 installations by using a common assembler. Nearly all computing center directors who had a 704 on order sent a representative out to California with less than two weeks notice. This indicates how much pressure they all felt to reduce their programming costs.[34]

Systems Programming

Producing a common assembler and a basic suite of programs for the IBM 704 defined the initial scope of Share's activities. But once this work was underway, Share quickly became a vehicle for further advances in systems programming. This followed from the computing center directors' commitment to operating efficiency and from the fact that systems programs contributed most directly to this chosen measure of efficacy. As if to reinforce this orientation, Share forbade the exchange of applications programs. While Share used this distinction to protect proprietary interests—application programs could contain important hints about a firm's engineering activities—this decision excluded end users and those who wrote application programs from an active role in Share. It also established systems programs and mathematical subroutines as legitimate areas for collaboration.[35]

Not all advances in programming originated with Share. Many member installations maintained independent coding efforts, and sites like Los Alamos and IBM's new programming group in Poughkeepsie continued to provide some of the most important innovations. However, Share's meticulous distribution of programs and program write-ups accelerated the diffusion of knowledge. Its meetings also provided a forum for discussing the most exciting developments, and this fueled further innovation.[36]

Just as many gains were made through the standardization of knowledge. The different words that Share representatives used to describe programs and operating procedures initially constituted an impediment to communication. The group resolved this difficulty by asking several West Coast members to produce a glossary. The exchange of programs also allowed many programmers to learn about common mathematical algorithms, as well as efficient ways to structure their computer operations. Meanwhile, the discussion about the initial coding effort had helped to establish what a basic suite of programs ought to be at all computing facilities. Share created a formal classification system for its programs, and it used this to organize its program library. While terms such as "compiler," "utility," and "mathematical subroutines" could already be found within the literature, these words gained further definition and stability through Share's activities.[37]

Share's efficacy, and another important example of standardization, can be found in its work on operating systems. Operating systems emerged as a stable concept during the group's first two years. Batch-processing systems were themselves an innovation. Although early schemes differed in their particulars, the basic idea was to load a set of jobs onto a reel of magnetic tape and to process them automatically. A central control program facilitated automatic operation by integrating all of the machine's systems programs. These integrated systems could eliminate "between-job" time, manage subroutine libraries, monitor output volume and computing time, and maintain an operations log. Referred to first as "automatic operators" or "monitors," these operating systems were firmly rooted in a computing center staff's commitment to production planning. Early discussions within Share were devoted to the underlying strategies for developing useful systems. Subsequent discussions compared different systems in terms of their operating efficiencies.[38]

Operating systems transformed the social relations within and surrounding computing facilities. Early computing centers differed substantially in their approach to such basic operating policies as who controlled and paid for computer time. No standards existed for what costs were passed on to the end user, even though there were clearly different costs associated with programming, machine time, resource use, and resource requirements. Share proceeded to organize various panels and surveys to establish common modes of operation. These were then encoded into the operating systems. An operating system, for example, could be optimized for throughput by giving exclusive control to the operator, or it could be optimized for programming through special debugging tools that could interrupt other computer operations. The different circumstances of each installation precluded initial agreement on a standard operating system. Nevertheless, Share altered the economics of the situation by distributing a preferred operating system quite early on. In general, collective action allowed computing center directors to secure greater control over the operation of their facility. Finding better ways to allocate costs also placed each center on a stronger fiscal footing.[39]

Clearly, Share was established with an eye toward productive efficiencies rather than the simple advancement of professional interests. The dissatisfaction with the programs offered by IBM, and the institutional dispersion of the relevant knowledge about computing and computer operations provided a technical rationale for collaboration; a wide-open postwar labor market and the computing center director's struggle to offer an efficient service provided a compelling cause. But the people who founded Share acted out of more than mere economic motives; they also sought to sustain a particular notion of efficiency. This in turn defined where the lines of solidarity were drawn among computing specialists. Share provided a way for computing center directors, systems programmers, and the programmers who wrote mathematical subroutines to work together to build this efficient facility. They did so by specifying the scope of Share's technical collaboration and by producing new knowledge and standards that reinforced their claims about efficiency. Systems programmers were the principal beneficiaries of the emphasis on production planning. More than anything else, Share's work on new utilities, assemblers, and operating systems helped give articulation to systems programming as a coherent body of expertise.[40]

The Workers and the Firm

Share's productive efforts also helped to redefine the relationship between computer programmers and their employers. Under different circumstances, the employers might have come to view Share to be some kind of labor organization. Computing had risen above the status of a technician's work only in the years after World War II. Even though the firms hired some PhDs, a computing center staff was sure to include men and women who had formerly worked as human computers and tabulating machine operators. Considerable ambiguities remained as to the social status of programming work.[41]

Share did not invoke the fear of labor organization among employers, largely because it was set up to represent the firms, not individuals. In addition to the formal rules of representation, the mere costs of travel tended to limit direct participation to computing center directors and the senior members of their staff. Employers could trust that, for these individuals, Share's first justification would remain the technical and economic benefits to their firms.[42]

Share in fact had little difficulty demonstrating the value of its work. While the latest advances in systems programming could be difficult to explain, expanding program libraries provided tangible evidence of the benefits of collaboration. Insofar as Share helped to standardize the functions of a computing center, computer services also became something that was much more familiar at each firm. At its best, Share did function as a trans-corporate association that allocated programming tasks across many different firms. But collaboration had its dangers. It was the lawyers from General Electric and Westinghouse, two East Coast manufacturers with a long history of concerns about anti-trust, who first raised the possibility of litigation in mid 1956.[43]

It was in reaction to these legal concerns that Share representatives began to make an explicit appeal to volunteerism. Share was already a voluntary association among its member installations, and it could have remained strictly within this model if the participating firms simply agreed to share some subset of the programs they wrote. But the group arguably transgressed a legal boundary by coordinating program development work. By asking each installation to take on specific "assignments," Share entered a gray area that could have brought on costly litigation.[44] Share representatives therefore began to stress the voluntary character of the actual work. This was an unusual and in some respects odd tactic, since every Share representative was a paid employee. But programmers and computing center directors put in long and unusual hours. They could, with some justice, portray part of their work as a voluntary effort over and above their salaried responsibilities. From a legal standpoint this remained a dubious claim, but the rhetoric of volunteerism began to pervade Share's letters and proceedings.[45]

Given the compelling technical and economic reasons for collaboration, employers could hardly curb Share's efforts without alienating their staff. They chose instead to look the other way. They allowed Share to draw on corporate resources while amassing a program library valued in the millions of dollars. The firms clearly knew that their staff was engaged in "Share work," and reimbursed their travel expenses. But for the most part they never formally sanctioned this work. Precisely because of their apprehensions about anti-trust, the firms willingly extended the privileged status of a profession to their new computing employees. It was the firms themselves that encouraged Share to conduct its affairs as if it were a professional society.[46]

Computer programmers, particularly systems programmers, benefited substantially from this situation. Computing remained a field dominated by electronics engineers and research mathematicians. In organizations such as the Institute of Radio Engineers and the Association for Computing Machinery, programmers were seen as pursuing derivative work or work outside the scope of established professions and disciplines. With neither a professional society nor academic discipline with which to align themselves, programmers found their path to professionalization blocked on many fronts.

Although Share's rules of representation had the effect of limiting direct participation to computing center directors and their senior staff members, the organization's influence extended down into the ranks of each facility. Many staff members of computing centers were involved in the coding of Share's programs and subroutines. Moreover, since all Share programs were attributed to their authors, the distribution of the programs served as a mechanism for garnering credit. In fact, during the mid 1950s Share provided many programmers with the most important forum in which to develop and demonstrate their skills. Meanwhile, the directors and their senior staff members found in Share's committee work and higher offices a path for advancement separate from those inside their firms. By patterning itself after a professional society, Share also helped to foster the sense of shared identity that emerged between computing center directors and their programmers.

The close association between computing center directors and their staff had significant consequences for the labor market. When Share was founded in 1955, a regional labor market for computer programmers did already exist in Southern California. However, Share helped extend this into a national labor market. Share made it easier to identify appropriate talent by establishing common measures of skills and abilities. Moreover, as computing center directors came to know one another, and became wedded to a common goal, they grew more willing to encourage their staff to accept a better position whenever an opportunity arose. The new postwar pattern of white-collar labor mobility grew mostly out of the boom-and-bust cycles of the defense industries. Still, Share helped to extend this pattern into the field of computing.[47]

An expanding job market became important to programmers as tensions began to flare back at the firms. Computing facilities were cast as a service to a firm's scientific and engineering divisions, and as noted elsewhere in this book, service work has its dangers. Disputes with upper management continued over operating costs and the burgeoning size of the staff. Meanwhile, end users complained of machine failures, inadequate programming support, and long turnaround times. Moreover, once a central computing facility successfully demonstrated the reliability of an electronic computer, powerful engineering divisions began to ask for their own computers. The threat of decentralized operations, which became increasingly common during the late 1950s, could clash with the central computing staff's sense of efficiency. Unfortunately, those promoting a central computing facility usually found themselves on the losing side of the debate. The details of such struggles varied from one installation to another, according to local circumstance. Nevertheless, by the end of the decade there were many good, disgruntled programmers ready to move somewhere else.[48]

Share came to serve as a surrogate professional society for many computer programmers. Still, most of Share's regular delegates remained loyal to their firms. This should hardly be surprising. These directors and senior staff members invested considerable energy building up their facilities. They also received steady promotions. While these individuals criticized the myopia of their superiors, they sought out other possibilities only when a situation grew intolerable. All the same, Share contributed to the professional interests of all those who were affiliated with the organization.

Share and IBM

If Share mediated the relationship between employer and employee, it also mediated the relationship between the member firms and IBM. Some of the most valuable interactions between Share and IBM were about computer hardware. In representing the users, Share conveyed information about IBM's equipment that helped to improve the firm's product line. While IBM sometimes seemed to be hard of hearing, Share representatives lost no time in realizing the benefits of a collective voice.[49]

IBM was quicker to grasp the value of Share's program exchange. After all, computer programs sold computers. When it became clear that Share lacked the resources to maintain its program library, IBM stepped in to help, contributing both staff and material resources. IBM also reproduced the group's administrative correspondence, transcribed its proceedings, and handled the local arrangements for its early meetings. Share drew as much upon the resources of IBM as it did upon its member firms.[50]

The strongest collaboration between Share and IBM occurred in the area of systems programming. When IBM announced its plans for its next major computer, the IBM 709, in late 1956, Share decided once again to produce a complete new suite of programs. This time, Share decided to also develop and distribute a standard operating system. Share drew IBM into these discussions. The IBM representatives agreed to supply the programmers, if Share could provide the specifications. In other words, IBM conceded that Share had better knowledge of systems programming and of the end users' requirements. It entrusted the design of the IBM 709's systems programs to Share.[51]

But this arrangement overtaxed the group's resources. Donald Shell of General Electric, who chaired the 709 System Committee, led Share's side of the effort. In a burst of initial enthusiasm many installations seemed willing to commit to a series of week-long meetings. However, uneasiness back at the firms forced Shell to scale back his plans to several three-day gatherings. At first he and the other committee members remained optimistic that they could still accomplish their task. But as technical discussions revealed the full scope of their work they were gradually forced to the opposite conclusion.[52]

This crisis revealed a serious limitation that was endemic to the organization. Share could not assemble the resources it needed to manage a significant collaboration. During the group's first year of existence, its elected secretary, Fletcher Jones, noticed that, despite the commitment to efficiency, there was substantial duplication in Share's program library. The organization's response was conservative, and telling. Jones first organized a bibliography in the hopes that each installation would reorient its work accordingly. When this failed, Share's general body established several committees to coordinate program development. But the sporadic and overlapping efforts of these committees drew specific criticism during the 1956 postmortem held by the Southern California installations. After the reorganization, the mathematics committee tried to collect formal evaluations of Share's mathematical subroutines so that the users could choose the best programs based on their needs. But this too was a cautious approach that chose to emphasize the autonomy of each installation. Moreover, this voluntary effort did not lead to a highly coordinated effort in programming, even in the limited domain of mathematical subroutines.[53]

Volunteerism had its limits. Coordinating computer programming work across the nation was something quite different from fostering an open program exchange. It required a strong executive body. Indeed, a corporate organization established to manage a coding effort on the scale undertaken by Share would have begun with a sizable central staff. In contrast, Share had neither the authority nor the staff with which to truly coordinate program development work. The systems programmers' emerging sense of professionalism only exacerbated matters by reinforcing their desire to produce innovative original work.

Southern California installations were already aware of this problem. During their 1956 postmortem participants had formed a subcommittee to discuss the need for greater coordination. This subcommittee mapped out two possible paths for Share. One was "an organization in which each installation contributes what he wishes in the way of programs and information"—a description of the current state of affairs. The other was to have some volunteers participate in a powerful new "working committee" that had the authority to coordinate program development work based on its assessment of all the member installations' needs. The subcommittee clearly preferred the latter path, but they were skeptical about its prospects. They expressed real concern about whether member installations could keep to their commitments of programming personnel, and whether physical separation might make coordination too difficult. Nevertheless, the subcommittee's recommendations provided the specific impetus for creating the 709 System Committee.[54]

Lack of coordination did not spell an immediate end to the 709 System Committee's work. Several Southern California installations assigned to design the input-output utilities expanded their charge to sketch out an overall system. North American's Owen Mock then proceeded to design the main control program. Rand Corporation's Thomas Steel led the effort to design a new compiler-assembler-translator (called SCAT). Both Mock and Steel coordinated their efforts with other firms located mainly in the region, thereby escaping corporate scrutiny about how they were spending their time. Once this work was done, IBM's new Applied Programming Group did most of the final coding work, as promised.[55]

But although the Share Operating System (SOS) was completed in mid 1958, it too proved to be of limited value. What was wrong, basically, was that the SOS was designed with a highly skilled programmer in mind. Mock's control program and SCAT offered an integrated programming language with several hundred different commands. Both Mock and Steel hoped to improve upon a computer programmer's productivity by relying on their ability to take advantage of a sophisticated programming language. But the SOS was quickly displaced by Fortran, a true high-level programming language that began to gain popularity at about the time the SOS was released. Fortran was developed by the IBM programming research group in Poughkeepsie, as carried out by John Backus. Fortran had a more limited set of commands, permitting a very different model of computing in which more of the coding burden was transferred to the end users, albeit with some expense to machine efficiency. It represented, in other words, a very different approach for dealing with chronic programmer shortages.[56]

The technical perspective that gave Share much of its initial impetus now became a hindrance. Share's emphasis on machine efficiency had steered its systems programmers away from a language like Fortran. But there were again the limits of volunteerism. And even the Southern California installations revealed a certain predisposition in this instance; in proposing to establish a powerful working committee, the California representatives expressed a common technical faith that a competent group of technical experts could cut to the chase and produce a strong technical solution. They implicitly asserted that there was no need for excessive planning or bureaucracy. As the use of Fortran increased, a different group of Share representatives organized Share's Fortran Standards Committee. This committee began to express reservations about the SOS design. However, in being denied—or rejecting—the option of more extensive discussions, the 709 System Committee donned its own blinders. It failed to envision the very different social arrangements that would come to dominate work in computer programming.[57]

The widespread use of Fortran rendered much of the SOS obsolete. This is not to say that IBM had greater foresight. Cost-benefit analysis counted for less in this instance than the multitude of decisions made by anxious end users. Many scientists and engineers, frustrated with the resource limitations of a central programming staff, asserted their own ability to acquire programming expertise. The general use of Fortran then proceeded to redefine the responsibilities of systems programmers. Although systems programmers continued to use assembly language to design highly efficient programs, their principal role became that of designing and maintaining a programming environment suitable for routine use by a large pool of inexperienced programmers. In other words, systems programmers were forced to adapt their knowledge to accommodate their own users' demands.

Share was not given the opportunity to make this last transition by itself. Owen Mock could not get the SOS to accommodate Fortran. It was soon replaced by a new operating system, written by IBM. The Applied Programming Group at IBM had learned a great deal about operating systems through its collaboration with Share. Although Share did contribute to the design of this new operating system, the limits of volunteerism kept it from retaining its early dominance in the field.[58]

The Limits of Voluntarism

For a number of years, Share also came to occupy an important middle ground between its member firms and IBM. Assuming a role as a technical intermediary, a new group of computing specialists learned to play one organization off the other in carving out a place for their expertise. As long as IBM could not understand all of its users, and as long as its end users had limited knowledge of computers, computing center directors and their staff occupied a privileged position. Share, a voluntary organization, provided one means of acquiring the resources needed to realize this privilege.

The general strategy of a technical intermediary has been described elsewhere. Carolyn Goldstein, for example, has detailed how home economists employed by gas and electric utilities established their professional identity by interposing their expertise between consumers and producers. But home economists built their reputation through the strength of their academic ties. In contrast, Share representatives had to articulate their expertise as employees of the customers rather than of a manufacturer. This arrangement also raised serious concerns about antitrust litigation, which forced Share to resort to a rhetoric of volunteerism. Institutional location clearly affected opportunities and outcomes. On the other hand, it was the home economists who in the end had a more difficult time retaining the status they had gained.[59]

Share representatives had to pursue many technical and organizational innovations to capitalize on the opportunity presented by their situation. To do so, they drew on their wide-ranging backgrounds—different traditions of engineering standards work, production planning, regional collaboration, voluntarism, and academic exchange all contributed to how Share conducted its activities, and in different ways. Some introduced norms of conduct. Others helped articulate a common goal. Yet others contributed directly to an ethic favoring the broad circulation of knowledge. Share representatives appropriated, combined and extended these traditions in forging a productive collaboration. In the process, Share created not only a body of knowledge, but an organization with a concrete body of practice for generating new expertise. All this helped to define computing, and systems programming in particular, as a new technical field. Novel and flexible organizations such as Share were an important part of Cold War research. It provided a valuable means of responding to the changing demands of a Cold War economy and the rapidly shifting grounds of technical knowledge. They were the norm, not an anomaly, in the context of Cold War research.[60]

In focusing on the issue of occupational formation and a group's earliest professional claims, I have attempted to show how Share did a lot more than simply coordinate the interests of different parties. The institutional dispersion of knowledge among a large number of computer installations, and the relevant technical and organizational traditions upon which the group chose to draw, helped to define a distinctive path for occupational formation and professionalization. Moreover, Share's technical efforts were themselves an important means of transforming the social relations surrounding computing facilities. Computing center directors and systems programmers were able to enhance their position by specifying different kinds of knowledge, delimiting Share's collective sphere of activity, and adding to the base of knowledge that reinforced their own views about efficiency. Some of these negotiations were built into the computer programs themselves. Systems programmers used operating systems to automate and appropriate the expertise of computer operators and to redistribute the control and costs of computing within their company.

This is not to say that those affiliated with Share could operate outside the politics of the workplace. The collective, technical efforts of Share altered the balance of these negotiations, but the group was not immune to external developments. The limits of volunteerism left Share vulnerable to the autonomous choices of the end users and the IBM programmers who, partly by Share's own choice, remained outside of the group's sphere of influence. The rise of Fortran forced systems programmers to redefine not only their role, but their expertise.

Nevertheless, the esoteric knowledge programmers created through Share gave these specialists the means by which to retain their emerging professional stature. Even though Share's new operating system was rendered obsolete by Fortran, only systems programmers possessed the knowledge needed to make the changes required by the end users. Here systems programmers had a distinct advantage when compared to home economists. Their new knowledge remained tied to the intricacies of a complex machine. Such expertise could not be readily absorbed by IBM's marketing staff or even its engineering staff. Nor was the knowledge rendered mundane for reasons of gender or any other social prejudice. Systems programmers could retain the status they gained because they could continue to construct credible claims about the efficacy of their knowledge in maintaining an efficient computing facility. This remained true even after considerable knowledge about systems programming was transferred to IBM. Just as tabulating machine operators in the aviation firms' old "IBM departments" professed a loyalty to one manufacturer, IBM programmers pledged their allegiance to an emerging profession. These overlapping loyalties were essential to the institutional integration required by Cold War research. They played a vital role in aligning the interests of different organizations, even as they created a defined sphere of technical authority that could harbor new technical occupations and professions. This, in a much more grounded sense, was the structure of institutional alignments that constituted the new "military-industrial complex."[61]

As I describe in the next two chapters, the knowledge assembled by systems programmers would merge with knowledge derived from other fields to produce new high-level languages, virtual memory, and computer time-sharing systems. Systems programmers, as we shall see, would also find a place within an academic setting. But as for Share, the organization was unable to sustain its role as a surrogate professional society. Apart from the limits imposed by its rules of membership, Share and other users' groups remained wedded to a single manufacturer. When an effort to form a joint users' group failed, the drive for professionalization shifted to other circles. Share did continue to exist as an organization, but it gradually became more of a vehicle for computer program exchanges among users with more fragmented needs.[62] Still, the group was an important stepping-stone. Share allowed computing center directors and their staff to forge a coherent body of knowledge and a shared identity. Moreover, in having already adopted all the trappings of a profession, the next step was clear. Share representatives were among those who pushed for new organizations that could better represent a computer programmer's point of view.[63]

Figure 8.1
Philip Morse in the MIT Computation Center. Courtesy of MIT Museum.

8 Research and Education: The Academic Computing Centers at MIT and the University of Michigan

In this chapter and the next, I return to the university as the site of Cold War transformations. I build both chapters around a comparative study of technical developments at MIT and at the University of Michigan. The present chapter describes how academic computing centers were first created at these institutions. The second extends the comparison by considering how both universities went about entering the field of computer-time-sharing research. By maintaining a sustained comparison of Michigan and MIT, it becomes possible to consider the differences between what occurred at the center and periphery of Cold War research. Not that the University of Michigan was located far out in the margins. Along with other Midwestern research universities, Michigan was part of an elite tier of state universities that benefited from federal patronage. Nevertheless, the difference between MIT and Michigan are stark. Researchers at MIT negotiated from a position of technical and institutional leadership. Those at Michigan remained accountable to regional obligations and educational commitments that emerged from their identity as a state institution.

Yet although circumstances drove efforts at MIT and Michigan in different directions, MIT did not produce an unambiguously superior program for academic computing or for time sharing. Michigan managed to introduce computers to far more students early on. And although MIT led other universities in laying the basic technical foundation for computer time-sharing work, internal tensions, organizational jurisdictions and technical choices at MIT enabled Michigan's researchers to beat them to the first large-scale facility for time sharing.

Once again, I am less interested in settling issues of priority than in providing a historical account of the unique institutional ecologies out of which new knowledge and organizations emerge. At MIT and Michigan, a different array of institutional pressures and opportunities shaped the strategies by which researchers and administrators went about introducing new academic computing facilities. These institutional contexts were far better defined than during Forrester's early efforts, reflecting the growing maturity of Cold War institutions. Nevertheless, the technical outcomes remained novel. This resulted partly from an underlying tension where computing facilities were seen simultaneously as a commercial technology, a research and educational facility, and an object of research. For a while, researchers at both universities sought to integrate these separate goals through novel administrative arrangements and technical solutions. Yet the final outcome reveals how MIT and Michigan had to resolve this tension in different ways.

In both chapters, I also heed Charles Rosenberg's observation that the differences between the professions and disciplines are as instructive as their similarities. They provide further evidence for tracing the complex trajectory of postwar technical occupations, as they unfolded at the intersection of different academic and professional traditions. In recapitulating the same kind of opportunism that drove the computing specialists at Share, it was the extension of the postwar technical labor market into the university itself that created a window for an academic computing staff to aspire to do research. But even more explicitly than for those affiliated with Share, these staff members ran up against the ambiguous status of computing as a science, a profession, and a vocational skill. This additional layer of tension had no simple one-to-one correspondence with the one described in the previous paragraph. The elaborate ways in which diverse institutional goals were woven into different technological designs and occupational identities helps to explain why computing never came to resemble the idealized model of the professions as represented by law and medicine—and why computers never became a single, standardized artifact. Seen from the standpoint of occupational and disciplinary formation, the story at MIT and Michigan would again be quite different.[1]

Computing centers were not the only or even the principal site for research on computing during this period. Much of the most respected work took place in the context of academic departments and special laboratories such as MIT's Lincoln Laboratories and Research Laboratory for Electronics. In part, I focus on academic computing centers precisely because of the numerous tensions between institutional, disciplinary, and occupational interests that could be found at this particular location. But the choice is also based on historical significance. Computer science emerged as an interdisciplinary field that transcended the interests of established departments and research programs. At many institutions, including MIT and Michigan, academic computing centers served as an important catalyst for new interdisciplinary programs in computer science.

The MIT Computation Center

Scientific and Engineering Uses of Whirlwind I

The MIT Computation Center was an outgrowth of Project Whirlwind. As described in chapter 5, when the Office of Naval Research announced its intent to sharply curtail Forrester's funding for fiscal 1951, Provost Julius Stratton offered a suggestion that helped to redirect the ONR's support to a more sustainable research program on scientific and engineering computing applications. This was the origins of MIT's new interdepartmental Committee on Machine Methods of Computation. In looking for a suitable candidate to head this committee, Stratton turned to Philip Morse.[2]

Morse was a noted member of MIT's physics department, and the former director of Brookhaven National Laboratory. It is difficult to capture any individual's biography in a single paragraph, let alone the postwar career of a physicist, like Morse, who traveled through many different technical and administrative circles. Morse had received his PhD in Physics from Princeton under Karl Compton before Compton became president of MIT. Trained as an experimentalist, Morse had spent his early career studying some of the practical implications of the quantum mechanics revolution through mathematical studies of electronic scattering and electro-acoustics. The latter work led him to wartime research in anti-submarine warfare, and the new field of operations research. After the war, Morse continued to work as a science administrator, both at Brookhaven and then as the civilian head of the Weapons Systems Evaluation Group, a new operations research unit established within the Department of Defense. Both positions enabled Morse to develop an appreciation for machine computation. Stratton appointed Morse both for his administrative experience, and his knowledge of computing.[3]

As was the case with the situation between Forrester and the ONR, disciplinary interests dictated the general orientation of Morse's committee.

While Morse convened his committee with the declaration that its directive remained undefined, he and his colleagues clearly saw the computer as a scientific instrument. At this point, however, this dovetailed with Forrester's present plans to reorient Whirlwind research around a study of computer applications. The mathematician, Charles Adams, who headed Forrester's Applications Study Group, was allowed to join the committee. Also present during the committee's deliberations was the head of MIT's Office of Statistical Services, Frank Verzuh.

This committee reaffirmed Stratton's suggestion to the ONR. Beginning with fiscal 1952, the ONR agreed to support technical computing at MIT at approximately $300,000 a year. The work consisted of three coequal parts. First, the ONR supported a new Scientific and Engineering Computing Group headed by Adams that would collaborate with other MIT researchers in attempting to use the Whirlwind I to solve real computing problems. Another third of the funds went toward the operation and maintenance of the Whirlwind I, with the stipulation that no funds go toward any new engineering work. The final one-third of the funds was written up as a new contract (DIC 6915) for supporting graduate research assistants who wished to use digital computers in their research.[5]

Morse and his committee agreed to administer this new contract. Despite the relatively modest level of funding, these assistantships played an important part in charting a new course for MIT's work in computing. The committee was charged with supervising thesis students with an interest in computing. Its most pressing issue was to design an appropriate curriculum for two doctoral candidates—one of whom was Verzuh—who already expressed their desire to work with the committee. Yet although the committee admitted that there was some need to adjust departmental degree requirements, they decided that it was in the degree candidate's best interests to proceed within a specific disciplinary tradition. Although Verzuh was granted some exemptions because of his "long service on the staff," his ScD thesis in Electrical Engineering, "Boundary-Value Problems of Automatic Computing Equipment," remained a mathematical study that fell within the disciplinary traditions of the Electrical Engineering Department.[6]

There was, moreover, a noticeable hierarchy. Originally, these assistantships were meant to be evenly distributed between the participating departments, namely Electrical Engineering, Physics and Mathematics.

However, of the four applications put forward in the first round by the Electrical Engineering Department, the committee accepted only one, namely Verzuh's. In the second round, nine assistantships were awarded to Physics, six to Mathematics, one to Chemical Engineering, and none to Electrical Engineering. These discrepancies reflected the status of different disciplines, as perceived by a committee charged with determining what the most interesting problems in computing were.[7]

Although Morse's committee was charged to oversee all computing activities on campus, historical records convey the sense that the committee felt more comfortable speaking about graduate students and their proper training, as opposed to advising other MIT faculty members or surveying their interest in computing. While this may have grown out of academic habits, it became something of a self-fulfilling prophecy. Morse and his committee members came to believe that it was easier to get a graduate student to take an interest in computers, rather than convincing an MIT faculty member to incorporate new analytical methods into their research programs. To be more precise, having a graduate student work on a technique that might advance their thesis advisors' research fit squarely within the experimental tradition in physics and other scientific disciplines. "These assistantships," Morse wrote in his subsequent recollections, "have been very fruitful in the spread of the understanding of machine capabilities throughout the institute, for not only does the graduate student and his immediate associates learn about computing machines and their use but, even more important, his faculty advisors in the various departments have a chance to learn of the usefulness of the new machines in their special fields."[8]

In exchange for funding, research assistants were asked to spend one day a week helping faculty members who needed assistance with their coding work.[9]

Academic traditions of hierarchy, advising, supervision, and discipline initially lent considerable structure to what was considered valid work within an academic computing facility—more so than the very open negotiations that determined initial task allocations within corporate computing facilities. But not entirely. Consider Fernando Corbato's experience. Corbato was one of the first research assistants supported by the ONR. He received his assistantship because of his proposed work in solid-state physics, and yet the service obligations of the assistantship brought him to spend considerable time assisting other researchers. One such benefici-ary was a visiting applied mathematician, Alec Delgarno, from Queens Uni-versity in England. Delgarno came to MIT to use the Whirlwind I, specifically to tabulate a set of integrals he hoped to use to calculate elec-tron energy levels. After his visit, Delgarno wrote to Morse thanking him for the "invaluable and very expert assistance of Corbato," and that, "I am afraid I wasted a very large amount of his time."[10]

But it was not simply time or coding expertise that Corbato had to offer. Delgarno had to leave before his work was completed, and he assigned the remaining coding work to one of the staff members in Adams's group. However, this staff member could make little headway because of his lack of knowledge about physics. A visiting physicist from the University of Chicago volunteered to take over the work, but Corbato doubted that this individual could complete the work any time soon because of his unfa-miliarity with programming. Even at MIT, where computing was cast as a scientific activity, Corbato and others like him appeared as valuable inter-mediaries who could span the gap between an existing discipline and the not-yet-articulated field of computer expertise. Morse would admit, later on, that "many of our best and most imaginative machine programmers have been introduced to computing equipment via these Assistantships."[11]

Origins of the Computation Center

At first, Morse viewed his position on the Committee on Machine Methods of Computation as a service obligation. But as his interest in computing grew, he found himself investing more and more of his energies in this direction. First a trickle, then a torrent, MIT faculty members began pre-senting interesting problems to the committee, with an accompanying request for computer time. D. P. Shoemaker, who had just arrived from Caltech, wanted to try Linus Pauling's new x-ray crystallography tech-nique. Quite a few requests came from MIT's engineering faculty. But the largest allotment of time, 20 of the 90 hours a month set aside for general use, went to John Slater, chair of MIT's physics department. By early 1955, Morse could claim that graduate students and their supervising faculty members from nearly all of MIT's academic departments had considered using machine methods of computation.[12]

As computing became a normal activity, not only at MIT but at research facilities across the country, the ONR began pulling back its general support of computing activities. In late 1953, the ONR expressed to Morse that it had no intention of supporting MIT's use of the Whirlwind I on an indefinite basis. Both the Univac I and the new IBM 701 could offer comparable performance with lower operating costs. In response, Morse declared that the next several meetings of his committee would be spent discussing how to sustain machine computation services at MIT. The committee estimated that Whirlwind I's minimum operating expense would be around $90,000 per year, exclusive of any programming support or funding for the assistantships. When the ONR announced that it would curtail support to $100,000 a year for the next two years, Morse was left wondering "whether an educational institution even as large as the Institute can afford to maintain a large computing facility solely for educational and free research purposes."[13]

Indeed, there was no fiscal solution in sight. Although researchers were desperate for computer time, nearly none of them could afford to buy an entire computer. Contract regulations, in any event, would have generally prohibited it. Yet the active use of the Whirlwind I generated many new converts, namely researchers who could no longer continue their research without the computer as a scientific instrument.

Even as MIT's senior administrators struggled to make institutional adjustments, both with postwar demobilization and then with the Korean War, MIT faculty members learned to push their own agenda forward through institute committees and summer studies that helped to articulate institute policies. Both practices had become integral to the postwar practice of faculty co-governance at MIT. As an experienced administrator, Morse was well aware of the methods by which MIT's faculty could compel MIT's administration to move in new directions. It was one of MIT's first summer studies, Project Charles, which created the rationale for brining the Whirlwind I into the fold of the Air Force's new continental air defense system. Morse joined ranks with the Whirlwind I's major users in convening a new Institute Committee on Numerical Analysis and Machine Computation (ICNAMC).[14]

This larger committee issued a report titled "Requirements for Electronic Computing Equipment for Training and Educational Research at the Institute." The principal aim of this committee was to find a way to finance computer operations at MIT. However, the committee's investigations and deliberations also served to clarify and articulate the institute's need for computing. For instance, it became clear during their discussions with Charles Draper, the head of MIT's Instrumentation Laboratory, that there were several large laboratories on campus that *could* afford to buy a digital computer. Especially with the arrival of smaller computers, Draper could justify a computer purchase in his research contracts. Indeed he already had on order an IBM 650 (a magnetic-drum computer). At the same time, Draper, who was both committed to education, and familiar with the latest developments in aeronautical engineering, could reveal to the committee how aircraft manufacturers were making heavy use of computers and that "they expected the men they hire to know how to use them."[15]

It was in assembling such information that the committee formed its recommendations. The most striking conclusion was that all engineers and scientists had to know how to use a computer in ten years time. This meant that MIT had to provide all its students with substantial computing experience, if the institute was to remain at the forefront of technical education. Of equal concern was the fact that many sponsored and unsponsored research projects could not afford to pay for a computer. Drawing an analogy to the facilities planning process by which MIT raised funds to build new laboratories, the committee argued that the computer was a major research facility that should be financed through institute planning, rather than financial arrangements tied to individual research contracts. The committee also pointed out that the current arrangement for free computer time on the Whirlwind I was merely a "by-product" of sponsored research, and that regardless of the solution, there was a need to find a more permanent arrangement for these "educational research" projects.[16]

The Whirlwind I continued to operate, even as Morse's new committee carried out its deliberations, forcing Morse and his original committee to deal with the immediate matter of finances. Since there was now a real need to raise revenue, Morse instituted a charge of $250 an hour for using the computer. This resulted in a 50 percent reduction in usage, confirming his suspicion that only sponsored research projects could afford to pay for computer time. The decline in usage also fueled the worry that it would be impossible to keep the Whirlwind I in operation.[17]

The crisis was resolved when IBM decided to install its largest machine on campus, and to make it available free of charge. In exchange, MIT was asked to operate a regional computing facility open to all New England colleges. It was also asked to train computer programmers and to share the programs it developed with IBM. Clearly, this was not just a philanthropic gesture. While computer sales had done very well in Southern California, New England, which by all indications should have been a good market, produced disappointing returns. Moreover, as IBM began to shift its emphasis toward business data processing, it became all the more important to enlist universities into training a large pool of programmers. A regional center, in which MIT (of all institutions!) provided computer training to students located at many different universities, was something of considerable value to IBM.[18]

Nevertheless, if it seemed like there was a simple convergence of interests, a close look at the detailed arrangements reveals the finer texture of negotiations in educational philanthropy. In part, it reveals the postwar authority of an academic institution like MIT. But these negotiations also make it clear that neither MIT nor IBM clearly understood their own interests at the outset, let alone the interests of the other party. Both were articulated through the course of their joint discussions.

Verzuh helped open the conversation between MIT and IBM. After completing his doctoral thesis, Verzuh had become the de facto operations manager for the Whirlwind I. Influenced by Morse's background in operations research, Verzuh produced an "operational analysis" of all computing activities at MIT. Focusing on the total volume of computation, he estimated that MIT was spending $856,120 per year on computing. Because it bundled together expenditures on machine design and machine use, as well as analog and digital computation, this was a somewhat spurious figure. Nevertheless, it drove home the point that there was sufficient activity on campus to justify a large, central computing facility. Verzuh then proceeded to offer his own interpretation of the ICNAMC requirements study, suggesting that a magnetic drum computer like the IBM 650 could satisfy all educational computing requirements. Hurd's Applied Science Department was already supplying Verzuh's office, the Office of Statistical Services with free use of some equipment, similar to IBM's earlier arrangements with Columbia University. In view of the concerns that were expressed about the new charges for the Whirlwind I, Verzuh wrote to Hurd in the fall of 1954 to ask whether IBM might grant the free or discounted use of an IBM 650.[19]

Hurd offered to pay a visit to MIT, at which point Morse stepped in to manage the negotiations. Morse instructed Verzuh to prime Hurd on MIT's computing activities, and the relevant details of their requirements study. Morse then arranged for Hurd to meet with the provost, during which Stratton presented Hurd with a formal request to donate an IBM 701, not the smaller IBM 650. Perhaps as a matter of posturing, Hurd claimed that IBM was more interested in training than research, and in data processing rather than scientific computing. Hurd advised Morse and Stratton that IBM's board members would evaluate MIT's request based on how well it fit with the firm's own objectives. Nevertheless, Hurd grew more interested in completing some kind of arrangement as the day wore on, especially after meeting with the dean of MIT's new School of Industrial Management. By the end of the visit, Hurd promised Morse that IBM would entertain a proposal from MIT, and that Morse should instruct Verzuh to carefully evaluate which of IBM's current machines MIT wished to have.[20]

With this expression of IBM's interest in hand, Morse set out to enlist his own administration. He also pressed for an even larger machine. In speaking with the administration, Morse offered his own interpretation that the ICNAMC's requirements study suggested that a large machine, such as a modified IBM 701, or preferably the newer IBM 704 or 705, was necessary to serve all their unsponsored research and educational computing activities. At the time, MIT was in the middle of a capital campaign for expanding institute facilities, and its current president, James Killian, was about to approach IBM with a request to contribute toward the new Compton Memorial Laboratory. Impressed by Morse's appeal, Killian agreed to redirect his request toward the support of a computation center "of considerable magnitude." Killian met with IBM president Thomas Watson Jr. the following June. It was apparently during this meeting that IBM introduced the idea of having MIT open this new facility to other regional universities. Killian expressed some initial reservations because of the service obligations that this would incur. Morse, however, convinced him otherwise, especially by making it clear that a regional facility would provide a more compelling reason for IBM to offer its largest machine. Morse explained that giving MIT's "sister" colleges in New England some access to the machine would make it "a little easier to explain why we got such preferential treatment."[21]

Morse was successful in leveraging the support extended by both organizations in expanding his proposal to fill, and articulate, the mutual interests of MIT and IBM. By the time the negotiations were finished, IBM had agreed to contribute an 18,000-square-foot extension to the Compton Laboratory, which would house the computer and the associated programming and training facilities. IBM also agreed to cover the costs of machine maintenance and most of the operating costs. It agreed to install its latest scientific computer, the IBM 704. In return, MIT agreed to provide a full-time director, to give designated New England colleges computer training and computer time, and to play host to an IBM programming research group that would be allowed to work on MIT's campus. MIT also gave IBM a non-exclusive paid-up license for any hardware improvements made by MIT researchers. As announced in a press release issued in December 1955, this agreement created "the largest and most versatile data processing facility yet to be made available primarily for education and basic research."[22]

There was one additional provision to this agreement, namely $30,000 per year for research assistants. The ONR had created a precedent for providing not only computer time, but programming assistance to anyone at the institute. Morse had insisted on continuing this arrangement. While the amount was small, it meant providing the new facility with about a dozen graduate students with solid knowledge about programming who could serve as "staff members." Moreover, the duties of the IBM RAs were defined a bit differently. Now that MIT's central computing facility was opened up to a much larger number of researchers and students, Morse declared that "the men would be expected to spend more of their time definitely helping out in the work of the Center." Corbato, whom Morse hired to supervise and train these new assistants, interpreted this to mean 36 hours of service *per week*. For this group of graduate students, it would be less clear whether their assistantship was about pursuing their thesis research, or about developing their computing expertise. Morse agreed to serve as the first director of MIT's new Computation Center.[23]

Computers and Engineering Education at MIT

It is necessary to take a step backwards into the history of engineering education to understand why education was such a central theme in the plans for the new computing center. As mentioned earlier, by the mid 1950s, and well before Sputnik, there were growing concerns about "technical manpower shortages" in the United States. There were broad calls for engineering educational reform across U.S. campuses, to which even MIT was not immune. Given that IBM explicitly touted the new arrangement as a "broad program to increase the number of computing specialists in the United States,"[24] it is worth considering to what extent the new facilities connected with MIT's plans to revise its undergraduate curriculum.

The postwar engineering education reform movement was quite complicated. While one part of this movement sought to respond to the national crisis in technical personnel, another sought to bolster the engineering profession in response to the postwar authority of the sciences. Even as technical personnel shortages helped to enhance the status of engineering, wartime technical accomplishments by scientists undermined the intellectual authority engineers could claim in the newest domains of technology. These were the countervailing tensions that pulled the engineering profession in different directions. Conversations about the professional stature of engineers were voiced at the national level, through organizations such as the Engineering Council for Professional Development (a major accreditation body for engineering schools in the United States) and the American Society for Engineering Education. The most infamous report to emerge from these discussions was the "Grinter Report," released by the ASEE in October 1953, which basically recommended a two-tier system of accreditation that placed "engineering science" above more traditional approaches to engineering education.[25]

The recommendations of this ASEE committee were quickly overturned amidst a storm of protest. However, if the proposed reform failed at the national level, it continued at MIT under Gordon Brown, the director of the MIT Servomechanisms Lab, who had become chair of MIT's Electrical Engineering Department in 1952. Brown drew on the impetus created by the so-called Lewis Survey, completed in 1949, which had recommended broadening engineering education in order to preserve the status of the profession.

After 1949, MIT became ever more popular among young prospective engineers and scientists. Its burgeoning enrollments, especially in physics and electrical engineering, became a matter of institute-wide concern. Brown came to chair a new institute Committee on Enrollment, which he used to push for the idea of engineering science. He and his committee recommended sweeping reforms that would give all MIT undergraduates "greatly increased competence in basic science and its exploitation," as well as "greater versatility to move across the fields of specialization." Significantly, Brown himself was trained in an interdisciplinary, analytical approach to engineering. His plan was clearly an expression of his own inclinations and experiences. But it was also a set of experiences that mapped all too easily onto MIT's recent history.[26]

Moreover, the strength of MIT's reputation among current and prospective students made it possible for the faculty to propose a general increase in course requirements. Yet even at MIT there was substantial, organized resistance to the idea of engineering science. The Dean of Engineering, Richard Soderberg, accepted that the School of Engineering needed to evaluate its educational policies and priorities. At the same time, this was a veiled protest that the engineering school should set its own curricular policies, rather than ceding the responsibility to an institute committee convened by MIT's physics-dominated administration. Uncomfortable with Brown's implicit claim about the higher status of scientific knowledge, Soderberg wrote an official response that suggested that engineering derived its strength from the independence of its fields of expertise, and their wide spectrum of educational philosophies. Responding with bureaucratic tactics, Soderberg appointed his own committee on engineering education, and charged the committee specifically to collect separate opinions from each department chair. In an informal report issued in October 1957, this committee labeled engineering science a popular catchword that lacked substance and had no meaningful definition.[27]

Even in the wake of Sputnik, Soderberg's committee concluded that they knew of no situation that required "emergency attention." Nevertheless, Sputnik was a watershed for engineering education. For one thing, it allowed Brown to act back at the national level through an ASEE follow-up committee to the Grinter Report. This committee proceeded to specify what a formal degree program in engineering science might look like. And while various members of MIT's engineering faculty remained quite dubious about the value of engineering science, Brown's initiative was picked up by the MIT administration—most notably by Stratton who stepped up to become MIT's acting president in 1957. Stratton was a physicist and the original director of the Research Laboratory for Electronics. Predictably, he emerged as a strong advocate of engineering science. Because of the strength of MIT's engineering faculty, the extent of the reform was still circumscribed. Even at MIT, the idea of a unified and more analytical engineering curriculum gained a strong foothold only in fields (e.g., electrical engineering and aeronautical engineering) where broader pressures for interdisciplinary training created an opportunity for more extensive reforms. Other departments changed more slowly. Still, the widespread concern for the quality of U.S. engineering education after Sputnik allowed individuals like Brown and Stratton to wail against the prevailing patterns of engineering education.[28]

If these educational reforms were cast only as a contest between professional and disciplinary interests, then this would miss an important element of the reforms, namely the undergraduates. It was the MIT student, after all, who stood as the object of this reform movement. MIT's administrators, faculty members, and students all had different ideas about who an MIT undergraduate was, and who they should become.

To begin with, from the standpoint of the MIT student, educational reform was not something that was always at the top of their agenda. Sure enough, even the students wrote about engineering education reform. They even reported on the unveiling of the new Computation Center in their undergraduate newspaper, *The Tech*. But daily life also consisted of news about the Inter-Fraternity Council, the latest hoax (or "hack") carried out atop the institute's great dome, and a new athletic facility opened on West Campus. There were occasional reports of a so-called student riot. In one incident, students took to the streets to protest an increase in dormitory rents and blocked off Memorial Drive by lighting several mounds of garbage on fire. In another, 47 students were arrested after a raucous water fight nearly doused a municipal judge. Nine paddy wagons and eleven fire engines were called out for this event.[29]

MIT administrators, beginning with Killian and Stratton, took these events seriously. The problem, as they saw it, was that their students were too slow in adjusting to the realities of the Cold War. MIT, once a good engineering school, now was one of the foremost institutions for scientific and engineering research. In their new world order, scientists and engineers assumed new social responsibilities that far exceeded their former purpose.

And yet it was the students who seemed fixed in their ways. In a series of talks in which they addressed the students, MIT's top administrators tried to get their students to see the opportunities—and responsibilities— associated with postwar science and engineering. In reality, the postwar expansion of the technical professions extended earlier patterns of class mobility, where many students who entered engineering, even at MIT,

came from working class backgrounds. Working class attitudes and practices were already a discernible aspect of the U.S. engineering culture. Nevertheless, MIT's administrators sought to instill in their students their own
image of an engineer, and proceeded to search for ways to encourage the
"moral, ethical, and social concepts essential to a satisfying personal philosophy." For them, the rigors of scientific methods and reasoning were
themselves a means of fostering valuable discipline, irrespective of the
immediate utility of such knowledge.[30]

Since these were still the ideals of the reformers, what were the attitudes
of the students themselves? No single description can capture the attitude
of a "typical" MIT student. The voices expressed in MIT's undergraduate
humor magazine, *Voo Doo*, for instance, were quite different from those
expressed in the undergraduate newspaper, *The Tech*. In some respects, a
publication like *Voo Doo* can provide a keener insights into how some of
the MIT students responded to the educational reform efforts at MIT. For
instance, a drawing of a physicist with his head up his ass, or of Norbert
Wiener as a "digital computer" counting his bare toes, constituted direct
assaults on the authority of physics and mathematics, along with the pur-
ported purity of science. Alongside other recurrent expressions of sexual
desire and self-deprecating images, this uneven publication posed frequent
and wishful inversions of disciplinary authority as well as the authority of
Cold War military institutions. Its clever combination of words and images
could deliver quite penetrating commentary on ongoing educational
reforms, and on the larger changes taking place at MIT. Such a study, based
on understanding the anatomy of humor, could lead to an honest understanding of many of the MIT students' thoughts and deepest anxieties. Yet,
while it is useful to note the existence of such forms of resistance, the
requisite analysis is beyond the means of the present study.[31]

In this account, I focus on the eager, literate, and often reform-minded
students who wrote for *The Tech*. This is not to suggest that *The Tech* was
either more or less representative of students' opinions. Yet so long as its
writers were MIT undergraduates, the words printed on the pages of *The
Tech* were hardly the product of individual opinion. Constructions of news-
worthiness, like constructions of humor, emerge through a degree of
hermeneutic closure with an intended audience. This would have been
especially true of recurring topics, as was the case with the articles pertaining to educational reform. To varying degrees, the text of these articles

Figure 8.2
"Hamus et Eggus." From *Voo Doo*, November 1956. Courtesy of *Voo Doo* and MIT
Institute Archives and Special Collections.

can be regarded as a reasonable representation of some subset of MIT students greater than its individual authors. *The Tech*, in any event, provides
a better insight into the relationship between engineering reform and computing at MIT.

From *The Tech* it is clear that MIT students were quite aware of the new
postwar prominence of physicists. These students welcomed the arrival of
the new 6-BeV accelerator and acknowledged that the new Compton
Memorial Laboratory would be a real tribute to research. They also recognized the new social authority of physicists. *The Tech* reported on MIT
physicists' involvement in the Atoms for Peace initiatives, and accepted as
authoritative their call to ban certain kinds of hydrogen bomb tests. Such
opinions no doubt echoed sentiments from mainstream media. Nevertheless, it is significant that postwar attitudes toward science could so easily
be reproduced within a body of young undergraduates. In a more local
reaction, MIT students celebrated the fact that their own president had
been appointed by President Eisenhower to serve as the new national

Digital Computer

Figure 8.3
Norbert Wiener as "Digital Computer." From *Voo Doo*, March 1956. Courtesy of *Voo Doo* and MIT Institute Archives and Special Collections.

advisor for science and technology. For them, Killian had been charged with "the task of leading the United States back to a position of technological supremacy."[32]

The pages of *The Tech* also make it quite clear that MIT students were quite aware of the changes that were being made to their own education. Students packed one of the dormitory dining halls to hear an address by Dr. Julius Stratton, then acting president, on the subject of curriculum reform. In an article titled "More Emphasis on Science Vitally Needed," the reporter understood Stratton to have suggested that, "the new order of civilization is founded in science, and those who hope to be influential in society must have a thorough understanding of its principles." MIT students accepted this responsibility. In an article that appeared shortly after Sputnik, *The Tech* noted:

Elsa Maxwell, famed party thrower, threw a barb into American education last week when she said, "Our young people are too serious. There's no lightness in their education. Everything is 'MIT-ism'. They try to beat the clock, beat the other fellow."[33]

While clearly reveling in the overall humor, the reporter nevertheless sided with MIT's Dean of Science, George Harrison, who retorted that it was public ignorance and apathy that was the real problem. According to Harrison, "what America is suffering from is too much Elsa Maxwellism." Opposite a picture of Maxwell, a fat woman, eating and wearing a funny hat, was a dry if respectable portrait of Dean Harrison.

With earnest concern, MIT students assembled their own Student Committee on Educational Policy. This committee raised the same questions as their faculty members, and asked whether an undergraduate engineering degree could still constitute a full professional degree. The committee organized a panel made up of MIT faculty members and administrators. Speaking to the students, Stratton fueled their worries with a flat declaration that a four-year undergraduate degree could no longer prepare students for a professional career. The article describing this event ran under the headline "MIT Graduate—Real Professional Man, or Merely Technician?"[34]

Stratton continued to push for the idea that science had to be integral to an engineering student's education. Both Stratton and Harrison also propagated the myth that the engineer's primary role was that of applying scientific knowledge. They allowed that there were important differences. Engineers had to "have a feeling for materials, a concern for cost." Scientists, by comparison, were supposedly trained to avoid such compromise—without which they could not do good scientific work—which meant that they were qualified to serve only as advisors to, rather than executors of human welfare. This article ran under the headline "Engineer Not 'Just Another Kind of Scientist.'"[35]

Since the digital computer was an analytical instrument, it might have been expected to play into the local plans for engineering education reform. However, the very distinctions, between scientist and engineer, and between undergraduates and graduate students, served to exclude most MIT undergraduates from a significant involvement with computers. In October 1957, when Sputnik was launched, *The Tech* reported that MIT faculty members were collaborating with Harvard astrophysicists in trying to compute the orbit of the foreign satellite. MIT's new computer was a powerful tool with which the United States could meet the Soviet challenge. The students marveled at the computer's abilities. And they worried when the researchers failed to obtain the required telemetry. This story, which ran for several days in the student newspaper, reveals how the computer remained a coveted tool for scientific research, instead of a resource generally open to MIT students. Other articles that reported on MIT's new Computation Center reiterated this general motif. Although a handful of courses did allow some undergraduates to make use of the Computation Center, the computer remained a remote entity for the vast majority of MIT students.[36]

This pattern of access was consistent with Morse's intentions all along. Morse remained a physicist and not a member of the administration when it came to matters of engineering education reform. He remained free to interpret, and thus define the policies of the MIT administration by choosing a specific direction for his Computation Center. When he presented his plans to the administration, Morse used the odd phrase "educational research." The real emphasis of the phrase was on the second word, and it was meant, with intentional ambiguity, to connote thesis research over changes in undergraduate education. Vannevar Bush would have been proud. The new Computation Center was one instance of the concern, expressed by a sincere member of the physics faculty, that MIT's undergraduate education was being made an appendage of its graduate schools. It was here that IBM wound up conceding the most to MIT's postwar authority. In the end, IBM accepted that MIT's Computation Center would be geared more toward research, not training, and science, not data processing. Through no coincidence, the second regional academic computing center IBM created was dubbed the Western Data Processing Center, a facility established in conjunction with UCLA's business school.[37]

This is not to say that MIT undergraduates had no access to the Computation Center. Students most fascinated with digital computers found their way past the center's doors. Morse was in fact quite proud of the fact that he had made IBM's biggest scientific computer available even to an undergraduate. Yet the limits of MIT's efforts appear in their sharpest relief when compared against the efforts at Michigan.

The University of Michigan Computing Center

The origins of the University of Michigan Computing Center parallel those of the MIT Computation Center in many ways. A military contract

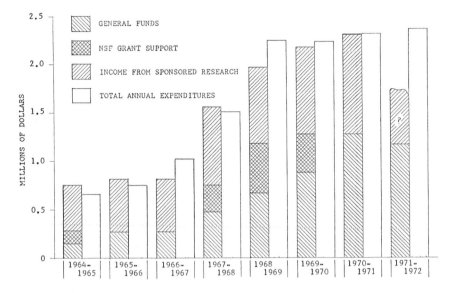

Figure 8.4
"Computing Center: Appropriations, Earning, Expenditures," Box 1/Annual Report 1958/59–1960/61, Computing Center Records, University of Michigan, Ann Arbor.

tangential to computing brought the first digital computer to Michigan. The university made this machine, Midac, available to members of its faculty. However, the machine was shut down because of a decline in Air Force sponsorship. IBM stepped in by making available an IBM 704. Michigan then turned to an applied mathematician knowledgeable about nuclear physics to serve as the director of its new academic computing facility.

Yet important differences were embedded within these similarities. Michigan lacked the postwar stature of MIT, and could not secure the same level of federal support, or the exact same arrangement with IBM. As a state institution, Michigan's obligations and priorities differed from those of MIT. All this would contribute to a rather different sequence of events, and a divergent course for computer education and computing research at Michigan and MIT.

Early Work at Michigan

Michigan's early work on digital computers grew out of a wartime aeronautical engineering facility. During the war, the Army Air Force set up a facility at the nearby Willow Run Airport to help the auto industry retool for aircraft production. This facility was transferred to Michigan at the end of the war. Beginning with a study of defenses against V2-type rockets, Willow Run Laboratories began focusing on continental air defense well before the work on the SAGE air defense system began at MIT. Michigan also created a new administrative unit, the Engineering Research Institute, to manage Willow Run along with several other major military contracts. MIT would pursue a similar move in 1953, in creating the Division of Defense Laboratories in an attempt to contain the threat of excessive military influence. But for Michigan, this action was more a reflection of the administration's desire to foster rather than restrain military research.[38]

As with Project Whirlwind, Michigan's Midac was built for a specific purpose. Willow Run began collaborating with Boeing on a new anti-aircraft missile, Bomarc, in 1950. Michigan was in charge of the guidance system. Given that the National Bureau of Standards was also engaged in guided missiles development, Michigan's researchers visited the NBS and became aware of Alexander's work on the SEAC. After running a set of guidance control simulations on the SEAC, Michigan convinced the Air Force that they needed a similar computer. The NBS assisted with its design.[39]

Michigan may have received help from Alexander, but was defeated by Forrester. By 1951, work on the Bomarc missile had progressed to the point where the Air Force authorized a preliminary study of an air defense ground reporting system to be used with the missile. Willow Run researchers had also begun to work on a larger, Air Defense Integrated System when they were confronted by Forrester's group at MIT. Historical accounts of this confrontation obscure the extent to which Michigan had begun to solve important technical problems pertaining to computerized air defense. Nevertheless, the computing capabilities at Michigan precluded a demonstration beyond that of the simulated interception of a single aircraft. In contrast, MIT could demonstrate the integration of multiple radars and provide a simulated command and control interface using the computational power of the Whirlwind I. Most important, Michigan could not envision Forrester's massive system of continental air defense—designed largely at the Navy's expense.[40]

While the termination of air defense studies at Michigan spelled an end to its machine-development work, technical computing using the Midac remained an important part of Willow Run's research activities. Building on its electronics expertise, Willow Run became a fairly diversified military research facility. The laboratory slowly shifted its focus to electronic surveillance and reconnaissance. This work justified the continued use of Midac.[41]

The use of the Midac was initially limited to military contracts. However, when Willow Run found itself in need of fostering better relations with the university, it opened Midac up to Michigan's faculty. Several members from Michigan's physics and mathematics departments immediately tried the kind of computational experiments being carried out elsewhere. This included studies of nuclear reactor transients, as well as Monte Carlo simulations of highway traffic congestion. This broader interest in computing prompted one of the younger staff mathematicians, John Carr, who had worked on the Whirlwind I, to develop a series of short courses on digital computing techniques. A two-week program in the summer of 1955 drew about 160 participants.[42]

However, there were limits to these early activities. As a secure military facility located some distance from the main campus, few researchers, let alone students, could gain access to the computer. More important, the Air Force took the Midac away from Michigan in 1958. Carr had also left Willow Run by then, having being given a regular appointment in Michigan's mathematics department. There were limits to how the military could influence, or support, academic activities at Michigan. However, there were also some lasting influences. Carr received his appointment partly because of his courses. He also went on to introduce a very popular course on campus called "Methods of High-Speed Computation" (Math 173). Most important, the early use of the Midac created an early nucleus of users with an avid interest in digital computing techniques.[43]

Origins of Michigan's Computing Center

The immediate antecedent of Michigan's first computing center was its Statistical Research Laboratory (SRL). Cecil Craig, a well-known statistician on the faculty, convinced the administration to set up a tabulating facility in the Horace Rackham School of Graduate Studies in 1947. Following the same technical trajectory being followed elsewhere, this evolved into a

computing facility used by many members of the faculty. Citing the interest the Midac had generated, Craig convinced the dean of the graduate
school that Michigan ought to get a computer for research and instructional use. In the meantime, the negotiations between MIT and IBM had
taken place. As an outgrowth, IBM announced a broader Educational Contribution Program that offered, in effect, a 60 percent discount to academic
institutions seeking to lease any of its smaller machines. This included
IBM's magnetic drum computer, the IBM 650, which was installed in the
SRL in March 1956.[44] About a year and a half later, IBM's new Applied
Science Director, Charles DeCarlo, approached Michigan's Vice President,
William Stirton. DeCarlo offered a deal similar to its new arrangement with
MIT. This would have been IBM's second regional academic computing
facility. However, Stirton rejected DeCarlo's offer. In part, the offer was not
as sweet as the one given to MIT. But more important, Stirton was unwilling to approve an installation open to other Midwestern universities. He
remained more guarded about committing Michigan's facilities and faculty
to other universities. Stirton may have also felt that such a service center
would produce unfair competition for regional businesses seeking to enter
the emerging computer services market. In any case, the offer was more
attractive when it originated with the institution in question, rather than
when tendered by a private firm.[45]

DeCarlo persisted in his efforts, recognizing that Stirton was an important person to enlist. It helped that Stirton, in turn, could not remain deaf
to the groundswell of interest in computers. Immediately after the IBM 650
was installed, eighty students from Carr's Math 173 besieged the new facility. Eighty-three percent of the machine's first 20,000 hours were spent on
instructional use or non-sponsored research, verifying that there was broad
interest in computers.[46]

In contrast with what happened at MIT, it was Stirton, an administrator,
who convened an ad hoc committee on computing needs. Working
through the committee and DeCarlo, Stirton approved a compromise plan
in which IBM would be asked to extend its Educational Contribution
Program to include one of its larger machines. This freed Michigan to set
its own policies regarding the machine's use. But it also meant that
Michigan, unlike MIT, would have to pay for computer time.[47] This raised
the matter of finances. One option was to simply extend the current
practice at the SRL. Like MIT's Office of Statistical Services, the SRL charged

sponsored research projects full commercial rates, while offering free use for classes and non-sponsored research. Craig was not altogether happy with this practice, and preferred that the main focus of the new facility be on education and non-sponsored faculty research, as was the case at MIT. Stirton agreed, and promised initially that the university's General Funds could be used to cover any unmet balance in Michigan's 40 percent rental obligation to IBM. However, budgetary cutbacks by the state legislature forced Stirton to withdraw his promise, making it necessary to raise real revenues. Craig thus turned full round, seeking now "to see if there were University units that would undertake to bring enough sponsored research to the 704." Craig approached the administrators at the Engineering Research Institute, now the University of Michigan Research Institute. Since they had lost the Midac, UMRI administrators stated that they could deliver $100,000 per year of computing use to the university's new computing facility. Craig also secured a three-year, $150,000 grant from a new National Science Foundation (NSF) program that specifically aimed to help universities create or expand their computing facilities. In August 1959, with these commitments in hand, the University of Michigan opened its new Computing Center. A member of the mathematics faculty, R. C. F. Bartels, agreed to serve as its first director.[48]

Things did not go quite as planned. During the first month of operation, sponsored research projects used only eight hours of computing time, producing paltry revenues of $2,500. Partly, the revenue was so small because of the speed of the IBM 704. Computations that were processed laboriously on the IBM 650 and Midac were solved so rapidly that it used very little CPU time, to which Bartels tied his revenues. But Craig and Bartels also failed to anticipate the difficulty of moving users over to a new machine. IBM's 704 and 650 were very different machines and had to be programmed in different ways. Even after Michigan closed down the IBM 650, users did not rush in to recode their applications to run on the new machine. Worse yet, Willow Run delivered almost no work to the Computing Center. Some of its researchers preferred to send their work to the IBM Technical Computing Bureau in New York, while others who were otherwise willing to transfer their problems found that the limited size of Michigan's 704's core memory precluded efficient computation. They too preferred to send their work elsewhere. It is significant that although UMRI administrators were eager to transfer their computing work to the new

Computing Center, they lacked the authority and technical judgment by which to obligate the choices of their own research staff.[49]

Although it appeared that the arrangements might unravel, the Computing Center had numerous allies, beginning with Stirton, who were wiling to shield the center and its new director from criticism. The Computing Center's Executive Committee, which was basically an extension of Stirton's ad hoc committee, also remained fully committed to the facility. With the assistance of the dean of the graduate school, Ralph Sawyer, who held a seat on the committee, Bartels began convincing various users to transfer their work to the Computing Center. By January 1960, revenues were up to $4,678—still well short of their mark. By March, revenues were up to $8,500, an amount high enough to sustain a balanced budget had they received this amount at the outset. Revenues then rose to $10,281 by May. Although revenues dropped off somewhat after the end of the semester, Bartels obtained an authorization to transfer unexpanded balances from the Center's salary account. He also obtained an additional $15,000 from a Ford Foundation grant a faculty member received to help bring computers into the engineering curriculum. At the close of the fiscal year, Bartels was able to claim that "With the assistance of the Ford Foundation, the Center will close the first year of activities without a deficit."[50]

The Ford Foundation Project

As is well known, the scale of postwar federal research expenditures rapidly eclipsed the influence philanthropic foundations had on science during the interwar years. Nevertheless, organizations like the Ford Foundation persisted in their interventionist practices, and sought to identify unique niches where they could make a difference. After World War II, the Ford Foundation initially focused on adapting the social sciences to the dominant concerns of the Cold War by funding studies of state formation, foreign relations and the larger genre of regional studies. However, in the wake of the Korean War, the foundation began to shift its mission to include technical domains that seemed to lack federal support. After Sputnik, this included improvements to engineering education.

It was within this historical and institutional context that a member of Michigan's Chemical and Metallurgical Engineering faculty, Donald Katz, submitted a proposal promising to bring computers into the engineering curriculum. Katz was a member of another of the ASEE follow-up committees, the Committee on Engineering Analysis and Design, established in the wake of the Grinter Report. It was no accident that a chemical engineering faculty member was on this committee. At the time, chemical engineering was undergoing substantial changes as control-systems engineering and the relatively new method of unit analysis began to come into general use. Following the same kind of logic that Gordon Brown pursued at MIT, Katz wished to install new analytical methods into his engineering curriculum. Significantly, Katz had succeeded not only in reworking his own courses, but a total of eleven courses in the Chemical and Metallurgical Engineering Department, where the students used the SRL's IBM 650 for their problem sets. With 225 students, Katz's department had been sending almost as many students to Craig's facility as the mathematics department.[51]

Building on this success, Katz had assembled a Committee on Computing in the College of Engineering, with the idea that he could extend his work to the other engineering departments at Michigan. Joined by an active and interested committee of faculty colleagues, Katz was able to put together a proposal of much greater scope than he could have advanced himself. His proposal to the Ford Foundation called for training the faculty themselves in computing. It paid for release time so that these faculty members could develop computing problems and exercises suitable for their curriculum. To broaden the impact beyond his own university, Katz also asked the foundation to support semester long leaves for fifty faculty members from other institutions. An even larger pool of instructors was to be invited to attend week-long summer workshops, as well as an annual conference on the use of computers in engineering education. As was the case at MIT, an important component of the project was its graduate research assistantships.[52]

Katz argued that the work of integrating the computer into the curriculum would force Michigan's faculty members to develop a better understanding of their own field. The unforgiving nature of computer programs required rigorous knowledge of engineering equations and their manipulation. Moreover, any broad curricular review provided an occasion for the faculty to reconsider the fundamental body of knowledge most relevant to their field. Katz had clearly learned a great deal, and picked up the appropriate language, in participating in curricular reforms at the national level.

It would have been clear to the Ford Foundation that Katz was proposing a general review of a university's engineering curriculum. Moreover, it was equally significant that from the standpoint of the Foundation, Katz was not merely improving engineering education. He was doing so with one of the most visible technologies in which the United States had a clear lead over the Soviet Union. Such work would bring attention to the Foundation itself. In 1959, the Ford Foundation gave Katz a three-year, $900,000 grant, an amount exceeding his original request.[53]

Katz's project was a resounding success. It did prove rather difficult to draw so many faculty members away from other institutions for sabbaticals and extended leaves. Nevertheless, this simply created a pool of money that Katz could draw on to address other things, like the Computing Center's deficit. The week-long workshops were immensely popular. During the first two years, 192 faculty from 87 colleges and universities received computer training at Michigan. Katz offered $500 stipends to faculty members who volunteered to draw up sample problems for use during the workshop. More important, 34 members of Michigan's engineering faculty developed a total of 37 undergraduate courses and 23 graduate courses that required students to use the Computing Center. By the end of the project, nine out of the college's fourteen engineering departments and programs, representing 75 percent of the university's engineering enrollment, required all their students to take Carr's programming course or an introductory digital computers course offered by the department during their sophomore year. These were changes that persisted after the grant ended. Moreover, with both faculty and students attuned to new computing techniques, this ensured a continuously growing demand for computing capacity at Michigan.[54]

Even with such numbers, it might be difficult to see just how rapidly computing entered into the undergraduate curriculum at Michigan. R. C. F. Bartels himself presumed that what Katz was doing was a natural outcome of the new technology and that similar work was taking place elsewhere. It was only during the first joint meeting of academic computing center directors, sponsored by MIT and IBM, that Bartels realized that his center was playing, "a more active role in the educational program of the university than does any other comparable facility in this country."[55]

Morse himself might have been puzzled by this discrepancy, but the difference stemmed from the different interests and resources of the two

institutions. As a state university located in the nation's preeminent industrial region, concern about undergraduate engineering education was more
broadly diffused among Michigan's faculty. It was no accident that Katz
could find other engineering faculty members excited about his project.
Michigan did not hold an undisputed lead in matters of engineering education, and yet local circumstances allowed Michigan to bring computers
into the undergraduate engineering curriculum more rapidly than any
other institution. It was difficult for Michigan to maintain its lead. The
reforms introduced by Katz were easily replicated at other universities. This
was indeed a major premise for his proposal. Yet what Katz did was a major
accomplishment.

Systems Programming in an Academic Computing Center

The scale, and more subtly, the type of demand for computer capacity at
Michigan also fostered technical innovation. As was the case for the corporate computing facilities described in chapter 6, academic computing
facilities also required a sizable systems programming staff.

Systems programming at Michigan began in earnest with the arrival of
the IBM 650. Bruce Arden, who ran Michigan's general tabulating service
(separate from the SRL), and Robert Graham, a machine operator for the
Midac, were hired to staff the new IBM 650 facility. Although they were
technically hired as operators, Arden and Graham were tasked to find and
install appropriate systems programs for the 650. Following the custom
established by Share, systems programs, subroutine libraries, and other programs for the 650 were freely available. These programs were distributed,
moreover, as source code,[56] so that they could be modified to meet the
needs of a specific facility. Much of Arden and Graham's work consisted of
maintaining the systems programs necessary for their IBM 650's efficient
operation.[57]

Initially, Arden and Graham followed the lead of other IBM 650 installations. However, as their programming proficiency increased, they began
to explore more advanced areas of systems programming. Again, by 1957,
Fortran was beginning to emerge as the standard language for scientific
and engineering applications. Many believed that high-level languages
such Fortran required large computers, such as the IBM 704. However, Alan
Perlis, then a mathematician at Purdue University, set out to demonstrate
that high-level languages could be built for the smaller IBM 650, which

was installed more often at academic campuses. Because the relative ease of a high-level language made it especially attractive for instruction, Arden and Graham obtained a copy of Perlis's Internal Translator (known simply as "It"). "It" became very popular at Michigan. Indeed, "It" was so popular that it threatened to max out their IBM 650's computing capacity, prompting Arden and Graham to see what they could do to improve "Its" computational efficiency.[58]

One difficulty presented by "It" was that *compiling* a high-level language into object code (that is, code that could be executed by a computer) was computationally intensive. Although the IBM 650 was a stored-program computer, its magnetic drum storage gave it the same programming properties as a computer built using delay lines memories in that it was necessary for programs to take into account the access delays associated with main memory. Especially after IBM made some modifications to the hardware (by adding an index register to the 650). Arden and Graham were able to make substantial improvements to the performance of Perlis's program. Although their General Algorithm Translator (GAT) was used only briefly at Michigan, being installed just shortly before the arrival of the IBM 704, it contributed to the decision to have Arden and Graham occupy the two senior staff positions in the new Computing Center.[59]

At this point, Arden and Graham were joined by Bernard Galler, a faculty member in the mathematics department. Galler had received his training in mathematical logic and topological algebra. First hired as an adjunct instructor, Galler was soon working for Paul Dwyer, a well-known mathematician at Michigan who pursued early work on linear programming in conjunction with the Air Force's studies on the optimization of military procurement. While Galler began by working on the Midac, Dwyer's problems quickly outgrew this machine, so Galler followed the problems as they migrated to Michigan's IBM 650, and then the IBM 701 and 704 at General Motors. In the process, Galler gained greater knowledge of programming, and of the differences between these machines.[60]

The original plans for the Computing Center called for a series of joint appointments, so that there would be an intellectual bridge spanning the Computing Center and the academic departments. Carr had left the University of Michigan when it became clear that he would be bypassed as the Computing Center's first director. As a consequence, Galler was given the second (and only) joint appointment, after Bartels. As a faculty member, Galler was free, under the terms of this appointment, to pursue his own interests. Nevertheless, he succumbed to the allure of computers and programming.

Michigan's IBM 704 was scheduled to arrive shortly, and it required an operating system. Galler was already visiting GM regularly because of his work for Dwyer, and he knew their systems programming staff well. He therefore volunteered to modify GM's operating system to serve Michigan's purposes. Through a special arrangement between IBM and GM, Galler was allowed to use as much as 15 hours per week on GM's IBM 704. He used it to rework the operating system to include such things as the detailed accounting system necessary for Michigan to charge its users and thus raise its revenues. In the end, Galler claimed to have rewritten 90 percent of the code in GM's operating system in developing the Michigan Executive System—known colloquially as a "Mess."[61]

Meanwhile, Arden and Graham occupied themselves with another high-level language, the Michigan Algorithm Decoder. Technically, this began as Michigan's attempt to implement a new language, Algol, which was gathering some momentum during the late 1950s. The specifications for Algol were written as a technical standard, leaving different institutions free to develop their own implementation. While Carr was still at Michigan, he had enlisted Arden and Graham to work on Michigan's own implementation of Algol-58.[62]

The Michigan Algorithm Decoder was supposed to be primarily an academic endeavor; Arden and Graham chose to install Fortran on the Computing Center's IBM 704 because of the demand for this language. However, when 80 students from Math 173 began using the Fortran compiler, this immediately occupied 30 percent of all available machine time. The more worrisome fact was that there were hundreds, if not thousands of students scheduled to arrive in conjunction with the Ford Foundation project.[63] Clearly, this would have to be addressed. As they had done before with Perlis's compiler, Arden and Graham, now joined by Galler, shifted their attention to the Fortran compiler. It was apparently Graham who in combing through the source code discovered that the main problem was that the Fortran compiler was making frequent references to translation tables stored on the 704's slower, secondary magnetic drum storage. Arden and Graham were familiar with optimizing programs to run on a machine with such a memory system, given their experiences with the IBM 650. Collectively, the three decided that a more efficient way to run a compiler was to develop all the translation tables in the faster core memory, and transfer portions of the table out to the magnetic drum only when the tables exceeded the space available in the core. Their solution also employed an abstraction, wherein a separate part of the compiler automatically kept the most commonly used portions of the tables in the machine's core memory, and automatically swapped in those sections on the magnetic drum when required by the compiler. Although Graham left the project before the Michigan Algorithm Decoder (MAD) was complete, he contributed substantially toward its implementation.[64]

Because of the rush to build a working compiler, MAD was neither a version of Fortran nor a version of Algol. Nevertheless, MAD drastically reduced the computing load on Michigan's IBM 704. With 13,016 of the 26,051 computing runs being dedicated to coursework during the spring semester of 1962, educational use accounted for only 13,511 out of 52,709 minutes, or 25 percent of the total machine time used. They had achieved between one and two orders of magnitude of reduction in compile time as compared to the Fortran compiler.[65]

In working to meet the needs of their university, Michigan's "Mad Men" (who had obtained permission from the publishers of *Mad* magazine to call their compiler MAD) wandered into some of the most advanced areas of computer programming. Yet although the Mad Men made the brash declaration at one of Share's 1960 general meetings that all current work on Fortran was unnecessary, MAD was not necessarily the most advanced compiler. To be specific, while most efforts were directed toward optimizing compilers for the object code's execution time, MAD was optimized for compile time. The former emerged from traditional concerns in scientific computing, where the efficient execution of a program was essential for the complex calculations found in many forms of scientific computation. Optimizing a compiler for execution presented formidable challenges, and it played into heated debates over the validity of using high-level languages in scientific and engineering computing. Yet, despite all this talk about optimized compilers, Michigan's computing center staff had demonstrated that there was no single standard to which compilers should be optimized to. Beginning with the same basic criteria of efficacy, at academic computing centers, where thousands of inexperienced students submitted simple programs that were run only once, a fast compiler contributed much more directly to a center's overall efficiency.[66]

Gaining Recognition for Research

The programming efforts of Arden, Graham, and Galler at Michigan followed a path quite similar to those within the corporate computing installations affiliated with Share. Much of their work on systems programs, operating systems, and high-level languages shadowed the work done in industry. Yet it is also clear that each of these systems had to be modified to suit the particular needs of an academic setting, and this accounted for much of the innovative programming work carried out by Michigan's Computing Center staff.

But in working in an academic setting, the systems programming work at Michigan also proceeded to take on more of the connotations of research. Whereas industrially employed programmers had to work to convince corporate executives of their professional stature, a university setting sanctioned, and even enticed programmers to cast their work as research. As a faculty member, Galler found it easiest to do so. He wrote a series of talks and publications based on his work. Arden, meanwhile, who had begun his career as a tabulating department supervisor, secured a slot in Michigan's doctoral program in electrical engineering. He would go on to have an illustrious career in computer science at Princeton.

There were tensions. So long as academic groups worked alongside industrially employed programmers, there were occasional misunderstandings about credit and the proper treatment of their work. The programmers at IBM, for instance, failed to give Arden and Galler credit for several of the features of the MAD compiler that they incorporated into Fortran IV. This was considered allowable in the context of the open appropriation of code as practiced by those affiliated with Share. Nor were the academics immune to the influences of the practical orientation of their center. For example, Galler and Arden had built a new "extensible language facility" into the MAD compiler. However, Galler failed to realize the broader implications this had for a formal understanding of programming languages, despite his formal training in mathematical logic. In later years, Galler was chastised by his colleagues for having failed to make this connection.[67]

Still, it was significant that Michigan's administrators recognized the intellectual merits of the work conducted within its computing center. Bartels and the members of his executive committee publicized the work on GAT, MAD, and other systems as academic achievements. The administration responded in kind, authorizing Bartels to augment the Computing Center's programming staff to expand on this work. IBM also contributed to the pot, in establishing a new fund for systems programming research at the center. For the moment, Michigan's administration accepted that the goals of research and that of providing a useful service were compatible, and that academic research on systems programming could, and indeed had to take place within the non-departmental context of the university's central computing facility.[68]

Interim Analysis

It is remarkable how quickly new forms of knowledge and new forms of work organizations emerged to meet the need for computing both at MIT and at the University of Michigan. A new hierarchy of skills and disciplinary authority was constructed at each site, drawing neither exclusively on academic traditions nor the precedent set by industrial computing facilities. Moreover different values and ideas, rooted in established practices for educational reform, academic research, and corporate and private philanthropy were embedded into the design of these new facilities. The central tension at both sites was between the goals of engineering education and scientific research. Morse opted for the latter by emphasizing the computational power of the electronic computer and its value to traditional scientific disciplines. The group at Michigan wound up emphasizing the former through their considerable, though not exclusive emphasis on undergraduate education. Different institutional ecologies generated different paths for technical development.

Simultaneously these academic computing facilities became yet another site for reworking occupational identities. Morse went from being a physicist to an administrator of an academic computing facility. Corbato went from being a physics student to computing supervisor. At Michigan, career trajectories crossed paths. Even as Graham and Arden went from being a machine operator and supervisor to researchers engaged in cutting-edge programming research, the same space was occupied by Galler, a faculty member who became intrigued by the practical aspects of a digital computer. In the end, there was a good deal of informality and contingency in the roles individuals came to assume in academic computing centers.

In fact, a look inside the work room of an academic computing facility would have revealed far more role improvisation. There, graduate research assistants, machine operators and experienced undergraduates—especially at Michigan—would have been found doing what they could to draw countless novices into the realm of digital computers even as they proceeded to refashion their own identities. Like Share, all this was driven by the need to maintain a productive computing facility.

Informality and the spirit of play has long been a self-ascribed characteristic of the computing field. In *The Soul of a New Machine* (1981), Tracy Kidder portrays a group of young programmers at Data General who develop an ad hoc game called "Tube Wars," where they set out to hack into their boss's computer. In doing so, they not only acquire knowledge and proficiency about systems programming but find their place at work. Such games are reminiscent of cultural anthropologist Clifford Geertz's well-known essay "Deep Play: Notes on the Balinese Cockfight" (1974/1972). There, Geertz describes the significance of play as a means of reflecting upon, and reinforcing the roles of different individuals within a community. But at the early computing facilities playfulness was more about reworking rather than reinforcing existing identities. In this respect, even something as simple as a systems programmer's playful use of language in acronyms like IT, MES, and MAD were indicative of deep ambiguities about the status and proper organization of computing work. This ambiguity was important. In a new domain of expert labor, where the work that had to be done remained unspecified and unspecifiable, it was more efficient to allow individuals from different backgrounds to enter a common work arena and to resolve their responsibilities through negotiations conducted *in situ*. This worked better than trying to formally allocate tasks before there was sufficient knowledge to do so.

Not surprisingly, it was at academic computing centers, and at Michigan in particular, where the meritocratic norms of academe, the resource constraints of a state institution, and the specific demands of academic computing ultimately worked together to create a highly dynamic environment for defining new occupational and disciplinary identities—

although clearly something quite similar occurred in the context of the IBM users' group Share. In fact, when viewed in the context of the past three chapters, allusions to informality in the computing field do in fact appear as coded references to the permission workers of different stature are given to enter a common arena of work. This is not to say that all the roles were worked out on a level playing field. Only faculty members and senior members of a corporate programming staff could directly receive the substantial rewards extended by their respective institutions. Nevertheless, metaphors of informality and play facilitated the rapid reconstruction of roles necessary to define the initial outline of new occupations, professions, and disciplinary specializations within the field of digital computing.

Figure 9.1
A caricature of a batch-processing facility. Courtesy of MIT Musem.

9 Discipline and Service: Research on Computer Time Sharing at MIT and the University of Michigan

MIT and Michigan each found a way to support the growing academic demand for computing. However, the ever-increasing popularity of computers soon overwhelmed both universities' large, central batch-processing facilities. Within just two to three years of operation, the computing centers at MIT and Michigan found that they could no longer support all the different demands that were being placed on their facility. Convinced, nevertheless, of the greater efficiency of a large, central facility, researchers at both universities launched a major initiative in computer time sharing. First conceived by electrical engineers with backgrounds in power systems engineering, this was a scheme for turning computers into a basic utility by relying on the same efficiencies of scale that applied to electric utilities. Computational power would always be online and on tap for hundreds if not thousands of users to use on demand.

In this chapter, I extend the comparison of the central computing facilities at MIT and Michigan by tracing their entry into this new domain of systems programming research. The course of events at the two universities would continue to differ. Moreover, they continued to revolve around the dual set of tensions surrounding computing as a science, a profession, and a vocation, and of computers as a commercial technology, a service, and an object of research.

This chapter revisits the account of computer time sharing presented by Arthur Norberg and Judy O'Neill in their 1996 book *Transforming Computer Technology: Information Technology for the Pentagon, 1962–1986*. Norberg and O'Neill's main thesis is that the Advanced Research Projects Agency and its Information Processing Techniques Office had a major influence on the development of computer time sharing and networking. Time sharing, in particular, provided a "neat fit" between IPTO director Joseph Licklider's interest in human-computer interaction, and ARPA's objectives in military command and control. This benefited the research program at MIT and elsewhere.[1]

I agree entirely with Norberg and O'Neill's interpretation that military interests gave shape to early time-sharing systems. Nevertheless, my goal is to trace a separate set of motivations that emerged from the expanding demand for academic computing facilities. The developments at MIT and Michigan described in the previous chapter fueled both universities' interest in supporting on-campus research in computer time sharing. Moreover the specific needs of academic computing centers continued to contribute to the technical articulation of agendas that remained distinct from those defined by military objectives and by Licklider's background in psychoacoustics and psychology.

The intent here is not simply to make a minor correction to the historiography. Only by taking into account this additional thread in the history of computer time sharing does it become possible to see the underlying pluralism of Cold War research. By taking into account the other academic pressures that drove computer time sharing, the story changes from that of a convergence of interests to that of sustained tension. By the end of this chapter, it will become clear how this tension contributed to technological innovation.

MIT and the Origins of Computer Time Sharing

Above all else, what drove universities to expand their academic computing facilities was the volume of sponsored research. At MIT, sponsored research was administered by its Division of Industrial Cooperation, the same office that assisted Forrester with Project Whirlwind. By 1953, sponsored research at MIT had risen to $13 million. While funding declined somewhat between 1953 and 1956, this figure rose to $20.3 million by 1961.[2] This translated into a rather sizable body of contract research on campus. The DIC supported 220 professors and instructors, over 500 research assistants, 1,100 research staff members, and nearly 2,000 technicians, secretaries and other "non-staff" personnel on its payroll even in 1953.[3]

Although Morse clearly favored non-sponsored, educational research use of his Computation Center, sponsored research still accounted for 65 percent of all use. Classroom use, unlike at Michigan, accounted for a mere 6 percent. At the same time, free use made machine usage rise quickly to the machine's capacity so that Morse had already authorized three-shift operation by mid 1958. MIT researchers also absorbed most of the machine time originally allocated for other regional universities.[4]

As Norberg and O'Neill have observed, one important motive for time sharing was that batch processing entailed delays. In 1960, the typical turn-around time at the Computation Center was $1\frac{1}{4}$ days. (See figure 9.2.) When students from an undergraduate computing course descended on the facility, this could grow significantly worse. The underlying concept of

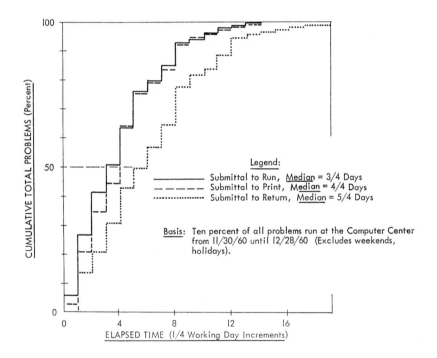

FIGURE NO. 2.12 TURNAROUND TIME BETWEEN USER SUBMITTAL, MACHINE RUN AND OUT AVAILABLE

Figure 9.2
"Turnaround time between user submittal, machine run and out available," MIT Computation Center, December 1960. From "MIT Computation—Present and Future," draft report [15 March 1961], Box 3/MIT Computation—Present and Future, Philip Morse Papers, Institute Archives and Special Collections, MIT Libraries.

computer time sharing emerged when John McCarthy, a junior faculty member with a joint appointment in the Computation Center and the Research Laboratory for Electronics (RLE), noted that while batch-processing facilities were created with routine computing jobs in mind, the most common use for computers was fast becoming that of running and debugging new programs. Although this was categorically false at commercial computing facilities, it was true for academic computing centers. The central computing staff at Michigan would soon realize this, albeit on a more pragmatic basis.[5]

In January 1959, McCarthy sent Morse a memorandum providing the first description of a computer time-sharing system that could provide several users with the illusion that a large, central computer was at their beck and call. This involved an operating system designed with a very different notion of efficiency, one more respectful of a systems programmer's time. Although this raised the underlying idea to a new level of abstraction, it was based on a practice firmly embedded in many computing facilities. Machine operators routinely allowed systems programmers, who stood as their superiors, to insert their jobs ahead of the other ones in the queue, since it was the systems programmers who did the important work of maintaining the efficient operation of a computing facility. With a computer time-sharing system, such insertions would occur automatically and in real time, in even interrupting a job in progress. Such a scheme could support a far greater number of privileged users even as the routine batch-processing jobs churned on in the background.[6]

McCarthy moved on to study artificial intelligence, and so the work on computer time sharing fell on the shoulders of a colleague, Herbert Teager. Teager also held a joint appointment with the Computation Center and the RLE. Working from McCarthy's reformulation of the notion of machine efficiency, Teager began by developing an interactive debugging utility that would support several programmers even as the batch jobs continued. Teager was allowed to do this work using the Computation Center's IBM 704. By 1960, Fernando Corbato was the Center's assistant director in charge of programming research. Insofar as he was in charge of systems programming, Corbato had every incentive to have a debugging utility that would ease the work of himself and his staff. When MIT decided later that year to move up to the new IBM 709, Corbato instructed his staff to rework Teager's utility into a general computer time-sharing utility. This was the origin of MIT's Compatible Time-Sharing System (CTSS), originally conceived of as a system that could support a small number of simultaneous users.[7]

While this provides a technical history of the origins of computer time sharing, there were broader issues of institutional context that drove Corbato to authorize an expanded program in computer time sharing. His decision came at a time when there were larger calls to decentralize academic computing at MIT. Long turnaround times were not problematic only for systems programmers. Vocal critics could be found in the Earth Sciences Department, in the Sloan School of Industrial Management, in the RLE, and in other units that made heavy use of the facility. The earth sciences faculty, in particular, had begun to complain of the Computation Center's "monopoly" on computer facilities, which resulted from the Center's privileged relationship with IBM. Others called for new real-time data analysis capabilities that could not be served by a central batch-processing facility. These groups all pushed for new arrangements where IBM's largess could be distributed more evenly across campus, with different facilities geared toward different ends.[8]

MIT administrators worked hard to secure the autonomy of their faculty during the immediate postwar years. As a consequence, MIT faculty members were free to negotiate directly with their sponsors, and most federal agencies remained willing to consider a request to lease or acquire a computer, if there was a compelling technical justification for doing so. Yet so long as computers remained large enough to require additional floor space, this decision remained partly within the administration's jurisdiction. All the same, the course of events at MIT continues to reveal the considerable strength of its faculty vis-à-vis the central administration.

Serious discussion about the future of computing services at MIT began with a formal request from the RLE's Center for Communications Sciences. The RLE's communications sciences faculty had set out to assess how important computers were to their research, and they reached the conclusion that they could not answer this question without an overall MIT policy toward computing facilities. However, instead of approaching the administration with a request for such a policy, several senior RLE faculty members approached Morse with the idea of creating an ad hoc committee to consider the issue. Once it became apparent to this core group that a large central time-sharing facility might meet most of the demands on campus, they decided to assemble a larger "Long Range Computation Study Group" by inviting other interested users of the MIT Computation Center. Only then did the ad hoc committee seek official sanction from the administration.[9]

Despite broad faculty representation on this study group, certain interests dominated MIT's new "long range" study. The Ad Hoc Faculty Committee on Computation was chaired by Albert Hill, a senior RLE faculty member and the former director of Lincoln Laboratories. Hill brought with him considerable knowledge about summer studies and systems engineering, both of which were becoming an increasingly familiar part of MIT's institutional landscape. In assembling the larger study group, Hill instructed its members to use summer study techniques in examining MIT's need for computing. He also instructed them to break down their work into separate stages for problem formulation, technical evaluation, and systems design. Yet, despite the apparent objectivity implicit to this systems design methodology, Hill's instructions carried with it an implied preference for large, integrated solutions similar to the air defense system developed by Lincoln Labs. Moreover, Hill prefigured the study group's conclusions by assigning Teager to chair it.[10]

Teager's evaluation is best understood as an effort to verify the scale economies of a large, central computing facility. At this point Teager specifically borrowed the technique of load balancing from the field of power systems engineering. After a preliminary discussion about an appropriate system of classification, Teager determined that 12 percent of the Computation Center's workload involved data reduction, 60 percent involved mathematical processing, 20 percent involved symbolic processing, and 31 percent involved educational use. (Total usage exceeded 100 percent because some of the use crossed categories.) After also considering future demands for real-time computation, which was not supported by the present facility, Teager ran these different usages against the different demands they placed on memory size, processing speed, and input-output facilities. He then considered different machine configurations and their ability to support this mix. The end result of this assessment was the conclusion that a single, large-scale computer time-sharing facility would provide the most cost effective solution for the future computing demands of MIT.[11]

There are several things to note about Teager's report. Despite the systematic nature of his investigation, there were occasional lapses where Teager placed the needs of computer programmers above others. Teager offered a defensible chain of reasoning why the needs of programmers were especially important for an efficient computing facility. Nevertheless, the basic premise that long turn around times forced "highly skilled personnel to remain idle for long periods of time," figured heavily into his cost comparisons. Teager's own assessment that 60 percent of the computing load at MIT still had to do with mathematical processing might have suggested otherwise. Second, Teager's assessment was really only designed to compare computer time-sharing facilities against a central batch-processing facility. It spent little effort considering other, more decentralized approaches to computation. Finally, Teager was clearly hoping, through this report, to fuse his own research interests with the current computing needs of other researchers at MIT. Adding a dash of urgency to his recommendation, Teager warned that, "MIT's major policy decision with respect to future computation if postponed, would effectively, and almost irreversibly, be made by default."[12]

Teager's recommendations were not left unchecked. Other members of the study group insisted that certain kinds of research may still require separate computing facilities. Meanwhile, Morse made sure to dampen the implied criticism of the Computation Center's current mode of operation. In a more private exchange with MIT's vice president, Hill wondered whether the MIT faculty would, irrespective of objective measures of efficiency, "agree to outside control over their own computer problems." Yet despite the fact that Hill's committee chose to support a revised report "reflecting more accurately the majority opinion" of the study group, even the revised report generally accepted the idea that a large, central computer time-sharing facility was the best solution to their current crisis.[13]

Norberg and O'Neill provide a good account of the rest of the history leading up to ARPA's sponsorship of Project MAC, a major computer time sharing project at MIT. Although MIT President Julius Stratton endorsed the long-range study, its $18 million price tag precluded immediate implementation. Stratton instead authorized Hill and his committee to look for outside support. Simultaneously, Stratton appointed another working group to begin developing a set of technical specifications. In August 1961, after consulting with a number of computer manufacturers, this working group produced its first set of specifications and wrote a $22 million proposal intended for the National Science Foundation.[14]

The NSF did not have the resources to support a project of this magnitude. However, MIT found the support through Joseph Licklider of ARPA's new Information Processing Techniques Office. Norberg and O'Neill describe Licklider as the "unofficial representative of the Cambridge community" inside ARPA. Licklider had served as a group leader at Lincoln Laboratories, had worked under Hill, and had spent some time as a vice president at Bolt Beranek and Newman, a small, specialized engineering research and consulting firm located in Cambridge. As has already been mentioned, Licklider was a psychologist and psycho-acoustician with a research interest in augmenting human intellect through "man-computer symbiosis." He wrote a seminal paper by that title while working at BBN. Licklider was chosen to head ARPA-IPTO in late 1962. Soon after, one of the members of Hill's Committee on Computation, Robert Fano, approached Licklider. Fano had been a Lincoln Labs group leader at the same time as Licklider, making it easy to approach him. Licklider immediately saw computer time sharing as not only fitting within his own research interests but also as an important foundation for military command and control systems. Rapid response and interactivity were necessary for both. Licklider proceeded to issue a $3 million study contract for Project MAC.[15]

Michigan's Shift to Time Sharing

The University of Michigan made its commitment to computer time sharing three years after MIT. Like MIT, Michigan turned to time sharing as a means of putting off the decentralization of its computing facilities. Michigan's initiative—even its long-range study—was modeled explicitly on the MIT effort, and drew directly on earlier work by McCarthy, Teager, and others. Still, the circumstances at Michigan were different. In particular, Michigan was driven by its need to produce revenues. This produced a different outcome.

It was out of their own conceit that Bartels and his executive committee first decided to expand their facility. Because of the success of their first year of operations, Bartels immediately submitted a letter of intent to lease an IBM 7090 (a transistorized version of the IBM 709), by then the most powerful IBM computer available. In the meantime, Willow Run Laboratories had acquired its own IBM 709, just as the Army was beginning to curtail its support for Willow Run's electronic surveillance research. Willow Run overextended itself, and began to negotiate for the transfer of the IBM 709 to the Computing Center's fiscal and administrative jurisdiction. Bartels felt he could utilize both the 709 and their current 704. But then Willow Run reneged once again on its promise to deliver a substantial volume of work. Compounding matters, university comptrollers advised Bartels of a discrepancy. The 704 and the 709 had relatively similar rental fees, so that federal regulations required him to charge $328 an hour for the larger machine, rather than the $450 an hour he was then charging. Given the current usage rates and revenues, Bartels had no choice but to release one of the two machines. Bartels chose to keep the larger IBM 709, in underestimating once again the programming costs associated with a machine transition. Bartels won few allies during this shift.[16]

A more serious crisis resulted from a subtle shift in IBM's educational discount policy. IBM's "standard educational contribution" amounted to a 60 percent discount on standard rental rates, but the contract Michigan signed with IBM was somewhat more complex. Michigan's contract stipulated that the university would charge no fees for educational or non-sponsored research use of the computer, while charging all sponsored research projects a "full commercial rate" set at $1/176^{th}$ of the normal monthly charges. Michigan was also required to limit sponsored research use to 70.4 hours per month. IBM included this provision to prevent the loss of business from sponsored research projects that could afford to pay for computer time. But as IBM came to dominate the market, it grew more concerned about how a complex agreement, irrespective of its legal standing, would play into accusations of market manipulation. In March 1963, IBM proposed to amend the agreement to provide Michigan with a flat 60 percent discount for all academic use. Seeing no obvious implications for his bottom line, Bartels agreed to the amendment.[17]

This time it was the university's legal counsel who advised Bartels that the Computing Center would have to reduce the rate it charged sponsored research projects from $460 per hour to $175 per hour because of the amendment. Since there was no longer any explicit provision for free academic use, federal contract regulations required that Bartels distribute the discount evenly across all uses of the computer. It was no longer permissible to charge sponsored research projects the full commercial rate in order to subsidize free non-sponsored uses of the computer. Bartels and his executive committee discovered a further irony. The more the students and faculty used the Computing Center for non-sponsored work, the more the hourly rate had to drop. Only by sharply curtailing educational and other non-sponsored use could Bartels bring his budget into line. This was obviously contrary to their objectives. Yet, with no other recourse, Bartels was forced to institute a system of time allocations that forced a sudden four-fold reduction in all non-sponsored work. This curtailed exactly those activities that were Michigan's claim to success.[18]

Such battles with accounting regulations were not an isolated event. As described by Larry Owens, serious discussions about contract regulations and academic overhead rates took place during World War II in conjunction with the NDRC.[19] Although many universities, including MIT and Michigan, had voluntarily accepted low overhead rates during the war, they were more adamant about fully recovering their costs during the postwar period. The 1946 revisions to the Armed Services Procurement Regulations launched what was generally heralded as the "blue book era," during which universities were able to pay for their expansion through the recovery of indirect costs. Nevertheless, disagreements continued, ensuring the build up of bureaucratic apparatus on both sides of the institutional boundary. Then, during the late 1950s, under the fiscal conservatism of the Eisenhower administration, there was specific discussion about separating research and educational expenditures to ensure that research allocations, including overhead rates, was not used to subsidize the educational activities of a university. Over the years, Michigan's sponsored research business office had grown quite attuned to issues of compliance. Moreover, in light of Bartels's perpetual fiscal crises, these watchdogs, whose responsibility it was to keep the university in good standing, were close at hand.[20]

Bartels realized the awkwardness of the situation. For some time, he and his committee had been devoting nearly all their energy to financial matters. They had had very few conversations about computer applications or computing research, which was their actual charge. Their enthusiasm for computers had carried them thus far. Nevertheless, with the Computing Center's annual budget exceeding half a million dollars, their role had become increasingly clear. Bartels and his executive committee were a group of faculty members trying to run a fee for service business within the confines of a research university. In a memo to the administration, Bartels abdicated all responsibility for the "business functions" of the Center.[21]

It was in the context of this administrative takeover that Michigan proceeded to carry out its own long-range study. Michigan's vice president for academic affairs, Roger Heyns, now joined Sawyer, who by then was the vice president for research, in their attempt to gain fiscal and administrative control over the Computing Center. But expertise about digital computers remained with the faculty. Because of the implied failure of Bartels and his committee, Heyns and Sawyer turned to the principal users of the facility. This ensured that there would be greater tension in Michigan's long-range study. Michigan's users had more serious grievances about the machine-transition difficulties there, and during the intervening years since MIT's study, academic users in general had grown more sophisticated about articulating their needs. In particular members of Michigan's powerful life-sciences faculty were quite adamant about creating a separate facility for real-time experiments. Heyns and Sawyer also interjected their interests. They instructed the new Ad Hoc Computing Advisory Committee to pay specific attention to the issue of the financial solvency of a central computing facility. Forging a consensus would be a somewhat more difficult task for Michigan's long-range study group.[22]

Fortunately for Bartels, Heyns and Sawyer asked Donald Katz, from the Ford Foundation Project, to chair the ad hoc committee. Katz remained an important ally of the Computing Center. Initially, Katz adopted an open and egalitarian approach, in which each of the ad hoc committee members were invited to contribute their own assessment. During the summer of 1964, Katz assigned each committee member to survey a specific set of schools, departments and other academic units, and asked them to describe these units' computing requirements. Since most members were appointed to survey their own schools and departments, this meant they were free to provide a direct statement of their interests and concerns. Nevertheless, insofar as each report dealt with a specific academic unit, these reports contained very little in the way of a general assessment or critique of the operations of the central computing facility.[23]

From this start, Katz was able to orchestrate general consent for a new, central time-sharing facility. Rather than focus on the Computing Center's failings, the ad hoc committee came to place much of the blame on Willow Run's autonomous purchase of a computer, which had incurred an unexpected financial burden on the university. This spoke against the wisdom of decentralized computing facilities. Moreover, in being conscious of Michigan's standing as a second-tier research institution, the committee concluded that the cost of maintaining a state-of-the-art computing facility made it imperative to perform all possible computation within a central facility. Although the committee held open meetings during which there were dissenting voices, the committee as a whole decided that it was necessary to override the concerns of those "who object to being controlled when they desire to acquire computing facilities." The position favoring centralization was also sustained by the very structure of the committee's final report, issued in December 1964. In reality, there had been considerable dissent, which could still be found in the individual committee members' surveys. These were included in the report. However, these remarks were overshadowed by the first two sections of the report, which offered an integrated plan built around a central time-sharing facility.[24]

Still, since there was opposition, it was all the more important for the systems programmers at Michigan to achieve social integration through a technical solution designed to meet all the different users' needs. In contrast to the MIT report, the proposed solution, as described in part I of Michigan's report, promised to support both batch processing and online use. It also allowed research groups to purchase smaller computers that could be connected as satellites to the central computing facility. Michigan's report was also more specific about classifying real-time applications, which they placed in categories requiring responses on the order of 10–15 seconds, 200 milliseconds, and less than 200 milliseconds. The committee conceded that they would not try to support real-time applications in the third category using a central facility. On the other hand, to support their claim that a central computing facility could support real-time applications down to 200 milliseconds, the committee secured the interest of an electrical engineering faculty member, Frank Westervelt, who agreed to undertake the research needed to make this possible.[25]

These technical discussions shifted, in effect, the boundary of the negotiation. By forming the abstraction of specified response times, this allowed those who favored time sharing to transform technical judgments about individual research projects into technical judgments about computer expertise. Teager's load-balancing calculations in MIT's long-range study had the same effect. This move also placed the burden of proof on the users to demonstrate why the proposed facility could not meet their needs, and on grounds with which they were not conversant.

While this was still all quite similar to MIT's study, there was one significant difference in Michigan's final report. Beginning with the premise that all universities had to have good research facilities to secure "large blocks" of funding from the federal government, and that computers were an important part of academic research facilities, the ad hoc committee reasoned that "a series of well-coordinated and integrated proposals [for computer equipment] should be prepared to seek the type of support which the University must have in order to progress as one of the leading institutions in the nation."[26]

Less confident about their own ability to secure the necessary support, Michigan's faculty members gave the administration the jurisdiction to oversee this coordination. Specifically, it asked the administration to create a permanent Committee on Computer Policy, with the power to review all research proposals that involved a request for computer procurement or computer-development research.[27]

The request to extend the administration's authority drew considerable interest from A. Geoffrey Norman, who replaced Sawyer as the university's vice president for research in 1964. Norman had also inherited the responsibility for overseeing the university's Institute of Science and Technology, an organization set up in 1959 along the lines of an increasingly popular formula for technology-led regional development. This was the scheme advanced by Frederick Terman at Stanford. Michigan's administrators convinced the state legislature that $1 million spent in state money could attract millions of dollars in federal support, which would foster, in turn, new technological industries of great value to the state. Initiatives like the IST gave sanction to centralized, administrative planning of academic research. Moreover, the ad hoc committee's proposal for a computer time-sharing facility, when interpreted in the context of the current plans for the IST, offered exactly the kind of project and investment in research infrastructure that would allow Michigan to leverage state funds into big federal research contracts.[28]

Norman agreed to bankroll the initial studies of time sharing by the current Computing Center staff. This was essential for Michigan's start in the field, for by the time Bartels approached ARPA, Licklider informed him that his office had already obligated all of the funds ARPA intended to spend on time sharing. Norman also appointed the Committee on Computer Policy, and instructed its members to pursue a carefully coordinated strategy for seeking external support. At Michigan, unlike at MIT, time-sharing research would be supported by a combination of state and federal funds.[29]

Racing to Beat Decentralization

Defining the Two Efforts at MIT

For both Michigan and MIT, research on computer time sharing was a race against time. So long as their central computing facility had no time-sharing capability, the university could not deny the effort by the departments and other academic units to acquire their own computers. Thanks to an early start, MIT clearly had the initial lead. However, owing to the greater autonomy of its faculty and the scale of its sponsored research, decentralization also proceeded apace at MIT. It began with the School of Engineering's decision to establish a Cooperative Computing Laboratory in 1962, followed by a quick succession of other facilities.[30]

In an effort to keep their other users from defecting, Morse authorized Corbato to extend the Compatible Time-Sharing System into a more general time-sharing utility. Though this was meant to be an interim solution that would last only until Project MAC delivered a more comprehensive system, it split MIT's efforts into two separate though not quite independent initiatives. Still, Morse was able to secure direct support from NSF for this work on CTSS, under the premise that the work would complement the long-term effort being undertaken by Project MAC.[31]

MIT's situation grew even more complicated when Fano decided to work with General Electric, not IBM. Fano had initiated his research through the use of yet another, 1963 summer study, paid for by ARPA-IPTO. Interested manufacturers were invited to these planning sessions. IBM was an active participant. But during the early 1960s, IBM was engaged in difficult conversations about an entirely new and integrated product line—what would soon be announced as the IBM System/360. Company engineers were already overcommitted in what many saw to be too big of a technical undertaking, even for IBM. At the time, GE, by comparison, was a new entrant to the field of digital computers. Fano found GE to be more open to his ideas about new machine architectures. In the ensuing design competition, GE advanced a proposal that Fano saw to be more compatible with "the broad goal of Project MAC." Because of the long working relationship between MIT and IBM, the MIT administration, beginning with Stratton, pressed Fano to justify his choice. Yet in the end, the administration bowed to the technical discretion of its own faculty. Morse was left to deal with the repercussions.[32]

Michigan and IBM

MIT's difficulties were Michigan's gain. After completing their portion of Michigan's long-range study, Bruce Arden and Bernard Galler, along with the electrical engineering faculty member, Frank Westervelt, moved forward with their technical evaluation of computer time sharing. They began approaching computer manufacturers shortly after Fano made the decision to go with GE. They therefore found a willing partner in IBM. IBM agreed to modify one of the machines in its IBM System/360 series, and to do so according to Michigan's specifications, so long as Michigan agreed to do all the systems programming for a computer time-sharing system. This plan gained the support of Michigan's new Committee on Computer Policy. It also drew renewed interest from NSF. Meanwhile, ARPA, in perhaps smarting from growing criticism that it was spending too much of its money at elite institutions, awarded a $1.3 million contract to Westervelt for the purpose of developing new graphics terminals and other equipment that would allow a central time-sharing system to support real-time experiments.[33]

One hurdle that remained was that of hardware modification. Here, Arden and Galler simply followed MIT's publications and MIT's lead. Nevertheless, in their desire to have their own sounding board, Arden and Galler enlisted Alan Perlis, along with prominent researchers from several other institutions. This led to a series of technical meetings between Michigan, IBM, and a number of "important" IBM customers, including General Motors, the Carnegie Institute of Technology, the Systems Development Corporation, and MIT's Lincoln Laboratories.[34]

The first of these meetings was organized by Michigan. But in forging this network, Arden and Galler helped to temporarily displace themselves from further work on time sharing. Once IBM's internal advocates of time sharing were able to document the broad-based interest in it, they convinced the firm to add a time-sharing computer to its System/360 product line. In the process, IBM reassumed all responsibility for systems programming, thereby killing NSF's interest in the work at Michigan. Arden and Galler felt that they were "left in the lurch."[35]

Difficulties at MIT

Back at MIT, the breach with IBM was not Morse's only difficulty. The same set of accounting regulations that got Michigan in trouble placed MIT in a similar but different bind. Having created a free computer facility for everyone at MIT, Morse had no revenue base with which to expand his facility. Yet free use ensured that usage quickly rose to and exceeded the Computation Center's capacity. While sponsored research projects were willing to pay for new facilities, accounting regulations prohibited them from paying into the pot. Their only option was to purchase their own equipment, thereby accelerating decentralization. In their own effort to stem decentralization, the MIT administration intervened in a fashion similar to Michigan, creating its own Committee on Computer Policy and Utilization with the authority to review all faculty proposals calling for a new computer installation. Dean of Engineering Gordon Brown was asked to chair this new committee.[36]

Meanwhile, IBM decided to sustain its relationship with MIT, if not with Project MAC. In response to the latest crisis over computer capacity, and the competitive pressures introduced by GE, IBM offered to make an outright donation of an IBM 360/67. Some MIT faculty members opposed the donation, seeing it as a continuation of a policy favoring a single, central computing facility. Brown and his committee also opposed the deal, but on different grounds. Citing Morse's assessment of the situation, Brown argued that MIT's computing capacity could no longer be tied so directly to IBM's philanthropy. Working together with Morse, Brown brokered a deal where IBM would require MIT to pay the same discounted rental as other universities, but then contribute $1 million per year to programming research, research assistants, and the non-sponsored use of the computer. This provided Morse with the resources he needed to subsidize free educational use. The money was also used to support Corbato's work on CTSS. IBM publicists were especially eager to receive assurances that there would be a "thick black line" between this work and Project MAC.[37]

Unfortunately, there was no real line between the Computation Center and Project MAC. Since CTSS was the best available time-sharing system, and an indispensable tool for systems programming, Project MAC installed an IBM 7094 that was identical to the one in the Computation Center. They installed CTSS for their own use. Moreover, Fano, acknowledging Corbato's knowledge of operating systems and of systems programming, gave him a joint appointment as the head of Project MAC's System Research Group. By this point, Corbato had also become the associate director of the Computation Center, and was in charge of all aspects of its operations.[38]

While in theory this could have been an opportunity to coordinate the two projects, in practice the arrangement proved detrimental to both. Corbato, in being drawn to the more ambitious technical goals of Project MAC, allowed the Computation Center staff to work on technical problems that served the needs of Project MAC. As CTSS became a test bed for Project MAC concepts, it lost the simplicity and stability required of a facility still burdened with a large batch-processing load. Meanwhile, the research staff associated with Project MAC, if it was even possible to make a simple distinction between the two groups, became caught up in the effort to make CTSS into a serviceable system. They directed very little of their energy toward GE's new computer, which stood idle for well over a year before there was any significant programming done on the system. In his 1966 annual report, Fano could say only that the work on CTSS "has thrown a great deal of light on the nature of computer utilities," and the deficiencies of existing time-sharing systems.[39]

It is possible to provide some sense of the technical challenges that Corbato and Fano faced. CTSS functioned adequately as an experimental system. However, placing CTSS into regular service at an overloaded facility meant placing it in the condition most conducive to system failures. Moreover, a computer time-sharing system's computational overhead exacerbated the overloading. Yet Corbato was being asked to extend CTSS's experimental instruction set into an operating system with a full complement of utilities and systems commands. Debugging the systems programs that made up a time-sharing system also proved to be a daunting task.

Precisely because a time-sharing system's failures tended to occur during overloaded conditions, this meant that many errors occurred under transient conditions that were impossible to reproduce. Moreover, overloaded conditions ensured that a system failure often left the machine in a state that was very difficult to diagnose.[40]

To offer at least one concrete instance of the kinds of situations Corbato faced on a regular basis, during April 1964, the Computation Center staff was busy completing three new commands, PRBIN, EXTBSS, and UPDBSS—utilities they desperately needed so that other programmers could easily install and retrieve subroutines from the machine's subroutine library. They then tackled the problem of creating a daily backup of their user's files. Partly because of the frequent system failures, it was important to have a good backup regimen. However, a significant technical challenge arose from fact that the secondary storage, a magnetic disk drive, was too large to permit daily backups without excessive use of precious computer time. Corbato's staff began to devise a work around, which involved producing a "new file" tape each day, which could then be followed by a complete backup once a week. Because of the pressure to provide continuous service, the staff also chose an aggressive scheme that would interleave the file backup operations in between other computations. But because of frequent disk failures, it was necessary to make sure that such a scheme was fully integrated with a disk salvage routine. The staff put all these requirements into a utility, but it failed to operate as planned. This left Corbato and his staff to wonder about the "philosophically profound" nature of the difficulty.[41]

Michigan's Success

Intractable difficulties also plagued the effort at IBM. Time sharing still was not one of IBM's top priorities, and designing a system that remained as compatible as possible with the rest of the System/360 series was an especially difficult challenge. In August 1966, IBM advised Michigan that it was unlikely that they could deliver their machine before December, whereas it had been scheduled to arrive early in the fall. Worse yet, IBM told Michigan that the accompanying software, Time Sharing System (TSS), would not be released until the following April, and with no guarantee as to its performance. In January 1967, this was followed with a more formal announcement that IBM had blocked further sales of the IBM 360/67, and was going to release TSS only for experimental and educational use.[42]

This brought Michigan squarely back into the midst of computer time sharing. Bartels had announced the inauguration of a new time-sharing service before he learned of IBM's difficulties. Feeling compelled to deliver on their commitment, Arden, Westervelt, and others began scrutinizing the situation. They came to the conclusion that IBM had lost confidence in their own system after a software simulation revealed that the IBM 360/67, when in time-sharing mode, would have $\frac{1}{10}$ the computational power of the older IBM 7094. From his knowledge of the hardware, Westervelt did not believe this could be true. He proceeded to carry out an independent evaluation of the IBM 360/67's performance. This led him to the conclusion that the source of the difficulty was IBM's time-sharing software, not the hardware. In the meantime, two of Arden's staff members had obtained the source code for the Lincoln Terminal System (LTS), an experimental time-sharing system developed separately at MIT's Lincoln Laboratories. Arden authorized his staff to port LTS first to their IBM 7090, and then their new IBM 360/67. At this point, Michigan also received a grant from NSF.[43]

Michigan was able to get a time-sharing system into operation, and to do so in short order. A rudimentary version of the Michigan Terminal System (MTS) was up and running by May 1967. The system attained an overall capacity for simultaneous users comparable to Corbato's CTSS by November. By the following August, Arden and his staff were able to release a more extensive and robust version of MTS that made full use of the IBM 360/67's hardware modifications. At the end of 1968, MTS was the only large-scale time-sharing system that was operating reliably.[44]

There was a technical basis for Michigan's success. One of the main technical challenges of time sharing had to do with the limited size of a computer's core memory. However, both Arden and Galler had already encountered a similar problem while working on high-level languages. They had developed a utility that automatically loaded sections of the MAD compiler's translation tables into core memory. Conceptually, by extending this idea from a single application to the operating system as a whole, it was possible to create an illusion that a machine had a very large core memory. This illusion was available to programmers as well as any other user, so that the abstraction simplified Michigan's systems programming work. The efficient implementation of this scheme required special hardware modifications—precisely the ones Arden and his colleagues requested of IBM. Technically, this involved having special circuits that made it possible to rapidly reallocate different sections of core memory. Through such work, Michigan became one of the sites that contributed to the development of "virtual memory."[45]

The ideas central to virtual memory had originated with MIT, not Michigan. It was Jack Dennis, one of MIT's electrical engineering faculty members, who first presented the important idea of "segmentation." Dennis was designing yet another time-sharing system for the Digital Equipment Corporation's (DEC) PDP-1, a tiny machine with very severe resource constraints. By Arden's own admission, Dennis's paper led to obvious questions of implementation. Drawing on the same work by Dennis, Fano's major disagreement with IBM had to do with IBM's initial unwillingness to incorporate similar dynamic reallocation hardware into its time-sharing system.[46]

Yet it was significant that only Michigan's Computing Center possessed both the requisite hardware and the pressing need to implement a large-scale computer time-sharing system. In comparison to the competing demands of MIT's Computation Center and Project MAC, Michigan was able to follow the more productive middle road.

The Disintegration of Research and Service

All the same, MIT received more recognition than Michigan for early work on computer time sharing. Despite, or rather because of their limited success, the Computing Center at Michigan became very unpopular. Its IBM 7090 was already fully loaded with jobs when Arden and his staff installed MTS onto this older platform. This hurt the performance severely. Some users complained that they had no use for a time-sharing service. Others complained about the delayed introduction of a more reliable time-sharing service. Even after MTS began operating reliably on the university's IBM 360/67, its capabilities were far short of what had been promised in the original long-range study. Researchers at Michigan could not deliver on the early promises of time sharing, any more than the researchers at MIT could. An administrator in Michigan's sponsored research office reported on the "continuous expressions of discontent" he had heard about the equipment in the Computing Center and about its overall systems philosophy. There was even a report that an impatient user had broken into the machine room and punched an operator.[47]

Underlying the users' discontent were broader changes in technology. The PDP-1 already existed at the time of MIT's original long-range study, but Teager dismissed the favorable cost-performance ratio of this machine as an anomaly. Yet the PDP-1 was the beginning of a trend toward smaller, less expensive machines, where costs were driven downwards by cheaper components and increased scales of production. While this included many minicomputers—small, stand-alone computers that fit within the existing space and budget constraints of a single research laboratory—it also included smaller time-sharing systems, such as the PDP-10, which could support the needs of academic departments and large university research labs.

Teager had used the analogy of load balancing and public utilities to justify the complete centralization of computing facilities. However, other MIT faculty members came to notice that a central facility offered real advantages only if it could pool together a computer's resources more effectively than a handful of smaller facilities. Clearly there were some users whose computational requirements called for a machine with a "large problem capacity." But for many others, this was not a necessity. Meanwhile, the standardization of computer programs, along with the universities' success in training larger numbers of computing specialists, made it entirely possible for an academic department or school to maintain a modest computer facility tailored to their specific needs. Proponents of computer time sharing were defeated on their own intellectual ground. Delays, poor service and technical limitations laid bare the false assumptions that had guided computer time sharing.[48]

Meanwhile, Michigan's Committee on Computer Policy had to admit that it was wrong to require computer users to "subsidize" programming research within the Computing Center. Judging that systems programming research had to be carried out separately from the routine computing services offered by the Center, the decision was made to limit the kinds of work that could be carried out on the campus's central computer.[49] MIT also made a similar decision. In 1968, MIT hired a new computing director, Richard Mills, who was charged with transforming MIT's Computation Center into its new Information Processing Service. As one of his first moves, Mill cancelled MIT's order for the IBM 360/67. "It is certainly an historical fact," he wrote, "that, as a *research* enterprise, MIT can produce software on its own that is without parallel. Unfortunately, we have also learned that research of this kind mixes poorly with computer-service objectives."[50] Reaching the same conclusion as Michigan's Committee on Computer Policy, Mills put at least an official end to the research aspirations of MIT's computing center staff.

The University of Michigan's work on computers and systems programming became scattered across its campus. Westervelt and his colleagues continued to pursue work on computer graphics, which grew independent of the work on time sharing. Others in the College of Engineering continued to build on Katz's work on computers and their use in engineering education. Meanwhile, the general eclipse of the Computing Center allowed the gestation of other initiatives. Michigan's business school set up its own program for research and instruction in data processing. The department of Computer and Communication Sciences, established in 1965 by ENIAC veteran Arthur Burks, proceeded to expand into the realm of systems programming. In other words, computing activities at Michigan became fragmented along the lines of its specific schools and colleges, each of which built upon some element of its past activities. With the technical disintegration of Michigan's time-sharing system came the social disintegration of the various forms of expertise that were bound together only temporarily during the course of this work.[51]

At MIT, however, Project MAC provided its researchers with an institutional foundation for creating a substantial program in computer science. Beginning with the exact same line of reasoning as the one presented by Mills, Fano redirected Project MAC in the opposite direction in favoring research over service. It did not take Fano long to discover this other option. The controversy over his decision to work with GE left Fano with little desire to remain entangled in the politics over institute resources, and he began to contemplate how his work could be transformed into a general program for computer science research. By the time he submitted his annual budget request to ARPA for fiscal 1968, Fano could confidently assert that Project MAC would soon turn to basic research and therefore complete its gradual evolution "from a task-oriented research and development effort to a long-term basic research laboratory in the computer sciences." In 1968, Joseph Licklider himself left ARPA to assume the helm of

Project MAC, and the research continued to diversify under his tenure. Project MAC was officially rechristened the Laboratory for Computer Science in 1975.[52]

It is important to temper any judgment about the failures of early computer time sharing. The technical difficulties time sharing presented were difficult to foresee. Neither MIT's original summer study, nor Michigan's long-range study, could produce an adequate assessment of the technical scope of the work. Moreover, amid the ever-increasing diversity of the demands that users wished to place on computer systems, it was unlikely that any group could have delivered a solution that would have met every user's needs. There was certainly a good deal of technical conceit in the early claims made by proponents of time sharing. Nevertheless, these researchers proceeded in a difficult context of institutional pressures, incompatible priorities, and uncertainties associated with the rapid pace of technological change. The projects at MIT and Michigan made important contributions, even though other organizations eventually came to dominate time-sharing work.

As for the administrators, the failure of computer time sharing did not put an end to their interventions. If their policies with respect to time sharing were a blemish on their record, administrators at MIT and Michigan continued to try to plan the overall course of their institution's research and research facilities. Even the policy of centralizing computing facilities was relaxed only gradually. Meanwhile, debates over defense-related research during the Vietnam War, along with the more recent calls for "reengineering" the university, would produce other occasions for revisiting issues of administrative authority. This I also leave to other scholars.

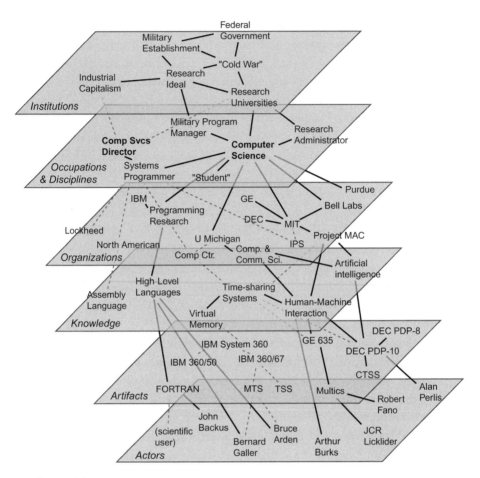

Figure C.1

The ecology of knowledge for computing expertise, 1968. Solid lines represent computer science. Dashed lines represent subordinated knowledge about computing and information-processing services.

Conclusion

This book makes a number of contributions to the history of computing. In revisiting various well-known episodes, from the wartime work on the ENIAC to the work on computer time sharing, it corrects or at least modifies historical interpretations of the origins of the stored-program concept, of IBM's entry into technical computing, and of the challenges associated with early research on time sharing. It brings fresh understanding to the roles of several historically neglected actors, including the National Bureau of Standards, IBM's Applied Science Department, and the University of Michigan. Moreover, by focusing on individuals in a way that makes them neither heroes nor villains, it introduces new perspectives on what the early "pioneers" did to establish computing as a dynamic discipline. In all these respects, the intention has been to bring additional nuance, if not closure, to some often-debated topics. The book's most important contribution to the history of computing is that is synthesizes earlier historical accounts by considering how computing as a whole was remade in relation to an emerging infrastructure for Cold War research in the United States. However, the book has broader purposes: to develop a better understanding of the social process of innovation during the early Cold War era, and to consider what an explicit focus on innovation could do to extend the academic tradition of constructivist scholarship. In this conclusion, I summarize what I believe to be the book's principal accomplishments in these two areas.

Innovation

This book could have been more strictly a study of innovation. Those familiar with the history of technology and its historiography will recognize that it incorporates many familiar themes about innovation. It echoes what is already well known about the historical shift from an era of independent inventors to one of institutionalized innovation. (Perhaps somewhat more originally, it suggests how this pattern may recapitulate itself with new technologies.) It considers alternative paths to innovation, the value of studying "failure," and the associated need to make the notion of failure relative if one's goal is to document the less linear paths of innovation. It also puts renewed emphasis on material culture, and on the distinct "epistemic cultures" of science and engineering more generally, as a way of peering into the process of innovation.[1]

With regard to innovation, the literature most relevant to this book lies outside the history of technology, in the literature on the "triple helix" thesis advanced by Henry Etzkowitz and Loet Leydesdorff (1995, 1997) in the broader realm of science and technology studies. I came across their work rather late in this project. Rather than try to recast my work relative to their thesis, I felt it would be more useful to review my findings in light of their observations.

Etzkowitz and Leydesdorff's "triple helix" thesis postulates that after World War II intensified interaction between universities, industry, and government accelerated the pace of technological innovation, especially in advanced knowledge-based economies. Their work therefore emphasizes institutional pluralism as a foundation for innovation. Etzkowitz (2004) also points to the creative tension that emerged between such norms and counter-norms as "capitalization," "interdependence," "independence," "hybridization," and "reflexivity," which came to characterize research within entrepreneurially minded universities. Etzkowitz and Leydesdorff both consider how organizational structures can facilitate innovation rather than simply institutionalizing (and thereby perpetuating) knowledge systems, as is traditionally argued in organizational sociology.

Without a doubt, this book confirms Etzkowitz and Leydesdorff's thesis. But in it I also extend their argument in several ways.

First, I take the resurgent interest among constructivist scholars in the "mutual shaping" thesis as indicating that the study of innovation is as much about institutional innovation as it is about technological innovation. Questions of organizational form and institutional change clearly are important to management and organizational theorists. Nevertheless, there is a tendency, even in Etzkowitz and Leydesdorff's analysis, to take the institutional configuration of Cold War research as given, rather than as a product of social processes conducive to analysis and description. By explicit use of constructivist and post-constructivist perspectives, I draw attention to the contingent and indeterminate nature of institutional change in ways that specifically counterbalance the functionalist assumptions that lurk beneath Etzkowitz and Leydesdorff's analyses.

This functionalist assumption is evident in Etzkowitz's 2002 study, in which he presents MIT as the institutional exemplar of "entrepreneurial science." Working in the mode of historical sociology, Etzkowitz uses historical sources. Nevertheless, what appears in Etzkowitz's study as a set of rational choices for MIT appears in my work as more contested and actor-driven. The contrast is also apparent in our respective conclusions. Whereas Etzkowitz suggests that the knowledge-based economy produced a political economy for science and technology funding that favored MIT's entrepreneurial model, my study points to the diversity of viable postwar institutional strategies. The paths chosen by the Institute for Advanced Study, the University of Michigan, the National Bureau of Standards, Share, and IBM all evidence the subtly different ways in which research universities and other institutions chose to situate themselves within the larger political economy of Cold War research. In this respect, I reject Etzkowitz's decision to cast MIT as an ideal type. What this suggests is that for Etzkowitz the institutional pluralism of Cold War research operated primarily within the confines of a single organization. If the goal is to understand the full diversity of Cold War research, I consider it important to study the varied arrangements that either survived or were created afresh during the Cold War. Thus, instead of casting MIT as the norm and Harvard and the Institute for Advanced Study as anomalous, I proceeded to explore how the underlying pluralism of Cold War research was a product of intentional choices, even while recognizing that historical currents for federal funding (and expanded industrial commitments to science and a vision of "applied science") tended to favor certain choices over others. The second difference, then, is that I used the empirical wealth of history to more fully historicize the diverse institutional arrangements that were created for Cold War research.

Third, I explicitly espoused a social historical perspective (and drew on an understanding of contingency and indeterminacy borrowed from the symbolic interactionists) to examine explicitly how local actors contributed to this diversity. I left the broadest changes in postwar science policy and research funding as an unexplored historical backdrop. Nevertheless, I focused on how individuals, sometimes acting individually and sometimes collectively, could interpret emerging policies and funding opportunities in ways that gave them substantial articulation. Moreover, as I showed in the later chapters, this remained true even as the institutional framework of science and technology funding grew more stable. Researchers, corporate executives, and administrators all continued to find fresh opportunities to manipulate not only knowledge but also the institutional arrangements for research.

To understand how this flexibility persisted, and how institutional pluralism contributed to the pace of innovation, it was necessary to trace how institutional objectives interacted with the most esoteric aspects of technological design (and, indeed, with all the layers represented in figure C.1). I pointed to this in the opening passage of the book. With the case studies behind us, it should be apparent how Cold War interests were given new articulation at every turn in the negotiations, vastly expanding the field of possibilities for technical and institutional innovation. This productive relationship between esoteric knowledge and institutional innovation was probably most evident in chapter 9, where multiple disciplinary and institutional interests were articulated and then integrated into a single artifact designed to uphold all the competing visions. However, such efforts to rework both knowledge and institutional objectives were noted in every chapter. They were present in John von Neumann's efforts to design digital computers and to redefine the postwar mission of the IAS, in Jay Forrester's efforts to reinterpret the computer's purpose and the purpose of MIT and the Office of Naval Research, and in the subtler efforts to rework knowledge and organizations in IBM's Applied Science Department and in Share. The fourth and final difference, then, is that I built my investigation around a specific technological artifact in order to study the mutually defining relationship between esoteric knowledge and its supporting contexts. Taken together, these four differences should bring further subtlety and historical understanding to Etzkowitz and Leydesdorff's basic thesis.

Constructivism and Post-Constructivist Analyses

It should also be clear that, by addressing larger questions about constructivist and post-constructivist scholarship, I went beyond the specific

genre of innovation studies. Quite important in this respect was my decision to espouse the notion of an ecology of knowledge as a governing
metaphor.

At one level, the notion of an ecology of knowledge functions metaphor-
ically in this book. It provides a way to think about multiple sites of practice, the different perspectives that they represent, and the nature of their
interrelatedness. An ecological outlook also allowed me to paint an overall
picture of a rather large domain of knowledge, even in recognizing that
such depictions remained accountable to and had to be anchored by an
understanding of the local circumstances that made up the whole. The
metaphor also pointed quite directly to institutional pluralism.

However, as a means of remaking constructivist scholarship into something more closely aligned with Anthony Giddens's structuration theory,
I also made a much more specific use of the metaphor that was tied to a
formal definition of metonymy and to a theory of practice. I take up this
discussion, as begun at the end of the introduction, first by returning to
the abstract typology of innovative practices described there and then
by considering how each of the chapters provided a much more situated
understanding of these practices so as to ultimately transcend the
abstraction.

The first category of innovative practice I defined was *syntagmatic extension*, by which institutions extend their research through the successive
reproduction of a stable assemblage of knowledge, artifacts, and instrumental practices. The mere fact that so many of the early figures in computing employed a language of proselytization and religious conversion
suggests that a cultivated body of practice for transporting knowledge from
one context to another was indeed an important part of an early computing specialist's identity. These practices, and their origins, were not hard
to identify. At North American Aviation and in Share, individuals drew on
a rich tradition of technical collaboration that already existed in Southern
California's aviation industry. IBM drew on the dual precedent of its sales
culture and the academic norms of open exchange. More narrowly, Jay Forrester drew on the hiring practices of the Servomechanisms Lab. Philip
Morse drew on the disciplinary traditions of physics education. Warren
Weaver used the contractual practices of business to replicate esoteric technical computing facilities during World War II.

Such actions strengthened familiar institutional forms. Thus, to the
extent that IBM (and, earlier, J. Presper Eckert and John Mauchly)

successfully replicated familiar sales strategies, this strengthening determined the extent to which the commercial development of computers became a viable approach to research during the Cold War. Von Neumann's early work and sociotechnical network pressed for a very different vision of computer development work that tied it to the idea of making fundamental advances in knowledge. Forrester's difficulties attest to the challenges an individual faced when trying to extend the mission-oriented activities of wartime research into a postwar context that clearly favored fundamental research, as was the case with the postwar direction chosen by the Office of Naval Research.

However, the point of taking a social historical approach is to acknowledge that such acts of "diffusion"—what Bruno Latour (1987) refers to as "translation"—are rarely so straightforward. It is often necessary to rework both knowledge and institutional arrangements incrementally through acts of *interpretive extension*. How this occurred through cultivated identities and disciplined practices is also discernible throughout this study. Mauchly adopted the identity of an inventor, which he had inherited from his father and from his contemporary social milieu; Forrester turned to new ideas and practices in the field of control-systems engineering to evaluate and extend the capability of digital computers; Teager borrowed the electrical engineer's technique of load balancing to assess the merits of computer time sharing, and proceeded to develop computer time sharing on the conceptual model of an electric power grid.

Nor were interpretive practices always applied to narrow technical concerns. Hurd remade academic conferences into a tool for selling Card Programmed Calculators; Forrester used rhetoric to redefine the mission of the Servomechanisms Lab; Morse used familiar academic rituals to remake digital computers into something that resonated with the educational mission of MIT. These latter examples simply reiterate a point that Bruno Latour (1987), John Law (1987), and Thomas Hughes (1986) all made: that scientists and engineers do not routinely distinguish the social from the technical in their daily work. My focus, nevertheless, has been on identifying the everyday practices that researchers *do* possess (or acquire) for working across the different layers and regions of an ecology of knowledge. Interpretive extension, in any event, speaks most directly to the traditional concerns of constructivist scholarship in describing how esoteric knowledge and artifacts come to assume a metonymic relationship with broader organizational and institutional entities.

Meanwhile, in building on what I have said in relation to Etzkowitz and Leydesdorff's thesis, the most striking thing about Cold War research was how often these interpretive extensions occurred at the intersection of multiple institutions and disciplines. This ought to be the underlying lesson for all Cold War research. Such acts of *recombination* occurred at the highest levels, as reflected, for instance, by IBM's efforts to both encourage and constrain the academic orientation of its staff. It also occurred at the most esoteric technical levels, as demonstrated by the integration and hybridization of different forms of disciplinary knowledge in all postwar computer-development projects.

Practices for working across institutions and disciplines clearly were cultivated during the Cold War. David Mindell (2002) has carefully demonstrated how World War II offered an important precedent. Nevertheless, all the individuals mentioned in this study, including Mauchly, Samuel Alexander, Paul Armer, and Fernando Corbato, created esoteric knowledge and remade research organizations by recombining prior knowledge and preexisting institutional forms. To the extent that these individuals became good at doing so, they prospered; to the extent that they were unable to make the adjustment, they failed. Irrespective of their individual successes and failures, it was clear that the field of computing depended on individuals who were willing to move beyond their disciplinary training and their institutional allegiances. Historically, interdisciplinarity emerged as an important part of the computing specialist's identity.

At the same time, new knowledge and new Cold War institutions depended as much on a process of metonymic *dissociation*—to put it more simply, a process of letting go. Thus, as is evident from Hurd's willingness to let go of some of his academic commitments, and from the Share representatives' efforts to see beyond their corporate commitments, working within the interstices of different institutions often meant forging new associations at the expense of other associations. Even von Neumann's work at the IAS demonstrates how military sponsors had to let go of some of their traditional expectations before they could support fundamental research.

At an institutional level, it is difficult to think about an "established" body of practice that facilitates dissociation. Institutions, by definition, entail commitment. It was more often the case that a particular institutional location, or else an irresolvable conflict that resulted from efforts to support what turned out to be incommensurable goals, presented recurring pressures to make certain choices about the proper course of research. Despite the pluralism of the Cold War era, institutional location did affect outcomes, as the cases of the machine-development program at the NBS and computer time sharing at MIT and Michigan show.

Dissociative practices were much more evident at the technical level. The periodic breaks between ideas, artifacts, and experiments may seem surprising in the realm of natural science because of our perceptions about the continuity of that body of knowledge.[2] However, as I suggested at the end of chapter 1, such breaks are quite commonplace with technology. Knowledge and artifacts can always be remade to suit new uses, and engineers and scientists often possess a cultivated set of habits—reading technical journals in a neighboring field, observing and drawing abstractions about the conduct of others, revisiting earlier hobbies and interests, attending interdisciplinary conferences, drawing on a disciplinary background in which they were previously trained, and so on—that allow them to selectively appropriate external knowledge and artifacts in ways that differ from syntagmatic or even interpretive extension. On many occasions, practices of this sort were at play. Notable examples were a systems programmer's ability to appropriate the skills of a computer operator and John von Neumann's appropriation of the knowledge and language of other disciplines.

This contingent system of classifying innovative practice has its merits. By developing an abstraction, we gain the opportunity to reflect on how these different (and non-exclusive) forms of practice construct and reconstruct metonymic relations so as to constitute an ecology of knowledge. As I suggested in the introduction, this can also lead to a more integrated approach to the study of language and practice, where these transformative practices are viewed specifically as a means of reworking the semiotic structures that span all the metonymic layers depicted in figure C.1. For instance, we can trace, through individual actions, how Mauchly came to understand "science" and "the weather," how Forrester came to define his principle of "computer importance," and how Hurd and his field men came to forge a hybrid identity based on the notions of "salesman" and "mathematician." Throughout the book, there were many subtle reconstructions of meaning that rebuilt the associations between the material culture of digital computers and the broad, institutional values and ideologies of the Cold War. An abstract focus on practice can provide an elementary understanding of the processes that drive such semantic reconstructions.

Still, from the standpoint of a well-grounded study of practice, it makes more sense, in the end, to move away from abstractions and to think about practice as situated within distinct epistemic cultures, even while recognizing that such cultures have become substantially fragmented with the prevalence of interdisciplinary research strategies. Only by doing so is it possible to view practice as the practical manifestation of social structure as consistent with Giddens's structuration theory.

In many respects, the case-study approach I took in this book does exactly that. Whether describing the rhetorical strategies Forrester used to sustain his machine development project or the different traditions of practice Hurd drew upon in building up the Applied Science Department, I described, to the extent that it was possible, a coherent body of practice that emerged from established traditions of practice. Admitting to the limits of historical sources, I drew on ethnomethodological perspectives in attempting to document the bodies of practice that constituted the "forms of life" that made up distinct cultures of knowledge production within the overall ecology of computing expertise.

These bodies of practice are not, in principle, reducible to any simple, abstract description. Nevertheless, insofar as I am using a study of practice as a way of thinking about constructivist methodologies, I find it instructive to look for structure *within* these aggregate bodies of practice and the narratives they produce. With this in mind, I redirect the reader's attention to titles of my chapters. Each title lent structure to the chapter's narrative. Each was also tied to a body of social theory. I used a highly narrative style in order to make the book accessible to a larger audience, and this was done at the expense of explicit references to theory. Here I would like to make the underlying theories of this study, and their relevance to rethinking constructivism, explicit.

Chapter 1 was about introducing the overall framework of an ecology of knowledge, and about what the "circulation of knowledge" looked like at a time when computing was not yet a coherent discipline. Here the focus was not on any single epistemic culture, but on the interaction among many related epistemic cultures located within the different kind of pluralistic institutional setting that existed before the end of World War II. Drawing on the symbolic interactionist literature in science studies, I noted how knowledge and artifacts sometimes traveled in "standardized packages"[3] during this period. For instance, organized practices for using mathematical machinery and labor were transported from one context to another. However, as often as not, this early period was characterized by the much more selective appropriation of knowledge and artifacts through practices of interpretive extension, recombination, and, most important, dissociation. The existence of robust bodies of practice for doing so, as complemented by underlying similarities in knowledge and purpose, had allowed ideas and artifacts to move easily across institutional and disciplinary boundaries. Such was the historically grounded configuration of knowledge and practice, and its relation to social theory that characterized the circulation of computing expertise prior to World War II. The sustained technical exchanges during World War II, as accelerated by the increased centralization of research, then produced the sociotechnical networks that began to remake computing into a coherent discipline.

In chapter 2, I used the narrative conventions of biography to describe a rather different approach to the circulation of knowledge. I opened the chapter by considering the theories and practices of socialization. I took a close look at John Mauchly's family correspondence, as complemented by other documentary sources, to consider how he was brought into an epistemic culture of scientific research and of technological innovation. Semiotic theory provided a silent foundation for this section, as it offered a means of thinking about the attitudinal structures that individuals develop in the course of constructing and reconstructing habituated identities. Thus, by knowing the value and the commitment Mauchly placed in certain objects, ideas, and accomplishments, I was able to explain how they came to define the constraints and affordances he perceived as he entered different institutional settings. These attitudes and perceptions propelled Mauchly into a peripatetic career.[4] The latter part of the chapter was devoted to the very different epistemic culture of the Moore School engineers, as transformed by wartime priorities, which took Mauchly's early work and ideas in a direction he could not master or entirely follow.

Biography remained an important theme in subsequent chapters, each of which addressed substantially different questions about the nature of the relationship between structure and agency. However, the principal theoretical aim of chapter 3 was to situate the competing metaphors of a loosely structured ecology, a trading zone, and a coherent network as applied to historical developments in computing. Building on the discussion begun in chapter 1 and followed through in chapters 4–6, I documented the very different patterns of conduct under these three epistemic regimes. I described the immediate postwar period as a time when computing as a whole operated as a trading zone. During that time, an extensive body of practice for exchanging and reproducing knowledge— mathematical correspondence, academic conferences, experimental replication, the hiring away of individuals—emerged to constitute the exciting new interdisciplinary domain of digital computing. But it was equally significant that many individuals initially chose to integrate this new knowledge through highly pragmatic decisions about their sponsors, their disciplinary and institutional priorities, and their technical capabilities, producing quite different approaches to the development and use of digital computers. Before long, this gave way to a more coherent network built around John von Neumann and his ideas. Engineering Research Associates, the National Bureau of Standards, Project Whirlwind, and IBM all came to adopt von Neumann's ideas (at least initially), each through a different process of evaluation. Though these processes varied substantially (for ERA, for instance, this involved conducting a survey and then "authoring" a report), they all remained anchored by some element of practice that acknowledged von Neumann's intellectual authority. That such authority could affect engineering practice itself was also evident in the context of the IAS's own Electronic Computer Project.

All these actions are specific instantiations of the abstract categories of practice described above. For ERA, to author a report that was substantially influenced by von Neumann's ideas meant replicating von Neumann's vision—an instance of syntagmatic extension. But all of these practices, insofar as they remained embedded within different epistemic cultures, also provided an opportunity to relate each practice to different bodies of social theory. For instance, chapters 4 and 5 provided a valuable occasion for thinking more explicitly about the metonymic relations that span the multiple layers of an ecology of knowledge, and to work with the social theories that speak to these relationships of metonymy. Thus, chapter 4 examined the relationship between competing institutions, disciplines, organizational structure, and esoteric knowledge. Here, common theories of organizational behavior were what could be extracted from, and lend structure in turn to, the narrative. I described how John Curtiss used common bureaucratic tactics to mitigate the influence of an external advisory committee. But this action, as complicated by the various contingencies of research, wound up invoking familiar practices of inter-organizational and intra-organizational accountability. In the end, this enabled the NBS's electronics engineers to utilize common organizational maneuvers and justifications to set up their own laboratory for designing digital computers. In another instance of mutual shaping, this work transformed the NBS's knowledge as well as its organization. Chapter 5 was cast in quite similar terms. There, however, I extended the analysis through a specific focus on rhetorical practice, drawing on and extending Latour's theories about the rhetorical practices of research in describing how Forrester gave definition to real-time computing, to MIT's postwar institutional identity, and (to a lesser extent) to the ONR's funding policies.

The subsequent chapters also drew on different elements of social theory. Chapter 6 drew on notions of identity formation, and on how it occurred in a situation characterized by two partially incommensurable institutions: corporate profit and academic exchange. Somewhat more subtly, chapter 6 drew on postmodern theory in studying the panoptic machinery of an intelligence network; it also served as a critique of a mechanical application of Foucault's concept by describing how identity, disciplinarity, and the complexity of esoteric knowledge could hinder the efficient circulation of knowledge and the simple subjugation of the scientific community. Chapter 7 revisited traditional narratives of professionalization. There I drew quite extensively from the symbolic interactionist literature in considering how the practices of intersubjective evaluation, of creating a "going concern," and of developing an identification with an occupation could create a space suitable for the proliferation of technical specializations as characteristic of the post-industrial era. Chapters 8 and 9 brought together many of the ideas presented in earlier chapters—ideas about identity, disciplinarity, organizational behavior, and accountability among them—to demonstrate how people within academic computing centers were able to manipulate social institutions and esoteric knowledge even as the larger context of Cold War research grew more stable.

I hope to have done this all in the best tradition of grounded theory.[5] Clearly, for social theory to stand as an expression of social structure, it must emerge out of everyday practice. All the theories to which I have referred conform to the fundamental phenomenological stance that sees the social analyst's role as that of formulating (giving articulation to) rules of conduct that are already part of everyday action. My study does so, first, through a direct account of the situated strategies for technical and institutional innovation that are entirely familiar to, if not explicitly articulated by the practitioners, and, second, through descriptions of more aggregate, self-organizing behavior that may not be visible to any single individual, as represented by concepts such as knowledge circulation and the ecology of knowledge.

This approach to social theory should serve as an important extension of constructivist methodology. Rather than taking knowledge and artifacts (or, for that matter, institutions) as being shaped by social interests through some vague notion of "interpretive flexibility," this approach sets out to describe the specific processes (and practices) of interpretation that remain embedded within distinct epistemic cultures. This should serve as a response to Ian Hacking's call to develop the underlying architectural metaphor of "constructivism" by thinking about the "structures" that emerge from acts of construction, while at the same time focusing on social processes and on what it means to "build" knowledge and society.[6] Stated in this form, this would be but one expression of Giddens's notion of the duality of structure. It is consistent with a historicized understanding of the structures of everyday action that, per Giddens (1984), are said to constitute society.

Still, the practical ramifications of such an approach to constructivist scholarship should be given some consideration. Jonathan Sterne and Joan Leach, the editors of a recent special issue of *Social Epistemology*, struggled to answer the question "What comes after social construction in the study of technology?"[7] But if their goal was to find a single unifying theory that would adequately characterize "post-constructivist" scholarship, their efforts were likely in vain. At least from what I have suggested, what seems necessary is a more eclectic approach to social theory, one that sets out to describe the very different kinds of social processes that become folded into the process of knowledge production. What was Cold War research but an agglomeration of such practices? Let me reiterate, however, that my training is in the history of technology. This book cannot be considered a formal treatment of the underlying theoretical issues. It merely stands as an invitation to those who are willing to use the empirical wealth of history (and ethnography) to continue the project of developing a more robust understanding of the social processes surrounding science and technology. Such an approach seems necessary for understanding a phenomenon as complex as Cold War research, and the era of "technoscience" that it engendered.

Notes

Abbreviations Used

Abbreviations Used

Archives

APS: American Philosophical Society Library, Philadelphia

BHL: Bentley Historical Library, Michigan Historical Collections, University of Michigan, Ann Arbor

CBI: Charles Babbage Institute for the History of Information Processing, University of Minnesota Libraries, Minneapolis

HUA: Harvard University Archives, Cambridge, Massachusetts

IAS: Institute for Advanced Study Archives, Historical Studies–Social Science Library, Princeton

IBM: IBM Archives, Somers, New York

LC: Library of Congress, Manuscripts Division, Washington

MIT: MIT Libraries, Institute Archives and Special Collections, Cambridge, Massachusetts

NA-II: National Archives II, National Archives and Records Administration, College Park, Maryland

NMAH: National Museum of American History, Archives Center, Smithsonian Institution, Washington

UCLA-UA: University of California, Los Angeles, University Archives

UPA: University of Pennsylvania Archives and Records Center, Philadelphia

UPE: ENIAC Museum, School of Engineering and Applied Science, University of Pennsylvania, Philadelphia

UPSC: Department of Special Collections, University of Pennsylvania Libraries, Philadelphia

Manuscript Collections

Aiken Papers: Howard Aiken Correspondence, UAV 289.2006. Computation Laboratory, Harvard University Archives, Cambridge, Massachusetts

AC 4: Office of the President, 1930–1958 (Compton-Killian), Institute Archives and Special Collections, MIT Libraries, Cambridge, Massachusetts

AC 12: School of Engineering. Office of the Dean, Records, Institute Archives and Special Collections, MIT Libraries, Cambridge, Massachusetts

AC 35: Office of the Vice President for Industrial and Government Relations, Records, 1957–1965, Institute Archives and Special Collections, MIT Libraries, Cambridge, Massachusetts

AC 62: Computation Center, Records, 1950–1962, Institute Archives and Special Collections, MIT Libraries, Cambridge, Massachusetts

AC 134: Office of the President, Institute Archives and Special Collections, MIT Libraries, Cambridge, Massachusetts

AC 268: Laboratory for Computer Science, Administrative Records, 1961–1988, Institute Archives and Special Collections, MIT Libraries, Cambridge, Massachusetts

AC 282: Laboratory for Computer Science Research Records, 1961–1985, Institute Archives and Special Collections, MIT Libraries, Cambridge, Massachusetts

AC 290: Whirlwind Computer Collection, Archives Center, National Museum of American History, Smithsonian Institution, Washington

AC 323: Paul Armer Collection, 1949–1970, Archives Center, National Museum of American History, Smithsonian Institution, Washington

AMD-EC: Records of the National Institute of Standards and Technology, Applied Mathematics Division, Electronic Computers 1939–54, Record Group 167, National Archives II, National Archives and Records Administration, College Park, Maryland

AMP: Records of the Applied Mathematics Panel, General Records, 1942–46, Record Group 227, Office of Scientific Research and Development, National Archives II, National Archives and Records Administration, College Park, Maryland

Astin Records: Records of the National Institute of Standards and Technology, Records of Dr. Allen V. Astin, Record Group 167, National Archives II, National Archives and Records Administration, College Park, Maryland

CBI 6: Francis V. Wagner Papers, Charles Babbage Institute for the History of Information Processing, University of Minnesota Libraries, Minneapolis

CBI 21: SHARE Records, Charles Babbage Institute for the History of Information Processing, University of Minnesota Libraries, Minneapolis

CBI 42: Herbert S. Bright Papers, Charles Babbage Institute for the History of Information Processing, University of Minnesota Libraries, Minneapolis

CBI 45: Margaret Fox Papers, Charles Babbage Institute for the History of Information Processing, University of Minnesota Libraries, Minneapolis

CBI 50: Edmund Berkeley Papers, Charles Babbage Institute for the History of Information Processing, University of Minnesota Libraries, Minneapolis

CBI 95: Cuthbert Hurd Papers, Charles Babbage Institute for the History of Information Processing, University of Minnesota Libraries, Minneapolis

ECP: Records of the Electronic Computer Project, Historical Studies–Social Science Library, Institute for Advanced Study Archives, Princeton

ENIAC Papers: ENIAC Patent Trial Collection, 1864–1973, UPD 8.10, Archives and Records Center, University of Pennsylvania, Philadelphia

ENIAC Trial: ENIAC Trial Exhibits Master Collection, 1864–1973, UPD 8.12, Archives and Records Center, University of Pennsylvania, Philadelphia

IAS Faculty: Records of the Faculty, Institute for Advanced Study, Princeton

IAS Faculty-M: Faculty Meeting Minutes, Institute for Advanced Study, Princeton

JvN Papers: John von Neumann Papers, Library of Congress, Manuscripts Division, Washington

Mauchly Papers: John Mauchly Papers, Department of Special Collections, University of Pennsylvania Libraries, Philadelphia

MC 75: Philip Morse Papers, Institute Archives and Special Collections, MIT Libraries, Cambridge, Massachusetts

NDRC-7: Record Group 227, Office of Scientific Research and Development. Division 7—Records of Chiefs and Members of Sections, 1940–46, National Archives II, National Archives and Records Administration, College Park, Maryland

RDB: Record Group 330, Records of the Office of Secretary of Defense, Research and Development Board. Records Concerning Organization, Budget, and the Allocation of Research and Development, National Archives II, National Archives and Records Administration, College Park, Maryland

RS 359: Administrative Files, 1935–1959, Chancellor's Office, University of California, Los Angeles, University Archives

UCCPU: University Committee on Computer Policy and Utilization, University of Michigan, Bentley Historical Library, Michigan Historical Collections, University of Michigan, Ann Arbor

UMich. Eng.: College of Engineering, University of Michigan, Bentley Historical Library, Michigan Historical Collections, University of Michigan, Ann Arbor

UM-CC: Computing Center Records, University of Michigan, Bentley Historical Library, Michigan Historical Collections, University of Michigan, Ann Arbor

VP-Development: Vice President for Development and University Communication, University of Michigan, Bentley Historical Library, Michigan Historical Collections, University of Michigan, Ann Arbor

VP-Research: Vice President for Research, University of Michigan, Bentley Historical Library, Michigan Historical Collections, University of Michigan, Ann Arbor

Notes to Introduction

1. Historical studies of U.S. Cold War research, which have tended to focus on academic institutions, include Leslie 1993 and Geiger 1993. Other studies of postwar science policy that are relevant to this book are cited in the text.

2. Akera, forthcoming.

3. See, for instance, Byerly and Pielke 1995.

4. See Star and Griesemer 1989, Star 1995, and other work discussed later in this introduction.

5. Winner 1986/1980; Pickering 1995.

6. Lynch 1993 and Schaffer 1991, both cited on pp. 41–43 of Golinski 1998.

7. For a more extended discussion of the critique of constructivist approaches in science studies, see Akera 2005.

8. This point is also made in Langdon Winner's frequently cited essay "Do Artifacts Have Politics?" (1986/1980). Calls for more attention to the "mutual shaping" thesis can be found in Wiebe Bijker and John Law's compilation *Shaping Technology/Building Society* (1992) and in Ron Westrum's *Technologies & Society* (1991).

9. As Hecht (1998, p. 9) notes, Ken Alder (1997) examines the mutual relationship between the French Revolution and military engineering, and Donald MacKenzie (1990) describes Draper's inertial guidance systems and their relationship to U.S. nuclear strategy. Hecht's study deals with the relationship between France's nuclear energy program and French technocratic ideology.

10. In the present work, I tend to use the phrase "mutual articulation," rather than "mutual orientation," to emphasize how the interchange between artifacts and institutions, and across different institutions, does not necessitate a simple convergence of discursive forms and practice. Nevertheless, I do not regard my choice of phrase to be original. Mutual orientation already presupposes mutual articulation, and the notion of articulation is already fundamental to the method of discourse analysis on which Edwards draws.

11. Ross 1996; Ashman and Baringer 2001. Perhaps the most important effect of the Sokal hoax was to bring a lingering question to the foreground: Where do we

go after social construction? (See Akera 2005.) Although this question is asked in earnest by those working in the constructivist tradition, when posed by critics it has tended to signal a desire to simply declare an end to constructivist scholarship. The argument generally advanced by these critics is not that the constructivist position is invalid. Rather, they suggest that the process of social construction can now be treated as a background assumption. This is argued most forcefully by Ian Hacking sin *The Social Construction of What?* (1999), where he notes that once science (and by extension, technology) is understood to be a form of social practice, the "social" in "social construction" loses all its meaning. Everything humans do is a social process. However, Hacking's argument is misleading. As with the difference between code and language, the real challenge of constructivist scholarship lies not in merely establishing that there is social influence, but to understand the historical and localized nature of the relationship between esoteric knowledge and its supporting institutions. To declare that science is a social process says very little about what kind of process this is. Rosenberg, in the second of his seminal essays laying out an ecological view of knowledge, suggests that the underlying question for our discipline is not that of pitting externalist and internalist accounts of science against one another, but to understand the "structure of their integration" (1997b/1988, p. 240). Hacking himself also offers a relevant insight when he calls on constructivists to fully develop the architectural metaphor embedded within the notion of social construction (1999, pp. 49, 50). In this study I turn instead to Rosenberg's own choice of metaphor, primarily because an ecological metaphor refers more directly to the metonymic relationship between knowledge and social institutions, as well as the elaborate, intersecting patterns of interdependence that can emerge between them. This is something I hope to have demonstrated with figure I.1.

12. This is another observation Rosenberg made, with obvious prescience (1997b/1988). Nevertheless, recent work at the intersection of history, philosophy, and sociology of science has produced more precise accounts of the relationships that exist within the "lower layers" of an ecology of knowledge. While the relevant studies are too numerous to list, they would certainly include Peter Galison's (1997) work on the material culture of science, Andrew Pickering's (1995) study of scientific practice, and the studies pursued by Michael Lynch (1993) from an ethnomethodological perspective. These are all works, and perspectives, on which I draw in the present study. Although it will be necessary especially to extend the notion of practice to deal seamlessly with matters beyond a narrow technical domain, any study that purports to investigate Cold War institutional changes through the historical lens of a technological artifact must of necessity engage with this literature.

13. Campbell-Kelly and Aspray 1996, chapter 5.

14. For a substantial discussion of these issues and a direct comparison to prior work in actor-network theory, see Akera, forthcoming.

15. See especially Hughes 1971a and Strauss 1993.

16. For an important exploration of similar issues, see Taylor 2005.

17. See my earlier note regarding the value of studying the "structure of integration" between knowledge and its supporting institutions.

18. Rosenberg 1997b/1988, p. 245.

19. Formally, drawing out specific parallels between an ecology of nature and an ecology of knowledge transforms the metaphor into an analogy. I continue to refer to an "ecology of knowledge" as a metaphor, as per Rosenberg's original usage.

20. Akera, forthcoming.

21. See especially Jakobson 1990b,c.

22. Jakobson 1990b; Chandler 2001.

23. Bourdieu 1977, pp. 4–15. A more dynamic approach to social practices can be found in Bourdieu's *Distinction: A Social Critique of the Judgment of Taste* (1984), and especially in his notion of reconversion strategies. These strategies would constitute the kind of cultivated practice that contributes to social transformation. Bourdieu attributes these strategies to be the product of a *habitus*, or the collectively embodied, phenomenological perceptions of a social group that produces concrete strategies of action (Fowler 2000, p. 3).

24. Bryant and Jary 1991.

25. Hacking 1999, pp. 49, 50.

26. Peirce 1991.

27. Israel 1998; Carlson 2000.

Notes to Chapter 1

1. MacLeod 2000, p. 6, as cited by W. Anderson (2002, p. 648).

2. Star 1995; Star and Griesemer 1989; Galison 1997, pp. 803–844; Law 1987.

3. See also Fine 1990; G. Anderson 1988. On banking, see Dohrn 1988.

4. Leslie 1983, pp. 19, 27; Cortada 1993, p. 93; Benson 1986.

5. Strom 1992, p. 34; Cortada 1993, pp. 99–101, 161; Rose 1995, chapter 2.

6. Geiger 1986.

7. Ibid.; Kohler 1991; Heilbron and Seidel 1989; Dupree 1986/1957; Cochrane 1976.

8. Bromley 1990; Goldstine 1993/1972, pp. 52–54.

9. Croarken 1990, pp. 17, 22–30; Goldstine 1993/1972, pp. 27–30, 107, 108.

10. Kidwell 1990. See also M. Wise 1995.

11. Carlson 1988.

12. Reynolds 1991; Williams 1985, pp. 111–118.

13. On Bell Laboratories, see Mindell 2002, p. 113; On GALCIT, see Culick 1983. On electrical engineering, see Owens 1986; T. Hughes 1983, pp. 376–385; Mindell 2002, chapter 5.

14. T. Hughes 1987, pp. 376, 377; Wildes and Lindgren 1985, pp. 62–72, 96–105.

15. Owens 1986; Wildes and Lindgren 1985, pp. 82–105; Mindell 2002, chapter 5.

16. Owens 1986.

17. Ibid.; Wildes and Lindgren 1985, pp. 67–75; Mindell 2002, pp. 153–161.

18. Owens 1986, pp. 64, 76.

19. Ibid., pp. 78–81; Genuth 1988, pp. 278, 279. The wordplay here is on the notion of centers of calculation advanced by Latour in chapter 6 of *Science in Action* (1987).

20. Buchwald 1985; Reid 1996/1972, pp. 113–116.

21. On the general direction of U.S. mathematical programs during the 1920s and the 1930s, see Parshall and Rowe 1994.

22. Owens 1989, p. 298; Parshall and Rowe 1994, pp. 35–40, 433–444.

23. Goldstine 1993/1972, pp. 75–81; Parshall and Rowe 1994, pp. 396, 436, 445.

24. Daston 1994; Schaffer 1994; Warwick 1995.

25. Porter 1995.

26. Pugh 1995; Austrian 1982.

27. Pugh 1995, pp. 67–69; Brennan 1971, pp. 3, 4.

28. E. C. Crittenden to Brehon Somervell, 4 February 1938, AMD-EC, Box 9/[Dr. Briggs—1938]; Raymond Archibald, "Foreword" [February 1940], AMD-EC, Box 9/Dr. Briggs—1940, NA-II.

29. "Dr. Arnold N. Lowan," December 1940, AMD-EC, Box 8/Dr. Lowan—Personal, NA-II.

30. Lowan to Lyman Briggs, 22 September 1939 and 19 December 1940, AMD-EC, Box 8/Dr. Lowan—Personal, NA-II.

31. "Status of Work on July 1, 1940." AMD-EC, Box 9/Dr. Briggs—1940, NA-II.

32. Lowan to Briggs, 29 September 1939, AMD-EC, Box 9/Briggs, LJ—1939. On Bethe's request, see correspondence in AMD-EC, Box 9/Professor A. H. Bethe.

33. Croarken 1990, pp. 23–32; Yates 1993.

34. Goldstine 1993/1972, pp. 108–110; Pugh 1995, pp. 71, 72; Brennan 1971, pp. 3–9.

35. Cohen 1999, chapters 2 and 3.

36. Ibid., chapter 3; [Aiken], "Proposed Automatic Calculating Machine" [4 November 1937], p. 5, catalog item HUG 4115.25.2, HUA.

37. Cohen 1999, chapters 6–8.

38. Schaffer 1994; Cohen 1999, chapters 9–11.

39. The stainless steel and glass exterior of the Mark I was designed by Norman Bel Geddes. See Campbell-Kelly and Aspray 1996, 71, 74; [Aiken], "Proposed Automatic Calculating Machine," p. 1; Cohen 1999, pp. 61–65.

40. Goldstine 1993/1972, pp. 127–129.

41. Ibid., pp. 96, 129; Burks and Burks 1988, pp. 107, 108. Goldstine's account of these events, and that of Burks and Burks, differ slightly from mine.

42. Reich 1985; Aitken 1985; Douglas 1987; Lukoff 1979.

43. Galison 1997, chapter 6.

44. Burke 1994, chapters 5–8; Bush 1945.

45. Burke 1994, pp. 42, 136.

46. Ibid., pp. 89–97.

47. Pugh 1995, pp. 81–86, 103.

48. Burks and Burks 1988, pp. 5–10.

49. Ibid., pp. 6–18.

50. Ibid., pp. 6–8, 18–21, 31.

51. Ibid., esp. chapters 1 and 5; Kidwell 1990, p. 39.

52. Dupree 1972/1970.

53. This point is raised by David Mindell. For a brief review of this literature, see Mindell 2002, pp. 185, 186.

54. On Bush and the organization of the NDRC, see Owens 1994; Genuth 1988; Kevles 1979.

55. Kevles 1979, p. 298. See also Mindell 2002, pp. 187, 188.

56. Owens 1994, Genuth 1988, Stewart 1948.

57. Mindell 2002, pp. 186–188.

58. Ibid., pp. 191–196.

59. Mindell 2002, chapter 8.

60. Ibid., pp. 197–200.

61. Owens 1989, p. 287; W. A. Wallis to Executive Committee, memo, 10 April 1944, AMP, Box 1/AMP Meetings, 1943–1946, NA-II.

62. Owens 1989, pp. 289, 290; "Mathematical Tables," Letter Circular 777, 26 January 1945, Aiken Papers, Box 1/Correspondence 1944–45, HUA.

63. Owens 1989, pp. 293–300.

64. Reid 1996/1972, pp. 236–245.

65. This work can be traced in AMP, Box 1/AMP Meetings, 1943–1946, NA-II.

66. Burke 1994, p. 206; Harrison to Weaver, 10 October 1941, NDRC-7, Box 2/Electronic Computers, NA-II.

67. Mindell 2002, p. 292; Stibitz to Weaver, 25 November 1941; Weaver to Zworykin, 23 December 1941 and 20 January 1942, NDRC-7, Box 2/Electronic Computers, NA-II.

68. Harrison to Weaver, 30 January 1942; Weaver to Zworykin et al., 16 February and 21 July 1942; "Conference on Electronic Fire Control Computers," 16 April 1942, NDRC-7, Box2/Electronic Computers, NA-II.

69. Mindell 2002, pp. 193, 300; Williams 1985, pp. 225–230.

70. Mindell 2002, pp. 300–302.

71. Ibid., p. 302.

72. Ibid., p. 303.

73. Ibid.; Williams 1985, pp. 225–240.

74. Burke 1994, pp. 208–253. Not described here are parallel efforts between the Army Signal Corps and IBM. (See Pugh 1995, pp. 98–106.)

75. Burke 1994, p. 275.

76. Ibid., pp. 275–302.

77. Goldstine 1993/1972, pp. 127–130.

78. Ibid., p. 128; "The Scientific Advisory Committee: Organization, Activities and Working Routine," draft [1940], JvN Papers, Box 11/1, LC.

79. Minutes, 3 May 1943, AMP, Box 1/AMP Meetings, 1943–1946, NA-II.

80. Cohen 1999, chapter 13; "The Computation Unit at Harvard" [27 August 1945], Aiken Papers, Box 1/Correspondence 1944–45, HUA.

81. Reid 1996/1972, pp. 241, 242.

82. Stibitz, "Final Report of Committee on Computing Aides to Naval Ballistics Laboratory," 28 April 1944, Aiken Papers, Box 1/Correspondence 1944–45, HUA.

83. Galison 1996; Galison 1997, pp. 803–844.

Notes to Chapter 2

1. Rosenberg 1997b/1988, p. 244.

2. Ibid., p. 242.

3. Published sources on Mauchly's biography include Stern 1981, Costello 1996, K. Mauchly 1984, and McCartney 1999. In this section, these are supplemented by family correspondence between John and his parents as filed with the Mauchly Papers, Series I, Box 1–2, UPSC.

4. Costello 1996, p. 45; James Ives and S. J. Mauchly, "A New Form of Earth Indicator," *Philosophical Magazine*, April 1911. Copy filed in Mauchly Papers, Box I/1/124, UPSC.

5. S. J. Mauchly to John Mauchly, 27 February 1915 and 28 September 1919, Mauchly Papers, Box I/2/19, UPSC.

6. See Rachel Mauchly's correspondence to her son, Mauchly Papers, Box I/1/13. See also Costello 1996, pp. 46, 47.

7. Costello 1996, pp. 46–48; "Accounts of J. W. Mauchly, from Sept 1, 1923," Mauchly Papers, Box I/10/166, UPSC.

8. Costello 1996, p. 48.

9. Mauchly, "Daily Record," diary [1921–], Mauchly Papers, Box I/10, UPSC.

10. Mauchly, "My Choice of a Vocation," 12 October [1925], Mauchly Papers, Box I/6/69, UPSC.

11. Rachel Mauchly to John Mauchly, 11 November 1925 and other letters, Mauchly Papers, Box I/1/13 and I/2/14, UPSC.

12. Rachel Mauchly to John Mauchly, 15 June 1926, Mauchly Papers, Box I/2/14, UPSC.

13. S. J. Mauchly to John Mauchly, 12 November 1925, 17 February 1926 and 26 October 1927, Mauchly Papers, Box I/2/20, UPSC.

14. Costello 1996, p. 48.

15. Ibid.

16. Envelope attached to Rachel Mauchly to John Mauchly, 26 November 1928, Mauchly Papers, Box I/2/17, UPSC.

17. Rachel Mauchly to John Mauchly, 20 November 1928, Mauchly Papers, Box I/2/17, UPSC.

18. Rachel Mauchly to John Mauchly, 8 November 1928, Mauchly Papers, Box I/2/17, UPSC.

19. Costello 1996, p. 49; Mauchly, The Third Positive Group of Carbon Monoxide Bands, PhD dissertation, Johns Hopkins University, 1932.

20. Costello 1996, p. 50; John Fleming to John Mauchly, 16 April 1932, Mauchly Papers, Box I/4/54.1, UPSC.

21. Costello 1996, p. 50; W. M. Gilbert to John Mauchly, 1 May 1933, Mauchly Papers, Box I/4/54.1, UPSC.

22. J. Mauchly 1982, p. 250; "Roster of Classes for 1934–35" and related correspondence, Mauchly Papers, Box I/4/54.1, UPSC.

23. K. Mauchly 1984, 119; J. Fleming to Mauchly, 9 July 1936, Mauchly Papers, Box I/7/86, UPSC; J. Bartels, "Statistical Methods for Research on Diurnal Variations," *Journal of Terrestrial Magnetism and Atmospheric Electricity* (September 1932). Copy filed in Mauchly Papers, Box I/9a/161.16.

24. J. Fleming to John Mauchly, 18 October and 10 November 1938, Mauchly Papers, Box I/9a/161.68, UPSC.

25. Indicative of a more general approach to the search for solar-related disturbances, Mauchly also turned to data on transmission disturbances collected by Bell Laboratories. Mauchly's collection of atmospheric data may be found in Mauchly Papers, II/5/39, II/7/105 and II/9/182, UPSC.

26. Mauchly to O. H. Gish, 24 March 1937; Mauchly to G. R. Wait, 24 March 1937. Both Mauchly Papers, Box I/7/86; Mauchly to Wilks, 25 November 1940; Mauchly to Hotelling, 25 November 1940. Both Mauchly Papers, Box I/7/90, UPSC.

27. John Mauchly, "A Significance Test for Ellipticity in the Harmonic Dial," *Journal of Terrestrial Magnetism and Atmospheric Electricity* (September 1932); J. Fleming to Mauchly, 16 March 1940, Mauchly Papers, Box I/7/89, UPSC.

28. John Mauchly, "Significance Test for Sphericity of a Normal N-variate Distribution," *Annals of Mathematical Statistics* 11(1940): 204–209. Mauchly to Larry, 17 August 1940, Mauchly Papers, Box I/7/90, UPSC.

29. J. Mauchly 1982, p. 250; "NYA Tentative Projects" [1940], Mauchly Papers, Box I/9b/161.27, UPSC.

30. [Mauchly], looseleaf design notes, various dates, Mauchly Papers, Box I/6/65, UPSC. See entry for 22 December 1937 for Mauchly's reference to E. C. Stephenson and I. A. Getting's description of a scaling circuit.

31. Burks and Burks 1988, p. 75; K. Mauchly 1984, p. 123; John Mauchly to John DeWire, 4 December 1940, Mauchly Papers, Box I/7/91, UPSC.

32. K. Mauchly 1984, pp. 122, 126; John Mauchly to Henry Tropp, 28 September 1978, published in *Annals of the History of Computing* 4 (1982): 245–247; Burks and Burks 1988, pp. 96–102.

33. Mauchly to Clayton, 15 November 1940, Mauchly Papers, Box I/7/91, UPSC.

34. Mauchly to DeWire, 4 December 1940.

35. Mauchly to Clayton, 15 November 1940.

36. Burks and Burks 1988, pp. 114–118.

37. Burks and Burks 1988, pp. 133–155, especially 145 and 152. Mauchly to DeWire, 28 June 1941, Mauchly Papers, Box I/7/91, UPSC.

38. Mauchly to Tom [Schonfeld], 23 June 1941, Mauchly Papers, Box I/7/91, UPSC.

39. [Mauchly], "Jan 15/41," record of household finances [January 1941], Mauchly Papers, Box I/6/80, UPSC.

40. K. Mauchly 1984, p. 128; J. Mauchly 1982, p. 251.

41. John Mauchly to Rachel Mauchly, 28 June 1941, Mauchly Papers, Box I/7/91, UPSC.

42. University of Pennsylvania, "Engineering, Science and Management War Training," 1942, Mauchly Papers, Box II/5/56, UPSC.

43. J. Presper Eckert was not related to Wallace Eckert.

44. K. Mauchly 1984, pp. 131–133; Burks and Burks 1988, pp. 155–168.

45. Mauchly to Tey [Teru Hayashi], 29 November 1941, Mauchly Papers, Box I/7/91, UPSC; John Mauchly to Mary Mauchly, 18 August 1941, as quoted in K. Mauchly 1984, p. 133.

46. K. Mauchly 1984, pp. 131–133.

47. Burks and Burks 1988, pp. 182–186.

48. Ibid., pp. 186–190.

49. Atanasoff to Mauchly, 31 May 1941; Mauchly to Atanasoff, 30 September 1941; and Atanasoff to Mauchly, 7 October 1941, all as quoted in Burks and Burks 1988,

pp. 127, 163–166. Mauchly to John [DeWire], 28 November 1941, Mauchly Papers, Box I/7/90, UPSC.

50. J. Mauchly 1942.

51. Ibid. For an etymology of the word 'program', see Grier 1996, p. 51.

52. J. G. Brainerd, handwritten note, as quoted in K. Mauchly 1984, p. 137; Mauchly to Tropp, 28 September 1978, op. cit. For a sense of the human computing work that took place at the Moore School, see Fritz 1996, pp. 14, 15.

53. Goldstine 1993/1972, pp. 133, 149.

54. University of Pennsylvania, "Report on an Electronic Difference Analyzer," 8 April 1943. Defendant's Trial Exhibit 3361, ENIAC Papers, Box 72/12, UPA.

55. The claim that the ENIAC was influenced by the design of the differential analyzer was originally made by one of the participants, Arthur Burks, and his wife, Alice Burks, in Burks and Burks 1981, pp. 334, 335, 344, 345.

56. "Report on an Electronic Difference Analyzer," p. 7.

57. Hazen, quoted in Stern 1981, p. 19. See also ibid., pp. 17–21; Mindell 2002, pp. 292–299.

58. Stern 1981, pp. 21–23; Goldstine 1993/1972, pp. 149, 150.

59. Stern 1981, p. 15.

60. For the best recent description of the ENIAC, see Burks and Burks 1981.

61. Burks and Burks 1981, pp. 340, 362–371.

62. Ibid., pp. 341, 360.

63. Owing to the ENIAC's innate parallelism, one of the most difficult aspects of programming it was the need to time each operation carefully.

64. Burks and Burks 1981, p. 341.

65. These early efforts are best discerned from the laboratory notebooks assigned to the project engineers, which are part of the ENIAC Papers, Box 22, UPA. For earlier comments about Mauchly's role in developing the ENIAC, see Stern 1981, pp. 37–39.

66. Sharpless, "Coupling Circuits," 26–28 July 1943, laboratory notebook Z14, ENIAC Papers, Box 22/5, UPA.

67. The idea of a resistor matrix, borrowed from RCA, made this substitution possible. The resistor matrix was also used in the ENIAC's high-speed multiplier.

68. H. R. Gail, 26 October 1943, 7 January 1944 and [March 1944], laboratory notebook Z4, pp. 5, 11, 25, 56, ENIAC Papers, Box 22/15, UPA; University of Pennsylvania, "The ENIAC (Electronic Numerical Integrator and Computer): Progress Report

Covering Work from 1 January to 30 June 1944," pp. IV-25–IV-26, UPE; Burks and Burks 1981, p. 346.

69. "Report on an Electronic Difference Analyzer," p. 18; Mauchly to Rademacher, 15 January 1944, Mauchly Papers, Box II/10a/209.18, UPSC.

70. Burks and Burks 1981, pp. 345, 346.

71. [Burks], "Positive Action Ring Counter," 17 August 1943; and "The NCR Thyratron Counter" [3 September 1943], Plaintiff's Exhibit 1630 and 1647, ENIAC Papers, Box 22/7 and 22/9, UPA.

72. Burks, notebook Z16, ENIAC Papers, Box 22/2; Stern 1981, pp. 21, 22; Burks and Burks 1981, pp. 351, 391; P. N. Gillon to Harold Hazen, 19 August 1943, Plaintiff's Exhibit 1633, ENIAC Papers, Box 22/7, UPA.

73. Burks and Burks 1981, pp. 345, 360; [Burks], "Function of Accumulator and Transmitter," 27 August 1943, draft report, as inset in notebook Z16 at p. 85.

74. T. K. Sharpless, "Mtg: Description of Programming Unit," 6 November 1943, notebook Z14, p. 19; Burks and Burks 1981, pp. 343, 359–362.

75. Notebook Z4, no title, no date, p. 135; "Programming," 20 November 1943, p. 144; no title, no date, p. 146; "Special Program Control," n.d., p. 181; Burks and Burks 1981, p. 372.

76. "Progress Report," 30 June 1944, p. IV-5.

77. Goldstine 1993/1972, p. 163; The quotation is from "Progress Report," 30 June 1944, p. IV-5.

78. James, 17 January 1944, notebook Z33, p. 95; "Summary of 6Y6 test of 99 tubes," 18 January 1944, p. 98; "PX tube checker," 5 January 1944, p. 88. See also early entries in Sharpless's notebook Z14.

79. James, n.d., notebook Z33, p. 120; 14 February 1945, p. 136.

80. "Progress Report," 30 June 1944, pp. III-1–III-5, IV-10; Burks and Burks 1981, pp. 352–356.

81. For further details on Zworykin's work at RCA, see Mindell 2002, pp. 292–296.

82. Here I also draw on arguments advanced by Mindell 2002, p. 7.

83. As I noted in the introduction, the perspective here draws on Anthony Giddens's structuration theory.

84. Stern 1981, chapters 5–7.

85. T. Hughes 1987, pp. 64–66.

86. For biographical information pertaining to Mauchly's later career, see Mauchly Papers, Series IV and V, UPSC.

87. Obituary, *New York Times*, 10 January 1980.

Notes to Chapter 3

1. This is a familiar position first advanced by the sociologist David Bloor (1991/1976).

2. For a more direct comparison of the metaphor of 'ecology' and 'network,' see Akera, forthcoming.

3. Aspray 1990, pp. 5–8; Macrae 1992, chapters 2–4.

4. Aspray 1990, pp. 8–10; Macrae 1992, chapters 5, 6.

5. Aspray 1990, pp. 10, 11.

6. Regis 1987, pp. 13–23; Aspray 1990, p. 12.

7. Aspray 1990, pp. 11–13; Aspray 1988; Macrae 1992, pp. 167, 168; B. Stern n.d.

8. Ulam to von Neumann, n.d., JvN Papers, Box 7/8, LC.

9. Ulam to von Neumann, 14 March 1941. Box 7, Folder 8, JVN Papers, LC; von Neumann to Ulam, 5 May 1941. Box 29, Ulam Papers, APS; Ulam 1976, chapters 4–5.

10. Regis 1987, chapter 5.

11. See, for instance, the Kurt Gödel Papers at Princeton University, Ulam's correspondence at the American Philosophical Society, and von Neumann's correspondence at the Library of Congress.

12. Among those who adopted this phrase or variants of it were Ulam, Nicholas Metropolis, and John Pasta. See correspondence files in JvN Papers, LC.

13. Reid 1996/1972, pp. 243–245.

14. The Scientific Advisory Committee met two or three times a year during the war.

15. Aspray 1990, pp. 25, 26; Macrae 1992, p. 190. For instance, Vannevar Bush, according to Regis (1987, pp. 127, 128), kept Einstein "out of the loop" with regard to the Manhattan Project because of his connections to Germany.

16. Aspray 1990, pp. 26, 27; Macrae 1992, pp. 200–215. Correspondence regarding von Neumann's NDRC project is in "Meetings of the Applied Mathematics Panel," minutes, AMP, Box 1/AMP Meetings, 1943–46 (hereafter AMP minutes), NA-II.

17. Aspray 1990, pp. 28, 29; Macrae 1992, chapter 10; Hewlett and Anderson 1962, pp. 124–136.

18. Aspray 1990, p. 29.

19. Aspray 1990, pp. 27–34; von Neumann to Weaver, 29 March 1944, JvN Papers, Box 7/12, LC.

20. Von Neumann to Veblen, 21 May 1943, as both noted and quoted by Aspray (1990, p. xv).

21. As indicative of this, von Neumann had already met with Leland Cunningham in July of 1944, shortly after Cunningham worked with Moore School engineers in determining the final makeup of the ENIAC (Aspray 1990, p. 31; Goldstine 1993/1972, p. 182; Macrae 1992, p. 281).

22. Goldstine 1993/1972, p. 205; Ceruzzi 1997, p. 5. For other discussions of the stored-program concept, see N. Stern 1981, pp. 74–81; Aspray 1990, pp. 36–42.

23. Eckert, Mauchly, and S. Reid Warren, "PY Summary Report No. 1," 31 March 1945, Mauchly Papers, Box III/C/4/70, UPSC; Aspray 1990, pp. 31, 32; University of Pennsylvania, "Report on an Electronic Difference Analyzer," 8 April 1943, p. 7. Defendant's Trial Exhibit 3361, ENIAC Papers, Box 72/12, UPA; Goldstine to Gillon, 2 September 1944, as quoted in Goldstine 1993/1972, p. 199.

24. Aspray 1990, pp. 37, 38; Goldstine 1993/1972, pp. 185, 186; "PY Summary Report No. 1"; S. Reid Warren, "Notes on the Preparation," affidavit, 2 April 1947, ENIAC Trial, Plaintiff's Exhibit 5764, UPA.

25. Aspray 1990, p. 40; von Neumann, "First Draft of a Report on the EDVAC," p. 1, UPE. However, William Aspray has suggested, in a personal communication to the author, that the entire content of the EDVAC report can be seen as an effort to specify digital computers in a simple, reductive form. This would be consistent with the practice of formal logic as performed within mathematical research circles. Von Neumann himself may have seen his effort in this way.

26. Mindell 2002, pp. 276–288.

27. Galison 1994, pp. 247–249; Aspray 1990, p. 183.

28. "First Draft of a Report on the EDVAC," pp. 4–6, 20, 21, 55–65, 68–73; For a good explanation of the tradeoffs associated with choosing a proper instruction set, see pp. 257–259 of Goldstine 1993/1972.

29. There are in fact surviving records that document the flavor of the collaboration. "PY Summary Report No. 1"; von Neumann to Goldstine, 10 December 1944 & 12 February [1945]; Burks, "Notes on Meeting with Dr. von Neumann," 23 March 1945, ENIAC Trial, Plaintiff's exhibits 2696, 2913, and 3040, UPA.

30. For similar arguments, see Aspray 1990, pp. 40, 41; N. Stern 1981, p. 86; Goldstine 1993/1972, p. 197.

31. Goldstine 1993/1972, pp. 255, 265.

32. Ibid., p. 266; Ceruzzi 1997, p. 6.

33. N. Stern 1981, pp. 79–81.

34. Ibid., pp. 48–52.

35. Ibid., pp. 96–99; Aspray 1990, pp. 42–47; von Neumann to Goldstine, 8 May 1945; Goldstine to von Neumann, 15 May 1945; Warren, affidavit, 2 April 1947; "Minutes of Conference Held at the Moore School," 8 April 1947, ENIAC Trial, Plaintiff's exhibits 3168, 3188, 5764, and 5824.5, UPA.

36. AMP minutes, 8 May and 13 November 1944, 19 February 1945, NA-II.

37. AMP minutes, 2 April 1945; Croarken 1990, pp. 83–87.

38. Kevles 1975.

39. Weaver to Rees, Fry, and Veblen, memo, 19 April 1945, AMP, Box 3/Organization—AMP, NA-II; AMP minutes, various dates, especially 22 July and 23 October 1944, 26 February 1945.

40. Sapolsky 1990, pp. 9–29; AMP minutes, 14 May and 24 September 1945.

41. Owens 1986, p. 83; Goldstine 1993/1972, pp. 219, 220.

42. Aspray 1990, pp. 50, 51; Leslie 1993, pp. 140–142.

43. Von Neumann to Harrison, 19 August and 20 November 1945, JvN Papers, Box 4/4, LC; Owens 1986, p. 83.

44. Owens 1986, pp. 83, 84.

45. Aspray 1990, p. 51; IAS, School of Mathematics, minutes of faculty meetings (hereafter IAS School of Mathematics minutes), 7 April 1945, JvN Papers, Box 13/8, LC; IAS faculty minutes, 22 May 1945, IAS Faculty-M, Box 1, IAS.

46. Aydelotte, "Report of the Director," appendix to minutes of annual meeting, 18 April 1947, IAS Board of Trustees Minutes, IAS; IAS faculty minutes, 2 June 1945; On the governance and power struggles within the IAS, see chapter 10 of B. Stern n.d.

47. IAS [Report of the Director], excerpts quoted in "Minutes of Regular Meeting," 20 April and 19 October 1945, IAS Board of Trustee Minutes, IAS.

48. "[Report]," 20 April 1945.

49. Von Neumann to Lewis Strauss, 20 October 1945, ENIAC Trial, Plaintiff's exhibit 3619, UPA; von Neumann to Aydelotte, 5 September 1945, JvN Papers, Box 12/1,

LC; "Minutes of Regular Meeting of the Board of Trustees," 19 October 1945, IAS, Board of Trustee Minutes, IAS; Aspray 1990, pp. 51–54.

50. IAS School of Mathematics minutes, 14 May 1946.

51. Morse himself would secure an ONR contract in 1947."Minutes of the Meeting of the Executive Committee of the Institute for Advanced Study," 18 February 1947, IAS, Faculty Meeting Minutes, Box 1, IAS; IAS faculty minutes, 29 January 1945.

52. For a broad description of von Neumann's accomplishments as related to computing, see Aspray 1990.

53. Weaver to von Neumann, 21 March 1946, JvN Papers, Box 7/12, LC; Aspray 1990, p. 54; von Neumann to Strauss, 4 May 1946, ENIAC Trial, Plaintiff's exhibit 4425, UPA; Goldstine to G. F. Powell, 14 February 1947, ECP, Box 3/27, IAS.

54. Aspray 1990, pp. 54–57; Goldstine 1993/1972, p. 242; Eckert to von Neumann, 15 March 1946, ENIAC Trial, Plaintiff's exhibit 4228, UPA.

55. Goldstine to Gillon, 8 December 1945, ENIAC Trial, Plaintiff's exhibit 3810, UPA.

56. War Department, Bureau of Public Relations, Press Branch, "High-Speed, General Purpose Computing Machines Needed," press release, 16 February 1946. ENIAC Historical Files, UPA.

57. N. Stern 1981, pp. 48–52. This was not a serendipitous gesture, but an action reviewed and approved by the university's Committee on Non-Medical Patents. McClelland to Mauchly, 15 March 1945, ENIAC Trial, Plaintiff's exhibit 3015, UPA.

58. N. Stern 1981, pp. 89–91; Mauchly to Harold Pender, 22 March 1946; Pender to Mauchly, 22 March 1946; ENIAC Trial, Plaintiff's exhibit 4255, 4259, UPA.

59. "Conference of Mathematical Computing Advisory Panel," 27 February 1946, CBI 45, Box 5/3, CBI.

60. Ibid., 27 February and 10 April 1946; J. H. Curtiss, "The National Applied Mathematics Laboratories of the National Bureau of Standards," 1 April 1953, CBI 45, Box 1/2, CBI.

61. Curtiss, "The National Applied Mathematics Laboratories," 1 April 1953, pp. 1, 12, 18.

62. Ibid.

63. Ibid., pp. 10, 21; E. W. Cannon, "Progress Report," 21 November 1946, CBI 45, Box 1/12, CBI.

64. Cohen 1999, pp. 177–183, 203.

65. Williams 1985, pp. 232–240.

66. Burke 1994, pp. 311–321; Tomash and Cohen 1979.

67. Pugh 1995, chapter 10; Brennan 1971, pp. 11–16.

68. Aspray 1990, p. 91; von Neumann to Bradbury, 7 October 1947; Bradbury to von Neumann, 24 June 1948, JvN Papers, Box 14/10, LC.

69. N. Stern 1981, pp. 101–104; [Bureau of Census], "The Development and Use of Electronic Tabulating Equipment by the Bureau of Census," May 1948, CBI 45, Box 8/8, CBI; and "Survey of Digital Computing Program," 1 August 1946, CBI 45, Box 8/34, CBI.

70. N. Stern 1981, pp. 103–106.

71. On Condon, see pp. 101 and 102 of N. Stern 1981. I am also indebted here to Pap Ndiaye, who has carefully reexamined the relationship between Condon, the military, and the Eckert-Mauchly Computer Corporation.

72. Aspray 1985, p. ix.

73. C. T. Joy, "Representing the Bureau of Ordnance," *Proceedings* 1985/1948, p. 5.

74. "Members of the Symposium," *Proceedings* 1985/1948, pp. xvii–xxix.

75. For an indication that Rees's concerns prompted the broader survey, see pp. 126 and 151 of Redmond and Smith 1980.

76. H. T. Engstrom, "Foreword," in ERA 1983/1950, p. xxxv; Arnold Cohen, "Introduction," ERA 1983/1950, pp. xi–xii; Tomash and Cohen 1979, pp. 89, 90.

77. Tomash and Cohen 1979, p. 89; Cohen, "Introduction," ERA 1983/1950, p. xiii.

78. ERA 1983/1950, pp. 193, 213–218, 251–261.

79. This consisted of three volumes published under the common title *Planning and Coding of Problems for an Electronic Computing Instrument* (Institute for Advanced Study, 1947–1948). Together these volumes constituted part II of *The Mathematical and Logical Aspects of an Electronic Computing Instrument.* Only volumes I and II are cited in the text.

80. This was a machine built for the Navy's cryptographic work in late 1950 and offered for commercial sales in modified form as the ERA 1101 in 1951. See p. 90 of Tomash and Cohen 1979.

81. As effectively summarized by Campbell-Kelly and Aspray (1996, pp. 99–104), the computer at the University of Manchester, completed on 21 June 1948, was a very limited machine designed to test the CRT-based memories developed by Williams. It had no input or output mechanisms, and was not designed for efficient computation. The EDSAC, completed in 6 May 1949, was based directly on von Neumann's draft report on the EDVAC, including its use of delay line memories.

82. "The Joint Research and Development Board of the War and Navy Departments," 30 June 1947, RDB, Box 429/1, NA-II.

83. Vannevar Bush to Grabau, 17 February [1948]; "4.3.4 Recommendations as to Personnel," n.d.; "Proposed Draft to Sec'y of Defense," 22 October 1948; Compton to L. R. Hafstad, 10 December 1948; Thomas Sherwood to William Houston, 12 November 1948, RDB, Box 587/6, NA-II.

84. W. V. Houston to Grabau, 5 July 1949, RDB, Box 587/6, NA-II. On the influence of the management and organization movement on the RDB, see, for instance, Vannevar Bush, "General Basis for Study of Programs," 6 November 1946, RDB, Box 428/1, NA-II; "Joint Research and Development Board," 30 June 1947. Bush chaired the RDB. For a general account, see also Redmond and Smith 1980, pp. 151–154.

85. Ad Hoc Panel on Electronic Digital Computers, "Report on Electronic Digital Computers," 1 December 1949, pp. 3–12, 25, 26, RDB, Box 303/5, NA-II.

86. Ibid., p. 44. The various responses can be found in RDB, Box 438/3, NA-II. See also Redmond and Smith 1980, pp. 152–154.

87. Grabau to Goldstine, 29 December 1949. Leslie Simon to Grabau, 21 January 1950; William Houston to Webster, 18 February 1950; Webster to Fink, 4 May 1950, RDB, Box 438/3, NA-II.

88. "Report," pp. 6, 7; Webster to Army, Deputy Asst Chief of Staff, 1 November 1950 and similar letters, RDB, Box 438/3, NA-II.

89. Aspray 1990, pp. 57, 74.

90. Ibid., p. 56. Burks took up a faculty appointment at the University of Michigan, but spent the subsequent three summers at the IAS.

91. A copy of the interim progress reports, along with the final report for the project, may be found in ECP, Box 5–8, IAS.

92. Aspray 1990, p. 57; First and third interim progress reports, January 1947 and 1 January 1948, ECP, Box 7/54 and 7/65, IAS. The influence of a generalized scientific culture and method is robustly visible in these documents. It shows up consistently in the reports' language as well as in their structure (where Bigelow clearly places the program in "Experimentation, Design and Construction" above "engineering belief," even while carving out a space for the epistemological validity of the latter).

93. Tensions within the project are visible in JvN Papers, Box 4/1, especially in Goldstine to von Neumann, 19 July 1947, IAS. See also ECP, "Monthly Progress Report." ECP, 4/40, IAS; and Aspray 1990, pp. 56, 57.

94. Aydelotte to von Neumann, 4 June 1946, JvN Papers, Box 12/1, LC; IAS School of Mathematics minutes, 14 May 1946.

95. Aydelotte to von Neumann, 5 June and 10 June 1947, IAS Faculty, Box 34/von Neumann, John, IAS.

96. Aspray 1990, pp. 241–247.

97. Daily log of visitors and phone calls, JvN Papers, Box 9/11–14, LC; Heims 1980, pp. 240, 369–371; Macrae 1992, pp. 373–380.

98. Aspray 1990, p. 251; Oppenheimer to von Neumann, 29 February 1956, IAS Faculty, Box 34/von Neumann, John, IAS. The crucial question posed by Dyson was "what is a proper role for the Institute to play in the fields of applied mathematics and electronic computing?" with the further comment that "The purpose of the Institute is to carry on long-term academic research in a scholarly atmosphere." Dyson and Deane Montgomery, "Memorandum to the faculty" [December 1954], IAS General File, Box 17/Electronic Computer Project—Considerations, Future of, IAS.

Notes to Chapter 4

1. Bijker 1995, pp. 122–127, 139–143.

2. Although the notion of closure has received extensive treatment in the science-studies literature, especially surrounding the question of how experiments end (see Galison 1987), the most relevant citation here would be Pinch and Bijker's (1987/1984) canonical essay that sought to transfer science studies approaches to the study of technology.

3. Committee for the Evaluation of the National Bureau of Standards, "First meeting," minutes, 29 April 1954, Astin Records, Box 5/[Kelly Committee], NA-II.

4. Harry Diamond, "Progress Report," 8 June 1946, CBI 45, Box 1/7, CBI.

5. Ibid.

6. N. Stern 1981, p. 105.

7. Stibitz, "Report on Computing Machines," 23 May 1946, CBI 45, Box 4/10, CBI.

8. Ibid.; Stibitz to Curtiss, 20 May 1946, CBI 45, Box 8/8, CBI.

9. E. W. Cannon, "Progress Report," 21 November 1946, CBI 45, Box 1/12, CBI.

10. Diamond, "Progress Report," 8 June 1946.

11. The ascent of Alexander's group is clearly visible by the fourth progress report, NBS, Electronics Section, "Progress Report," 8 October 1946, CBI 45, Box 1/7, CBI.

12. NBS, Electronics Section, "Progress Report," 11 September and 8 November 1946, CBI 45, Box 1/7; W. J. Weber, "Conference on Extending 'Improved Vacuum Tube Program of the Airlines,'" 22 September 1947, CBI 45, Box 5/42, CBI.

13. "Progress Report," 16 December and 31 December 1946, CBI 45, Box 1/7, CBI.

14. J. H. Curtiss, "The National Applied Mathematics Laboratories," 1 April 1953, CBI 45, Box 1/7, CBI.

15. Ibid.

16. Ibid., p. 67; Cannon, "Progress Report," 21 November 1946. N. Stern 1981, p. 113.

17. NBS, Electronic Computers Laboratory, "Automatic Computing Machines," annual report, June 1950, CBI 45, Box 1/7; Mary Stevens to A. V. Astin, memo, 6 October 1955, CBI 45, Box 8/37; Curtiss, "The National Applied Mathematics Laboratories," pp. 67, 68.

18. Curtiss to von Neumann, 11 August 1947; Harry Huskey, "Relations with Eckert-Mauchly," 18 March 1948. Both CBI 45, Box 8/8, CBI. For an account based more or less on the NBS's subsequent reconstruction of the events, see N. Stern 1981, pp. 107–112.

19. Subcommittee, "Memorandum to the NRC Committee" [March 1948], p. 5, CBI 45, Box 1/13, CBI.

20. Ibid., pp. 7, 8.

21. Von Neumann to Curtiss, 16 March 1948; Curtiss to Chief, Division 13, 18 March 1948; Curtiss to von Neumann, 11 August 1947, CBI 45, Box 8/8, CBI.

22. Huskey to Members, Division 13, 18 March 1948, CBI 45, Box 8/8, CBI; Curtiss, "The National Applied Mathematics Laboratories," p. 67.

23. For a somewhat different account, see N. Stern 1981, pp. 113, 114.

24. See also N. Stern 1981, pp. 106, 112–115.

25. Huskey to Members, Division 13, 18 March 1948; NBS, "Statement by the Director," 28 September 1949, CBI 45, Box 8/8; A. L. Leiner to Huskey, "Acceptance Tests for Univac," 24 June 1948, CBI 45, Box 8/17; "Schedule A. Checkpoint Program" [1948], CBI 45, Box 8/8, CBI.

26. Electronic Computers Laboratory, "Annual Report," June 1949, CBI 45, Box 1/7, CBI.

27. Stevens to Astin, 6 October 1955.

28. "NBS Interim Computer" [May 1948], CBI 45, Box 2/14; Alexander to Astin and Cannon, "Research and Development Projects for Office of Air Comptroller," 9 September 1949, CBI 45, Box 1/13, CBI.

29. "Meeting of High-Speed Circuits Technical Planning Group," minutes, CBI 45, Box 7/25. See also ECL, "Quarterly Progress Report," and "[Annual] Progress Report." CBI 45, Box 1/7.

30. For some of the relevant records, see meeting minutes for 1 and 16 March 1949, 3 May 1949; and "[Annual] Progress Report," 19 July 1949.

31. A. W. Holt, "Technical Discussion Meeting," 14 September 1949, CBI 45, Box 8/9, CBI.

32. ECL, "Automatic Computing Machines," June 1950.

33. Wang 1999, pp. 130–145.

34. Again, I am indebted to Pap Ndiaye for his work and insights pertaining to the political economy of research at the NBS. NBS, "Scientific Program Plan, Fiscal Year 1952" [Spring 1951], CBI 45, Box 1/4, CBI.

35. Ibid. The provision for the funds transfer was established through Section 601 of the Economy Act of 30 June 1932. On Condon's subsequent struggles with HUAC, see Wang 1999, pp. 144, 145, and his unpublished autobiography in B:C752 Edward Condon Papers, Folder: "Autobiography." APS.

36. "SEAC," brochure, 20 June 1950; Condon, "Address at the Dedication," 20 June 1950. Both CBI 50, Box 8/26A, CBI.

37. NBS, Applied Mathematics Advisory Council, minutes, 11 May 1951, CBI 45, Box 7/34; For an actual record of the use of the SEAC by its sponsors, see the monthly reports by R. J. Slutz in CBI 45, Box 4/7, CBI. The AEC's maximum use was 190 hours per month, or roughly one full shift of operation.

38. ECL, "Annual Report," 30 June 1951, CBI 45, Box 1/7, CBI.

39. NBS, "SEAC Facilities Expanded," 8 January 1952, CBI 45, Box 1/8. See also Alexander's lecture and talks facilitated by the SEAC's popularity, CBI 45, Box 8/33 and 6/45, CBI.

40. NBS, Applied Mathematics Executive Council[sic], minutes, 14 February 1950; NBS, Applied Mathematics Advisory Council, minutes, 8 November 1950. Both CBI 45, Box 7/35, CBI.

41. Curtiss, "The National Applied Mathematics Laboratories," pp. 72–74.

42. Minutes, 14 February 1950, p. 19; 8 November 1950, pp. 4, 10; 11 May 1951, p. 3; Leslie Williams to Curtiss, 2 November 1951, CBI 45, Box 7/34.

43. Committee for the Evaluation of the Present Functions and Operations of the National Bureau of Standards, minutes, 29 April 1953, Astin Records, Box 5/Kelly Committee, NA-II. A basic history of the controversy can be found in [Churchill Eisenhart], "Chronology of the Battery Additive Controversy and Its Aftermath," which is filed with the finding aid for the NIST Records (RG 167) at NA-II.

44. Ad Hoc Committee on Evaluation, "A Report to the Secretary of Commerce," 15 October 1953, Astin Records, Box 5/Kelly Committee, NA-II. The overall NBS

budget fell from \$52.2 million to \$20 million between 1952 and 1955, while the size of the staff fell from 4,781 to 2,800. See also Cochrane 1976, pp. 559, 563.

45. Alexander, "Division 12—Data Processing Systems" [1955, Astin Records, Box 40/Visiting Committee, 1935, NA-II]; On the negotiations to transfer INA to UCLA, see RS 359, Box 26/[INA], UCLA-UA. The details pertaining to the NBS's response may be found in its annual reports to the NBS Visiting Committee, Astin Records, Box 40 & 41/Visiting Committee, NA-II.

46. Alexander, "Division 12" [1955]; and Cannon, "Division 11—Applied Mathematics" [1955], Astin Records, Box 40/Visiting Committee, 1955, NA-II.

47. C. C. Hurd, "Visit to Computation Laboratory," 17 October 1951, CBI 95, Box 1/2. CBI.

48. Applied Mathematics Advisory Council, minutes, 8 November 1950; "Notes for Course on 'Introduction to Automatic Data Processing,'" [1955], CBI 45, Box 8/22, CBI.

49. Powell and DiMaggio 1991.

50. Akera 2000.

Notes to Chapter 5

1. See, for instance, Edwards 1996, pp. 99, 100; T. Hughes 1998, p. 40; Campbell-Kelly and Aspray 1996, chapter 7.

2. I draw here on Paul Edwards's notion of "mutual orientation" (1996, pp. 81–83). Though I use a slightly different term here, I do not regard it as original; mutual orientation already presumes articulation.

3. Latour 1987, chapter 3; Agar 2003.

4. On Brown, see chapter 8 of Mindell 2002.

5. Redmond and Smith 1980, pp. 1–3.

6. Ibid., pp. 3–9.

7. Ibid., pp. 9–15.

8. Ibid., pp. 14–22.

9. Ibid., pp. 22, 23; Sapolsky 1990, pp. 23–26.

10. Mindell 2002, p. 291; Redmond and Smith 1980, pp. 26, 27, 67.

11. Redmond and Smith 1980, p. 33.

12. Mindell 2002, esp. chapter 8; S. Dodd and L. Bernbaum, C-1, 22 October 1945, AC 290, Box 9/2; Kenneth Tuttle, "Conference on Techniques of Computation," C-6, 14 November 1945, AC 290, Box 9/3, NMAH.

13. Forrester, "Outline of Discussion on Digital Computation," C-9, 28 November 1945; Stephen Dodd, "6345 Conference," C-8, 29 November 1945, AC 290, Box 9/3, NMAH.

14. Forrester, C-9, p. 9.

15. Tuttle, C-6, p. 3.

16. Redmond and Smith 1980, pp. 41–44; Sapolsky 1990, pp. 26, 27, 39–47.

17. Owens 1994, p. 565; Stewart 1948, p. 322; MIT, "Policy Relating to Governmental Contracts," Special bulletin, 1 November 1944. In F. L. Foster, "Sponsored Research at MIT," vol. II, part II, p. 45–3; Compton 1944, p. 26.

18. Kleinmann 1995; Geiger 1993.

19. Leslie 1993, pp. 22–26; James Killian to Compton, 30 August 1945, AC 4, Box 59/4, MIT.

20. Redmond and Smith 1980, pp. 8, 39–43.

21. Redmond and Smith 1980, pp. 41–44.

22. Redmond and Smith 1980, pp. 46, 48; Forrester, C-10, 11 February 1946; Forrester, "General Meeting," C-12, 10 June 1946, AC 290, Box 9/7 and 10/2, NMAH.

23. Owens 1986); Mindell 2002, chapters 5 and 8; Redmond and Smith 1980, pp. 10, 36, 46, 79; Forrester to John Slater, 17 December 1945, AC 290, Box 9/4, NMAH.

24. Redmond and Smith 1980, pp. 49–51; Forrester, E-9, 4 March 1946; Everett, E-13, 3 April 1946; other "E-series" memoranda, AC 290, Boxes 9/8–10/1, NMAH.

25. Redmond and Smith 1980, pp. 51–54; L. Bernbaum and J. Bicknell, "ASCA Equations," R-61, 4 April 1946; Everett, E-13, 3 April 1946; Hugh Boyd, "[Report of the] Mathematics Group," E-14, 12 April 1946. All AC 290, Box 9/8, NMAH.

26. Mabel Latin, C-13, 17 June 1946, AC 290, Box 10/1; Forrester, C-14, 9 October 1946, AC 290, Box 10/6; Harris Fahnestock and Everett, "Coding of Aircraft Flight Equations," C-15, 9 October 1946. Box 10/6, NMAH.

27. Forrester to Dodd et al., M-26, 22 October 1946; Dodd to Forrester, C-19, 23 October 1946. Both AC 290, Box 10/6, NMAH.

28. Everett, E-13, 3 April 1946; David Brown, "Specifications for the Computer, Switching, and Engineering Program," E-26, 16 September 1946; Fahnestock and Everett, C-15, 9 October 1946; Everett, "Discussion of a Parallel Computer," C-22, 6 November 1946, AC 290, Box 9/8, 10/5, 10/6 and 10/7, NMAH.

29. Redmond and Smith 1980, p. 55; Forrester, C-12, 10 June 1946; Fahnestock and Forrester, M-33, 20 November 1946; Fahnestock, C-25, 2 December 1946, AC 290, Box 10/2, 10/7 and 10/8, NMAH.

30. Fahnestock and Forrester, M-33; Fahnestock and Forrester, "Prototype Computer Meeting," M-37, 22 November 1946, AC 290, Box 10/7, NMAH.

31. Everett, "Notes and Block Diagrams for the Pre-Prototype Computer," M-52, 27 December 1946, AC 290, Box 10/8, NMAH.

32. Ibid.; Fahnestock and Forrester, M-37; Fahnestock to Forrester, M-47, 17 December 1946, AC 290, Box 10/8, NMAH.

33. Fahnestock, C-25, 2 December 1946, AC 290, Box 10/8, NMAH.

34. Sage to Forrester, 30 January 1947, AC 290, Box 10/9, NMAH. Redmond and Smith 1980, pp. 63–66.

35. Redmond and Smith 1980, pp. 69–73.

36. Weaver to Rees, 26 June 1947, as cited in Redmond and Smith 1980, p. 73.

37. Forrester to Sage, "Warren Weaver Visit to Laboratory," 19 February 1947, AC 290, Box 10/10, NMAH.

38. Weaver to Rees, 26 June 1947.

39. Forrester to Sage, 19 February 1947.

40. Forrester, "Electronic Digital Computer Development at M. I. T.," 17 February 1947, AC 290, Box 10/10, NMAH.

41. Ibid.

42. Ibid.

43. Redmond and Smith 1980, pp. 62–66.

44. Forrester to J. C. Hunsaker, 1 April 1947, AC 290, Box 9/8; Fahnestock, M-60, 13 March 1947, AC 290, Box 10/11, NMAH. Redmond and Smith 1980, p. 77.

45. De Florez became the director of technical research at CIA near the end of World War II. Different naval officers rotated through the position of SDC Director, and Forrester generally sent his correspondence as addressed to the office rather than the specific individual.

46. Redmond and Smith 1980, pp. 79, 80, 107, 108; Fahnestock to Forrester, M-47, 17 December 1946, AC 290, Box 10/8, NMAH.

47. Redmond and Smith 1980, pp. 79, 92; "WWI Computer—Material Cost," 30 July 1947, AC 290, Box 11/4, NMAH.

48. Forrester to SDC Director, 5 September 1947, AC 290, Box 11/6, NMAH.

49. Forrester to SDC Director, M-102, 15 September 1947, AC 290, Box 11/6; Redmond and Smith 1980, pp. 107, 108, NMAH.

50. Anonymous memorandum, 15 September 1947, quoted in Redmond and Smith 1980, p. 68.

51. Redmond and Smith 1980, pp. 66, 73, 101–107.

52. Ibid., pp. 73, 74.

53. Ibid., p. 74; Murray collaborated with von Neumann during World War II, and was a member of IAS in AY 1946–1947, IAS School of Mathematics minutes, 14 October 1946, JvN Papers, Box 13/8, LC.

54. Redmond and Smith 1980, p. 75; Forrester to Sage, "Analysis of Whirlwind Program," L-10, 13 January 1949, AC 290, Box 17/2, NMAH.

55. Murray to SDC Director, "Report on Mathematical Aspects of Whirlwind," 21 November 1947, AC 4, Box 236/12, MIT.

56. Ibid.; Redmond and Smith 1980, pp. 80, 81.

57. These difficulties are meticulously documented in the E- and M- series memoranda in AC 290, Box 10–15, NMAH. See for instance, Fahnestock to Forrester, M-49, 28 February 1947. Box 10/10; John Hunt, "Performance Chains of Gate Tubes," E-137, 5 August 1948, Box 15/3. See also Redmond and Smith 1980, pp. 92, 129, 130.

58. Dodd to Forrester, C-19, 23 October 1946, AC 290, Box 10/6; Dodd et al., "Storage Tube Summary," M-159, 10 November 1947, AC 290, Box 12/5; Everett, "Storage Tube Organization," M-687 [November 1948], AC 290, Box 16/2, NMAH.

59. Fahnestock to Nat Rochester, 24 October 1947, AC 290, Box 12/2; "Bi-Weekly Report," M-267, 5 March 1948, AC 290, Box 5/3; Forrester to D. M. Rubel, "Information of Whirlwind I," L-11, 15 February 1949, AC 290, Box 17/3, NMAH.

60. "Non-Staff Organization Chart," D-32267, 10 June 1948, AC 290, Box 14/6; Forrester, "Talk Delivered by Jay Forrester," R-142, 31 August 1948, AC 290, Box 15/2, NMAH.

61. Redmond and Smith 1980, pp. 108–110; Sage to SDC Director, 4 August 1948, AC 290, Box 15/3, NMAH.

62. Redmond and Smith 1980, pp. 95–101, 111; Forrester, C-10, 11 February 1946, AC 290, Box 9/7; Forrester, "Professor Bryant's Memorandum on Report Writing," A-22, 4 April 1947, AC 290, Box 11/1, NMAH.

63. Redmond and Smith 1980, pp. 93, 110; "Staff Indoctrination Program," M-626, 29 September 1948, AC 290, Box 13/2; Forrester, "Outlook for Electronic Digital Computers," R-154, 16 April 1949, AC 290, Box 17/3, NMAH. On systems integration in the context of the NDRC's Division 7, see chapter 10 of Mindell 2002.

64. Redmond and Smith 1980, p. 58; Forrester and Everett to SDC Director, "Digital Computation for Anti-Submarine Program," M-108, 1 October 1947; Forrester and Everett to SDC Director, L-2, 15 October 1947, AC 290, Box 12/1, NMAH.

65. Solberg to Compton, 2 September 1948, as quoted in Redmond and Smith 1980, pp. 111, 112.

66. Redmond and Smith 1980, p. 112; Forrester, "Visitors to Project Whirlwind," L-7, 2 November 1948. Box 16/2, NMAH. The other members of the committee were Hugh Boyd and Robert Nelson.

67. Redmond and Smith 1980, pp. 116–118; Forrester, "A Plan for Digital Information Handling Equipment in the Military Establishment," L-3, 14 September 1948, AC 290, Box 15/4, NMAH.

68. Redmond and Smith 1980, p. 118; Forrester, Boyd, Everett, Fahnestock and Nelson, "Alternative Project Whirlwind Proposals," L-4, 21 September 1948, AC 290, Box 16/1, NMAH.

69. Forrester, L-3, p. 2; The idea to include staff levels originated with Compton, who specifically requested that the report include information about "time, money and staff." Compton to Sage, 8 September 1948, as quoted in Redmond and Smith 1980, p. 112.

70. Compton's confidence in Forrester derived in part from an internal evaluation carried out in the summer of 1948 by Ralph Booth of Jackson and Moreland, a consulting firm with deep ties to MIT. Booth was the third representative sent to this meeting. See Redmond and Smith 1980, pp. 108–114.

71. Redmond and Smith 1980, pp. 119, 120. For an alternative account of the root causes of the ONR's discontent, see ibid., pp. 101–107.

72. Redmond and Smith 1980, p. 122; Forrester and Everett, "Comparison between the Computer Programs at the Institute for Advanced Study and the M. I. T. Servomechanisms Laboratory," L-6, 11 October 1948, AC 290, Box 16/1, NMAH.

73. Redmond and Smith 1980, pp. 123, 124, 136, 137; Forrester, L-5, 11 October 1948, AC 290, Box 16/1, NMAH.

74. Forrester, L-5, p. 2.

75. Ibid., pp. 2, 3.

76. Ibid., p. 2.

77. Redmond and Smith 1980, p. 127; Forrester to Sage, L-10, 13 January 1949.

78. Redmond and Smith 1980, pp. 128, 145, 146; Fahnestock to C. V. L. Smith, 17 August 1949, AC 290, Box 19/3, NMAH.

79. Pinch and Bijker 1987/1984. Though I also cite this work in chapter 4 in relation to the concept of sociotechnical closure, Pinch and Bijker also introduce the notion of rhetorical closure in this article.

80. Vannevar Bush to Grabau, memo, 17 February [1948].

81. Redmond and Smith 1980, pp. 112, 121; Compton to L. R. Hafstad, 10 December 1948; Sherwood to William Houston, 12 November 1948, RDB, Box 587/6, NA-II.

82. Forrester, "Discussion on the Comments on Project Whirlwind," L-16, 13 January 1950; Forrester, "Comment on the Report of the Ad Hoc Panel," L-17, 13 January 1950. Both AC 290, Box 20/3, NMAH.

83. Ad Hoc Panel, "Report on Electronic Digital Computers," 1 December 1949, pp. 39, 52–54, RDB, Box 303/5, NA-II. See also Redmond and Smith 1980, pp. 59, 60, 153, 154.

84. Ad Hoc Panel, "Report"; Redmond and Smith 1980, pp. 153, 154, 166, 167.

85. Forrester, L-16, p. 7; Forrester, L-17, p. 5.

86. Redmond and Smith 1980, pp. 154–157; 2000, pp. 67–69.

87. See also Hacking 1999, pp. 68–80.

88. Pickering 1995, pp. 169–176.

89. Noble 1984, chapter 11.

90. MIT administrators would reach a similar conclusion: "The Institute has been at fault in permitting a man so young and relatively inexperienced to assume such an enormous burden of responsibility without closer guidance and supervision." Stratton to Killian and Sherwood, 3 February 1950, AC 4, Box 236/14, MIT.

91. Redmond and Smith 1980, pp. 113, 121–128, 155. Redmond and Smith (2000) also revisit this issue and point to the conciliatory stance taken by MIT and ONR administrators. See especially chapter 5. meanwhile, the best source for tracing the development of MIT's research policies and practices of research administration is F. L. Foster, "Sponsored Research at M. I. T., 1900–1968" (unpublished manuscript), MIT.

92. Formally, SAGE was an acronym for Semi-Automated Ground Environment, although it has been argued that the acronym was also meant to be a tribute to DIC Director Nathaniel Sage.

93. See especially Sapolsky 1990, pp. 49, 50.

94. C. V. L. Smith, "Summary of Conference . . . 6–7 March 1950"; Waterman, diary entry, 6 March 1950 (both as cited in Redmond and Smith 2000, p. 69).

95. Redmond and Smith 2000, chapter 4. See also Edwards 1996, chapter 3; T. Hughes 1998, chapter 2; Valley 1985.

96. Forrester's principal works in the field of system dynamics, as applied to different realms of policy, are *Industrial Dynamics* (1961), *Urban Dynamics* (1969), and *World Dynamics* (1971).

Notes to Chapter 6

1. The following works already proceed in this direction: Hounshell 1992; Hounshell and Smith 1988; Graham 1986.

2. For an earlier account of Hurd's activities, see Weiss 1997. On IBM, see Pugh 1995; Fisher, McKie, and Manke 1983.

3. Strasser 1993; Friedman 1998; Goldstein 1997; Zunz 1990.

4. Pugh 1995, pp. 1–57. On the office equipment industry as a whole, see Cortada 1993. On NCR, see Leslie 1983. On the history of industrial research, see Hounshell 1996, Israel 1992, Hounshell and Smith 1988, Reich 1985, and G. Wise 1985.

5. Friedman 1998; Campbell-Kelly and Aspray 1996. On the rationalization of the U.S. sales force, see Spears 1993.

6. On Edison and system selling, see pp. 105–128 of Passer 1953.

7. Pugh 1995, pp. 42–44, 57–60, 62–64. Goldstein's 1993 article describes a similar strategy employed by electrical utilities.

8. Campbell-Kelly and Aspray 1996, pp. 110–117; Pugh 1995, pp. 117–129; Weiss 1997, p. 66.

9. Pugh 1995, pp. 123, 124, 152–157.

10. The Clinton Laboratory, operated by Union Carbide, became the Oak Ridge National Laboratory in 1948. For biographical information on Hurd, see Weiss 1997. Alston Householder to Hurd, 29 March 1949, CBI 95, Box 6/11, CBI.

11. Weiss 1997, p. 66; Truman Hunter to author, 8 August 1999 (letter in author's possession); See Hurd's correspondence with Householder at Oak Ridge and Cecil Hastings at the Rand Corporation in CBI 95, Box 6/11 and 6/23, respectively, CBI; Hurd 1985.

12. Hastings to Hurd, 31 October 1949, CBI 95, Box 6/23, CBI.

13. Pugh 1995, p. 157; Weiss 1997, p. 66; T. V. Learson to R. W. Hubner, 1 November 1949, CBI 95, Box 3/11, CBI. On trips made by Ralph Hopkins in late 1949, see CBI 95, Box 1/11, CBI.

14. Hurd to Hastings, 8 November 1949, CBI 95, Box 6/23, CBI.

15. Weiss 1997, pp. 66, 67; On women and accounting, see Strom 1992. On women and mathematics, see Green and LaDuke 1988. On postwar retrenchment and gender-based segmentation, see Milkman 1987. On women's mathematical work during and immediately after the war, see Grier 1997 and Light 1999.

16. Weiss 1997, p. 67.

17. On salespersons and their identities, see Spears 1993.

18. S. W. Dunwell to Pendery, 2 March 1950; G. A. Roberts to Hurd, 14 April 1950 (both CBI 95, Box 6/9, CBI).

19. Edwards 1996, pp. 83–90.

20. Wolanski 1957. See also chapter 7 of the present work.

21. L. A. Ohlinger to T. J. Watson Jr., 26 June 1950; [Hurd], memo [c. Summer 1949]. Both CBI 95, Box 6/9, CBI. See also Ceruzzi 1997, pp. 5–12; Pugh 1995, pp. 152–155.

22. See various letters from Pendery to Hurd in CBI 95, Box 2/9, CBI.

23. Carolyn Goldstein's observations about home economists in light and power (1997, pp. 135, 136, 142–146) are relevant here. Home economists used their familiarity with domestic work to gain a special rapport with housewives. However, even though the utilities made a specific appeal to science during their better lighting campaigns, the particular distribution of knowledge ensured that science in this instance would be used more as an instrument of persuasion rather than of a mutual exchange. On sales representatives, image, and the art of social improvisation, see Spears 1993.

24. Evidence of how Pendery became enmeshed in a regional and national system of technical exchange can be found throughout his correspondence with Hurd (CBI 95, Boxes 2/9 and 2/10, CBI). See, for example, Pendery to Hurd, 11 October 1951 and 20 May 1952.

25. Pendery to Hurd, 10 August 1951, CBI 95, Box 2/5, CBI. See also Pendery to C. A. O'Malley, 8 January 1951, CBI 95, Box 2/9, CBI.

26. W. A. Johnson, "Call Report," 22 May 1951, CBI 95, Box 2/1. Helen Dunn to Pendery, 2 February 1951, CBI 95, Box 2/9; [Pendery], "Prospects for Technical Computing in Southern California," 25 August 1950, CBI 95, Box 2/8; Pendery to Hurd, summary report, 7 January 1951 [sic, 1952], CBI 95, Box 1/4, CBI.

27. Although Northern California was in District 10, Hubbard had to cover for Pendery after Pendery's reassignment to Los Angeles. Pendery was first assigned to District 10 because of his work at Boeing. See Hubbard's early correspondences to Hurd in CBI 95, Box 1/12, CBI.

28. Hurd's correspondence with Los Alamos can be found in CBI 95, Box 5/8, CBI.

29. Hubbard to Hurd, 9 February, 31 May and 30 July 1951, and 11 January 1952, CBI 95, Box 1/12, CBI.

30. See, for example, Hubbard to Hurd, 28 May and 30 July 1951, CBI 95, Box 1/12; Hubbard to Hurd, personal and confidential memo, 2 October 1951, CBI 95, Box 5/9; Hurd, "Visit to Los Alamos Scientific Laboratory, Oct. 29–30, 1951," CBI 95, Box 1/2, CBI. Zunz (1990, 165–169) also describes how McCormick sales agents established themselves as important members of their local communities.

31. Hammer to [Hurd], 24 July 1951, CBI 95, Box 5/9; Hubbard to Hurd, 31 May and 30 July 1951, CBI 95, Box 1/12; R. B. Thomas to Hurd, personal and confidential memo, 10 September 1951, CBI 95, Box 5/9; Hubbard to Hurd, 25 October 1951, CBI 95, Box 1/12, CBI.

32. Mason 1983.

33. Mason to Hurd, 29 June 1951, CBI 95, Box 2/5, CBI.

34. Mason to Hurd, 26 September 1951, CBI 95, Box 2/5, CBI.

35. See, for example, Mason to Hurd, call report, 24 June, 4 September and 26 September 1951, CBI 95, Box 2/5, CBI.

36. Mason 1983, p. 176; On correspondences from Petrie, Hunter, and Chancellor, see CBI 95, Box 2/11, 1/14 and 1/8, respectively, CBI.

37. On the value of a hard and soft sell, see Goldstein 1997, pp. 135, 158. Perhaps not surprisingly, the quote itself was from Pendery. Pendery to C. A. O'Malley, 5 March 1951, CBI 95, Box 2/9, CBI.

38. Pugh 1995, p. 154; [Hurd], memo [c. Summer 1949]; and L. A. Ohlinger to T. J. Watson Jr., 26 June 1950. Both CBI 95. Box 6/9, CBI.

39. Ohlinger to Watson Jr., 26 June 1950; Dunwell to Pendery, 2 March 1950. Both CBI 95, Box 6/9, CBI. The CPC's IBM 417 was a modified IBM 402, and is often referred to as the latter in the literature.

40. Pendery to Dunwell, 5 March 1950, CBI 95, Box 6/9, CBI.

41. Ohlinger to Watson Jr., 26 June 1950. The general character of IBM's engineering department is described in some detail on pp. 150–152 of Pugh 1995.

42. Dunwell to Pendery, 2 March 1950; Pendery to Dunwell, 5 March 1950; Hurd to O'Malley, 27 March 1950; and also Ohlinger to Watson Jr., 26 June 1950. All CBI 95, Box 6/9, CBI.

43. See, for instance, Ohlinger to Watson Jr., 26 June 1950; G. A. Roberts to Hurd, 29 June 1950; Hurd to Ohlinger, 7 July 1950; and Ohlinger to Hurd, 17 July 1950. All CBI 95, Box 6/9, CBI. Goldstein (1997, pp. 151, 152) writes more directly about the work technical intermediaries do in redefining the respective roles of the consumer and producer. On organizational responses to crisis, see Vaughan 1996).

44. Toben to [Hurd] [c. December 1949], CBI 95, Box 6/9; Hurd to C. E. Love and T. V. Learson, 27 February 1950, CBI 95, Box 1/3; Ralph Hopkins to L. H. LaMotte, 16 December 1949, CBI 95, Box 1/11, CBI; Pugh 1995, p. 326.

45. The principal accounts of these events are Pugh 1995, pp. 167–169; Hurd 1981; Birkenstock 1983; and Weiss 1997, pp. 67, 68.

46. For two of the more balanced accounts, see Pugh 1995, pp. 117–129, 145–182; and Campbell-Kelly and Aspray 1996, pp. 105–130.

47. Hurd and Birkenstock received an initial authorization to proceed with the project in December of 1950. This included preliminary engineering work, so that by February 1951, Hurd had a rough schematic of the Defense Calculator and an approximate price with which to approach prospective clients. See Pugh 1995, pp. 167–170.

48. Both men would claim later on that they had "independently" arrived at the same conclusion regarding the Defense Calculator design (ibid., p. 170; Weiss 1997, pp. 67, 68).

49. The records of Hurd's trips in early 1951 are filed with the correspondence pertaining to individual installations, in the files of individual representatives, and in CBI 95 Box 1/2.

50. Pugh 1985, p. 180; Campbell-Kelly and Aspray 1996, pp. 116, 117, 123, 124.

51. Pendery to O'Malley, 8 January and 26 February 1951, CBI 95, Box 2/9, CBI. At least several dozen reports on data reduction are scattered throughout the Cuthbert Hurd Papers.

52. Hubbard to Hurd, 9 February and 28 May 1951, CBI 95, Box 1/12; Pendery to Hurd, 1, 17 and 29 October 1951; and Hurd to Pendery, 9 October 1951. All CBI 95, Box 2/9, CBI.

53. IBM did have a magnetic drum computer in engineering development since 1948. See Pugh 1995, pp. 178–182; also Glaser 1986, pp. 30, 31. To provide some sense of Hurd's intelligence mechanism and the rapport he enjoyed with the users, Hurd obtained the CRC's specifications from at least three different sources.

54. Ralph Harris to John McPherson, 20 July 1952; H. R. Keith to Hurd, 22 July 1952. Both CBI 95, Box 6/9, CBI. There were larger politics within IBM about the magnetic drum computer, most of which had to do with the computer's potential in commercial data processing. Hurd was deeply involved in the matter, and his concerns at the time were certainly much broader than that of the IBM 701's sales. See Pugh 1995, pp. 178–182; the special issue on the IBM 650 of the *Annals of the History of Computing*, 8/1 (1986); J. B. Greene to Hurd, H. R. Keith, S. W. Dunwell, and J. H. Vincent, "Intermediate Computer," 12 August 1952, CBI 95, Box 1/4, CBI.

55. Hurd to R. Harris, 9 September 1952, CBI 95, Box 6/24, CBI.

56. Hurd to W. A. Johnson, 20 December 1951, CBI 95, Box 1/4, CBI.

57. T. V. Learson to H. R. Keith, 31 March 1952, CBI 95, Box 1/1, CBI. In fact, the tension here was amplified by the fact that the field men reported to Hurd on "technical" matters, but to Learson on all "administrative" matters. What was under contention was how these two words were defined.

58. Ibid.

59. Campbell-Kelly and Aspray 1996, pp. 151–153; Weiss 1997, pp. 67–70.

60. Weiss 1997, pp. 70, 71; Pugh 1995, pp. 232–237.

Notes to Chapter 7

1. Armer 1980/1956. The other general account on the history of Share is Mapstone and Bernstein 1980. "Share," although always styled as 'SHARE' by its participants, was not an acronym.

2. For other accounts on the history of programming, see Mahoney 1997, Light 1999, Ensmenger n.d., and Grier 2005. I am also indebted to Ensmenger for his observations about the degree of autonomy and authority eventually enjoyed by programmers. In addition to Leslie 1993 and other works previously cited, and for more direct background on the defense industries as relevant to this chapter, see Markusen et al. 1991 and Lotchin 1992.

3. For broad sociological and historical studies of the professions, see Abbott 1988, Haber 1991, and Perkin 1996. On occupations, see Abbott 1989, Trice 1993, and Barley and Kunda 2001.

4. On the role of technical intermediaries, see Goldstein 1997 and Zunz 1990.

5. Layton 1971.

6. My emphasis on professionalization as a process is based on a symbolic-interactionist approach to the study of work and occupations, for which the canonical work is Bucher and Strauss 1961.

7. Rosenberg 1997a/1979, p. 227. Another important, classic interactionist study here is Becker and Carper 1956. In contrast to Becker and Carper, who examine how individuals are socialized into an existing occupation, this chapter describes how the skills and norms of an occupation, as well as the means of their reproduction, come into existence in the first place.

8. This may have been the Fort Worth division of Convair. "Engineering Computing History" [1961], CBI 6, Folder 12, CBI; Pugh 1995, pp. 71, 72.

9. The Rand Corporation, a not-for-profit "think tank" in Santa Monica, was created by the Air Force in 1947. See Kaplan 1983.

10. [C. L. Davis], 21 May 1946; F. M. Sigafoese to Engineering Facilities, 5 September 1946; and P. E. Bisch to R. L. Schleicher and C. L. Davis, 24 August 1950, CBI 6, Folder 12, CBI. Fred Gruenberger, "A Short History of Digital Computing in Southern California," 11 November 1958, CBI 6, Folder 6. On the Aircraft War Production Council, see R. Ferguson 1996.

11. Bisch to Schleicher and Davis, 24 August 1950; "Engineering Computing History" [1961]; Gruenberger, p. 10; Mapstone and Bernstein 1980, pp. 367, 368. Northrop's contributions to the CPC were described in the previous chapter.

12. Armer 1980/1956, pp. 124, 125; Mapstone and Bernstein 1980, pp. 367, 368. For historical recollections on the development, and more important, use of the IBM 701, see *Annals of the History of* Computing 5 (1983), no. 2 (special issue). The full name for PACT was Project for the Advancement of Coding Techniques. Frank Engel to J. T. Ahlin and W. G. Bouricius, 8 April 1957, AC 323, Box 7/Share Correspondence, NMAH.

13. Although I often use the term "computing center directors" in this chapter, firms defined their most senior position for technical computing activities in different ways. Paul Armer was the head of the numerical analysis department at Rand, Lee Amaya the head of technical computing at Lockheed, and Jack Strong and Frank Wagener were the heads of central computing and engineering computing at North American.

14. Mapstone and Bernstein 1980, p. 363; Armer 1980/1956, pp. 124–126; Fletcher Jones, 19 August 1955, CBI 21, Box 1/1, CBI.

15. C. E. Hughes 1971b, p. 54.

16. I use the term "men" to indicate the tenor of the early meetings, although some women were involved in the organization.

17. Technically, the computer installations, not the firms or individuals, were the members of Share. Initially, any computer installation that had one or more IBM 704s installed or on order could join. This criterion was later expanded to include IBM 709s. I use the terms "participating firm" and "member installations" to distinguish between the interests of the corporation and those of individual computing center directors and their staff members.

18. "Synopses of Proceedings," in *Reference Manual for IBM 704,* pp. 10.01-01–10.01-20, CBI 21, Box 3/1, CBI. See also Mapstone and Bernstein 1980, pp. 370, 371; Armer 1980/1956, p. 126.

19. See, for example, W. A. Ramshaw to Fletcher Jones, 6 April 1956, CBI 21, Box 1/2, CBI.

20. "Share Programs" [August 1955], CBI 21, Box 1/1, CBI; "Synopses of Proceedings," pp. 10.01-01–10.01-02.

21. "Share Membership and What It Entails," 10 February 1956, AC 323, Box 6/Share Correspondence, NMAH.

22. I borrow here from sociological and ethnographic accounts of institutional development, where shared rules, norms, and practices, whether codified or not, are regarded as central to creating social identities. See Clarke 1991 and Barth 1969.

23. *Reference Manual for the IBM 704*; "Synopses of Proceedings"; Armer, "Discussion of SHARE Agenda, Committee Item #1–4, Tape Reliability," 3 May 1956 and related correspondence; North American Aviation, "704 Machine Usage Statistics," September 1956, CBI 21, Boxes 1/3, 1/4, and 1/7, CBI.

24. Jones to Engel, 21 September 1956; "Revision to Procedure for Electing Officers," 25 September 1956, CBI 21, Box 1/5, CBI; D. E. Hart to Engel, 20 November 1956, AC 323, Box 7/Share Correspondence, NMAH.

25. Jones to Engel, 21 September 1956; "Government of Share," 18 September 1956, CBI 21, Box 1/5; *Reference Manual for the IBM 704*; "Synopses of Proceedings"; Strong, "Indoctrination Committee to Share," April 1957, CBI 21, Box 1/11, CBI.

26. Jones to Engel, 21 September 1956; "Share Committees" [December 1956]; and John Greenstadt, "Report by the Chairman of the Share Mathematics Committee," 14 December 1956, CBI 21, Box 1/8, CBI.

27. Davis, "Tabular Calculations"; Bisch to Schleicher and Davis, 24 August 1950.

28. Strong, "Indoctrination Committee to Share"; "701 Symposium Notes," 17–18 August 1953, pp. 11–17, CBI 6, Folder 6, CBI.

29. Liston Tatum to Hurd, 2 September 1952, CBI 95, Box 1/4, CBI; "701 Symposium Notes."

30. Charles Davis, "A General Description of the Current and Immediately Projected Work," 19 October 1954, CBI 6, Folder 12, CBI; "Engineering Computing History" [1961.]

31. "701 Symposium Notes"; Armer 1980/1956, pp. 124, 125; Mapstone and Bernstein 1980, p. 365; "Rand Survey of Salary Structures," 1 October 1956, CBI 21, Box 1/9, CBI.

32. H. N. Cantrell to Fletcher Jones, 1 May 1956, CBI 21, Box 1/3; "Publicity on the Need for Computer Personnel," 16 January 1956, CBI 21, Box 1/2; "Debugging," n.d., CBI 21, Box 1/4, CBI.

33. Armer 1980/1956, pp. 124, 125; Francis Wagner to members, "PACT IA Compiler System," 31 July 1957, CBI 21, Box 1/11, CBI.

34. Fletcher Jones, 9 August 1955, CBI 21, Box 1/1, CBI.

35. H. S. Bright to B. B. Fowler, 4 May 1957, CBI 42, Box 3. 19, CBI.

36. *Reference Manual for the IBM 704*; "Synopses of Proceedings."

37. "Share Programs"; "Share Classifications," 17 November 1956, CBI 21, Box 1/7, CBI. "Utilities" referred to common programs such as those for input-output, file maintenance, and binary-to-decimal conversion. This term preceded "systems programming."

38. Francis Wagner to Richard King et al., 11 March 1958, CBI 6, Folder 12; 709 System Subcommittee to members of Share, "Progress Report on Supervisory Control," 13 September 1957, CBI 21, Box 1/13, CBI.

39. Strong, "Indoctrination Committee"; "Debugging"; Wagner to King. et al., 11 March 1958.

40. This analysis draws upon Bucher and Strauss's (1961) observation that the relative status of different specializations within a profession is constantly changing in response to extensions of knowledge.

41. On the ambiguous status of early programmers and its relation to job segmentation within the workplace, see Light 1999.

42. "Obligations of Share Membership" and "Share Procedural Standards," in *Reference Manual for the IBM 704*, 01.03-01 and 03.2-01; On loyalty to the corporation, see Layton 1971.

43. Arthur Strang to H. N. Cantrell and Allen Keller, 19 June 1956, CBI 21, Box 1/3. Albert Pitcher to M. J. Rodgers, 25 September 1956, CBI 21, Box 1/9, CBI.

44. T. M. Kerr to Frank Engel, 1 November 1956, CBI 21, Box 1/7; "Share Programs"; Fletcher Jones, letter regarding nonmember distribution [January 1956], CBI 21, Box 1/2, CBI.

45. See, for example, H. S. Bright to Frank Wagner, 14 February 1958, AC 323, Box 7/Correspondence, NMAH; "Meeting to Form Agenda Items," minutes, 20 November 1956, CBI 21, Box 1/7, CBI.

46. Kerr to Engel, 1 November 1956; "Share Membership and What It Entails."

47. See Akera 1998, pp. 437–443.

48. Walter Bauer, "The Informatics Story," 15 November 1962, CBI 6, Folder 10. Tensions could lead to the wholesale departure of the central computing staff, as occurred at North American Aviation between 1961 and 1962. See "Engineering Computing History" [1961] and Bob Patrick, "This is a backbone" (n.d.), CBI 6, Folder 12, CBI.

49. "Synopses of Proceedings," pp. 10.01-02–10.01-04; Fletcher Jones to Charles DeCarlo, 29 August 1955; Armer to DeCarlo, 7 September 1955; and DeCarlo to Jones, 12 September 1955, CBI 21, Box 1/1, CBI.

50. "Programming Library," *IBM Research News,* May 1963, AC 323, Box 7/Share Correspondence, NMAH; L. Joanne Edson to J. Hunter White, 16 November 1956, CBI 21, Box 1/7; H. S. Bright to Executive Board of Share, 15 December 1958, CBI 21, Box 1/14, CBI.

51. D. L. Shell to members, 7 January 1957, AC 323, Box 7/Share Correspondence, NMAH; "Verbatim Transcript of the 9th Share Meeting," 1 October 1957, p. 47, CBI 21, Box 3/13, CBI. This arrangement involved neither a contract nor direct subventions from IBM.

52. Shell to members, 7 January 1957; Frank Beckman, "Minutes of the Meeting on February 21, 1957"; and [Irwin Greenwald to Paul Armer], "709 Meetings," handwritten notes, 1 March 1957, AC 323, Box 7/Share Correspondence, NMAH.

53. Lewis Ondis, "Share Routines Received Up To April 1, 1956," n. d, CBI 21, Box 1/2; "Proposed Share Questionnaire" [c. September 1956]; Fletcher Jones to Frank Engel, 21 September 1956, CBI 21, Box 1/5, CBI; and "Verbatim Transcript of the 9th Share Meeting," pp. 8–10.

54. "Share Subcommittee Report," 1 October 1956, CBI 21, Box 1/6, CBI.

55. 709 System Subcommittee to Members of Share, 13 September 1957, CBI 21, Box 1/13; *Share 709 System (SOS) Manual,* circa 1958, with sections revised periodically, CBI 21, Box 3/8 and 3/9; "Verbatim Transcript of the 9th Share Meeting," pp. 30–32, 120–125.

56. Charles Swift, "The MockDonald Multiphase System for the 709," 9 September 1957, CBI 21, Box 1/13; "Introduction to and the Philosophy of Modify and Load," in *Share 709 System (SOS) Manual;* F. V. Wagner to T. W. Alexander, 8 April 1960, CBI 6, Folder 22.

57. "Share Subcommittee Report"; James Fishman to Donn Parker, "Report of Share Subcommittee on 709 Fortran Operating System," 17 March 1959, CBI 6, Folder 6, CBI.

58. "Engineering Computing History" [1961]; F. V. Wagner to T. W. Alexander, 8 April 1960, CBI 6, Folder 22, CBI.

59. Goldstein 1997.

60. The new systems engineering organizations, science advisory groups, and other hybrid organizations described by T. Hughes (1998), Leslie (1993), Kaplan (1983), and others were also ad hoc arrangements established to accommodate competing interests over how to conduct Cold War research.

61. The significance of multiple allegiances is described more formally in sociological accounts of how "people typically participate in a number of social worlds simultaneously" (Clarke 1991, p. 132).

62. Share still exists—see www.share.org. In addition to a software exchange, one of Share's principal roles continued to be that of providing IBM with feedback about its software and hardware.

63. Paul Armer, joined by others from Share and from the Univac Scientific Exchange, first tried to transform the Association for Computing Machinery's charter to encompass the computer programmer's point of view. Upon failing to change ACM to his satisfaction, Armer went on to help establish the American Federation of Information Processing Societies. See Akera 1998, pp. 567–602.

Notes to Chapter 8

1. Rosenberg 1997a/1979, p. 228.

2. Redmond and Smith 2000, pp. 67, 68; Committee on Machine Methods of Computation, minutes (hereafter CMMC minutes), 22 November 1950, AC 62, Box 1/CMMC, MIT.

3. Morse 1977.

4. CMMC minutes, 22 November 1950.

5. CMMC minutes, 3 January and 7 February 1951.

6. This was discussed in a series of meetings between November and December 1950, see CMMC minutes.

7. CMMC minutes, 21 March 1951, 9 April 1952 and 5 November 1952.

8. Morse to Carl Floe [c. December 1954], AC 62, Box 1/Computation Center, MIT.

9. CMMC minutes, 21 March 1951.

10. Delgarno to Morse, 30 September 1954, AC 62, Box 1/Machine Project.

11. Morse to Floe [December 1954]; Corbato to Delgarno, 11 October 1954, AC 62, Box 1/Machine Project, MIT; Morse 1977, p. 308.

12. CMMC minutes; Morse to Joachim Weyl, 17 January 1955, AC 62, Box 1/Computation Center, MIT.

13. This was discussed between October 1953 and January 1954. CMMC minutes.

14. Morse to Floe [December 1954].

15. ICNAMC, "Requirements for Electronic Computing . . ." [1 May 1954], AC 62, Box 1/Computation Center, MIT; CMMC minutes, 11 February 1954.

16. ICNAMC, "Requirements."

17. CMMC Minutes, 7 January 1954, 3 November 1954, 1 December 1954; F. M. Verzuh, "MIT Digital Computing Equipment Requirements," 1 December 1954, AC 62, Box 1/Computation Center, MIT.

18. C. C. Hurd to T. J. Watson Jr., 7 April 1955. C. Hurd Papers, Watson, T. J. Jr. & Sr., 1955–56, Box 18/6, IBM.

19. [Verzuh], "Operational Aspects of Present MIT Computing Facilities" [1 March 1954], AC 62, Box 1/Computation Center; Verzuh to Morse, 17 November 1953, AC 62, Box 1/Statistical Services Center, MIT.

20. Verzuh, "Minutes of Meeting with Dr. Hurd," 17 December 1954, AC 62, Box 2/Computation Center—Conferences, MIT.

21. Morse to Stratton, "High Speed Computing Equipment for the Institute," 28 December 1954; Stratton to Morse, 25 January 1955; J. R. Killian to Thomas Watson Jr., 2 June 1955; Morse to Killian, 25 July 1955. All AC 62, Box 1/Computation Center, MIT.

22. Press release, 11 December 1955; "Operating Agreement," 20 January 1956; Morse to Stratton, 28 December 1954. All AC 62, Box 1/Computation Center, MIT.

23. CMMC minutes, 25 January and 3 April 1956; Corbato to IBM Research Assistants and Associates, 17 September 1956, AC 62, Box 2/Morse, Philip, MIT.

24. "IBM to Install Giant Computer," *The Tech*, 13 January 1956.

25. E. Ferguson 1992, pp. 159–161; ASEE, Committee on Evaluation of Engineering Education (CEEE), "Preliminary Report of the Committee on Evaluation of Engineering Education," 10 October 1953, AC 12, Box 2/ASEE Committee, MIT. See also Seely 1993.

26. Ad Hoc Committee on Enrollment, "Report to the President," 16 January 1956; Brown, "Report of the Committee on Enrollment," 17 February 1956, AC 12, Box 5/Committee on Engineering Education, MIT.

27. Soderberg to E. R. Gilliland, 24 October 1956; Gilliland, "Informal Oral Report," 26 October 1957, AC 12, Box 5/Committee on Engineering Education, MIT.

28. Brown to R. L. Bisplinghoff et al., 26 November 1957, AC 12, Box 2/ASEE Committee; Minutes of Engineering Education Committee, 3 February 1958; "Meeting with Stratton and Soderberg," 15 March 1958, AC 12, Box 2/ASEE Committee, MIT.

29. "IBM to Install Giant Computer in Compton Laboratories," *The Tech*, 13 January 1956; "Student Riot Rocks Campus," 5 March 1957; "47 Students Arrested in Tuesday Fracas," 6 May 1957.

30. ASEE, CEEE, "A Summary of Institutional Committee Comments," 28 August 1953, AC 12, Box 2/ASEE Committee, MIT.

31. The two specific references appear in the March and June 1956 issues of *Voo Doo*, which can be found in the MIT Humanities Library.

32. "Laboratories, Lecture Hall," *The Tech*, 26 April 1955; "Six Billion Volt Accelerator," 24 April 1956; "Ban At Least Some H-Bomb Tests, Says MIT Physicists," 19 October 1956; "Killian Becomes Science Chief," 8 November 1957.

33. "Harrison Sees Apathy as 'Real Problem,'" *The Tech*, 10 December 1957.

34. "MIT Graduate—Real Professional Man or Merely Technician?" *The Tech*, 18 January 1958, p. 3.

35. "Engineer Not 'Just Another Kind of Scientist,'" *The Tech*, 4 March 1958.

36. "Compton Lab '704' Will Compute Orbit of Satellite Sometime Today," *The Tech*, 8 October 1957; "Moon Orbit Search Keeps On: Insufficient Data for 704," 11 October 1957; "Institute Leading Computer Center," 13 January 1956, pp. 5, 6.

37. Morse to Stratton, "High Speed Computing Equipment for the Institute"; "Mueller Calls Teaching Here Inadequate," *The Tech*, 18 March 1958.

38. Institute of Science and Technology (IST), "Willow Run Laboratories," March 1970, VP-Research, Box 12/[Willow Run], BHL.

39. "Midac," 15 November 1960, UM-CC, Box 1/Historical Files, BHL.

40. IST, "Willow Run Laboratories"; Edwards 1996, pp. 96, 97.

41. IST, "Willow Run Laboratories."

42. "Computer Research Facilities at the University," *ERI Newsletter*, August 1955, p. 2; "John W. Carr, Computing Specialist," *ERI News*, December 1955, p. 1, UM-CC, Box 1/Historical Files, BHL.

43. "Midac," 15 November 1960; "John W. Carr," *ERI News*.

44. "Computing Center" [1972], UM-CC, Box 1/Historical Files, BHL; "Midac," 15 November 1960.

45. DeCarlo to Stirton, 9 September 1957, UM-CC, Box 1/Historical Files, BHL.

46. DeCarlo to Stirton, 11 September 1957, UM-CC, Box 1/Historical Files; "Annual Report of the Statistical and Computing Laboratory" [1959], p. 3, UM-CC, Box 1/Annual Reports, BHL.

47. "Midac," 15 November 1960.

48. Computing Center, Executive Committee, Minutes (hereafter ExC minutes), 15 July 1959, VP-Research, Box 4/Executive Committee for the Computer Center, BHL. Bartels, "Educational Allowance Supplement" [May 1960]; C. C. Craig to B. W. Arden et al., "Meeting Held 27 February 1959," VP-Research, Box 4/Computing Center, BHL.

49. See the Executive Committee minutes from this period. Also Bartels to R. A. Sawyer, 30 March 1960, VP-Research, Box 4/Computing Center, BHL.

50. ExC minutes; Bartels to Sawyer, 28 April 1960, VP-Research, Box 4/Computing Center, BHL.

51. ASEE, "Report of the Committee on Engineering Analysis and Design" [1957], AC 12, Box 2/ASEE Committee, MIT; College of Engineering, "Upgrading Instruction in Engineering Education," 30 June 1959, UM-CC, Box 6/Ford Grant, BHL; "Annual Report" [1959].

52. "Ford Foundation Project," 21 October 1959, UM-CC, Box 5/Ford Grant.

53. Ibid.; "Upgrading Instruction," 30 June 1959; "Conference of Project on Computers in Engineering Education" [August 1961], VP-Research, Box 4/Executive Committee for the Computer Center, BHL.

54. "Use of Computers in Engineering Education," *Research News*, 1 November 1961, pp. 4, 5, UM-CC, Box 1/Historical Files, BHL; Katz et al., "The Use of Computers in Engineering Education: Final Report," 1963, UMich. Eng., Box 62/Project on the Use of Computers, BHL.

55. Bartels to Sawyer, 25 July 1961, UM-CC, Box 1/Annual Reports, BHL.

56. That is to say, in the original language in which they were programmed, as opposed to the compiled, executable version of the program.

57. Bernard Galler, transcript of an oral history interview by Enid Galler [1980]. Tape 1, pp. 11, 17. Accessioned but unprocessed item, BHL.

58. "Midac," 15 November 1960.

59. Galler interview, Tape 1, pp. 13–17.

60. Ibid., pp. 2–4.

61. "Annual Report" [1959], pp. 1–4; Galler interview, Tape 1, pp. 15–17.

62. Galler Interview, Tape 1, p. 18.

63. Arden, "Computing Center," *Michigan Technic*, February 1963, p. 47, UM-CC, box 1/Historical Files, BHL.

64. Galler interview, Tape 1, p. 20.

65. Arden, "Computing Center," pp. 47, 50.

66. Galler interview, Tape 1, p. 18.

67. Galler interview, Tape 1, p. 21 and Tape 3, p. 10; Donald Thackrey to Doug van Houweling, 29 May 1985, UM-CC Box 1/Historical Files, BHL.

68. "Annual Report" [1959]; "President's Report," 1960, UM-CC, Box 1/Historical Files, BHL; ExC minutes, 10 July 1962, IBM's grants for programming research from 1962 to 1966 totaled $500,000.

Notes to Chapter 9

1. Norberg and O'Neill 1996, pp. 5–12, 28.

2. This was separate from the work conducted under the Division of Defense Laboratories, which in 1961 included $35 million for Lincoln Laboratories, $21 million for Charles Draper's Instrumentation Laboratory, and $2.1 million for a military operations research group. These facilities had their own computing facilities, and therefore are not included in this discussion.

3. For the general records of the DIC, see F. Leroy Foster, "Sponsored Research at M. I. T., 1900–1968," unpublished manuscript, MIT.

4. Verzuh to P. V. Cusick and C. F. Floe, "Summary Report," 9 April 1959, in Foster, "Sponsored Research at M. I. T.," volume IV/I.

5. Norberg and O'Neill 1996, pp. 79, 82–83.

6. Committee on Computation, "MIT Computation—Present and Future" [April 1961], p. 16, AC 268, Box 13/33, MIT.

7. Norberg and O'Neill 1996, pp. 83–86; Morse to Corbato, 1 July 1960, MC 75, Box 3/Computation Center.

8. M. Loren Bullock to John Slater, 27 June 1960, AC 62, Box 2/[B]; S. M. Simpson to R. R. Shrock, 21 January 1960; Howard Johnson to Robert Atkinson, 23 September 1960. Both AC 62, Box 2/Computation Center, MIT; Committee on Computation, "MIT Computation," pp. 21, 22.

9. Norberg and O'Neill 1996, p. 84; Long Range Computation Study Group, "First Organizational Meeting," minutes, 21 April 1960; Robert Fano, Philip Morse and Albert Hill to Carl Floe, 30 March 1961. All MC 74, Box 1/Hill Committee, MIT.

10. Teager to Fano et al., 12 September 1960, AC 62, Box 2/IBM Correspondence, MIT.

11. [Teager], "MIT Computation—Present and Future," draft report [15 March 1961], esp. pp. 24 and 71–81, MC 75, Box 3/MIT Computation, MIT.

12. Ibid., pp. 5 and 8; Long Range Computation Study Group, minutes, 21 April 1960.

13. Fano, Morse, and Hill to Floe, 30 March 1961; Hill to Floe, 26 April 1961, AC 62, Box 2/Long Range Computation Study, MIT.

14. Norberg and O'Neill 1996, p. 85; MIT, "Advanced Computation System" [August 1961]; Working Group to Computer Committee, 31 August 1961. MC 61, Box 1/Hill Committee, MIT.

15. Norberg and O'Neill 1996, pp. 27–29, 89, 94–96.

16. Institute of Science and Technology (IST), "Willow Run Laboratories," March 1970, VP-Research, Box 12/[Willow Run]; Computing Center, Executive Committee, minutes (hereafter ExC minutes), 15 July 1959 through 5 July 1961.

17. E. A. Cumminskey to F. E. Oliver, 13 October 1960; H. J. Gibson to Bartels, 7 march 1963; Bartels to W. K. Pierpont, 28 June 1963, VP-Research, Box 4/Computing Center, BHL.

18. Bartels to Executive Committee, 31 October 1963; Bartels to Sawyer, "Computing Center Budget for 1964–1965," 13 January 1964; "Financial Status of Computing Center," 23 April 1965. All UM-CC, Box 3/Administrative Committees, BHL.

19. Owens 1994, pp. 555–558.

20. McCormack and Vincent Fulmer, "Federal Sponsorship of University Research" [1960], pp. 3-35–3-40, AC 35, Box 2/F, MIT.

21. [Bartels], "A proposal for Supporting a Centralized Computing Facility" [9 April 1964], UM-CC, Box 3/Administrative Committees, BHL.

22. Heyns and Sawyer to distribution, 5 June 1964; Ad Hoc Computing Advisory Committee, minutes, 14 July 1964. Both UCCPU, Box 1/Ad Hoc Computing Advisory Committee.

23. "Survey Assignments to members of Ad Hoc Computing Advisory Committee," n.d., UCCPU, Box 1/Ad Hoc Computing Advisory Committee, BHL.

24. Ad Hoc Computer Committee, "A Report on Machine Computation at the University of Michigan," 7 December 1964, VP-Research, Box 4/Ad Hoc Computing Advisory Committee, BHL.

25. Ibid., pp. 21, 22.

26. Ibid.

27. Ad Hoc Committee to Heyns and Norman, 7 December 1964; Heyns and Norman to Members, 8 February, 1965, VP-Research, Box 4/Ad Hoc Computing Advisory Committee, BHL.

28. Robert White, "The Next Ten years of Research in Michigan" [1959]; and IST, "Progress Report to the Governor, November 1960," n.d. Both VP-Development, Box 5/IST, BHL.

29. Committee on Computer Policies, minutes, 23 February 1965, UCCPU, Box 3/Minutes; "Computing Center: Appropriations, Earnings, Expenditures," n. d, UM-CC, Box 1/Annual reports, BHL.

30. Michael Barnett, "MIT Cooperative Computing Laboratory," 7 May 1962, AC 134, Box 13/9, MIT. The other requests for decentralized facilities can be found in MC 75, Box 3/Computation Center (1 of 4), MIT.

31. Morse, "Proposal to the National Science Foundation," 12 March 1963, p. 14, MC 75, Box 3/Computation Center (3), MIT.

32. Norberg and O'Neill 1996, pp. 98–102; Fano to Stratton, 29 June 1964, "Commentary on IBM Proposal," n. d, AC 268, Box 14/31, MIT.

33. See ExC minutes from March 1965; "Report on the Computing Center's Evaluation," n.d., filed with ExC minutes; Press release, 11 February 1966, VP-Research, Box 4/Computing Center, BHL.

34. Arden to Bartels, 9 August 1965, UM-CC, Box 5/MAC Project Controversy; Committee on Computer Policy, minutes, 6 July 1965, UM-CC, Box 3/Minutes. A direct record of these technical meetings may be found in UM-CC, Box 9/IBM 360 System, BHL.

35. Interview with Bernard Galler, *Computing Center Newsletter*, 12 November 1980, 1, UM-CC, Box 1/Historical File, BHL.

36. Morse, "Position Paper on the Future of the MIT Computation Center," 15 August 1964, MC 75, Box 3/Drafts of Position Paper; Stratton to Members of the Faculty, 13 August 1965, AC 268, Box 14/32, MIT.

37. Brown to Committee on Information Processing, 13 December 1965; "Master Agreement for Research Program," 4 August 1966, MC 75, Box 3/Computation Center. Both AC 268, Box 13/37; Brown to Stratton, 12 August 1965, AC 134, Box 13/9; Robert Scott, minutes, 19 July 1965, MC 75, Box 1/Committee on Computation, MIT.

38. Norberg and O'Neill 1996, p. 98; "Preliminary Proposal to the NSF," 27 March 1964. MC 74, Box 3/Computation Center (2), MIT.

39. R. M. Fano, proposal to ARPA, 20 December 1966, esp. pp. 2, 3, 13, and 14, AC 268, Box 13/36, MIT.

40. Brown, "First Progress Report to the International Business Machines Corporation," 11 May 1967, AC 268, Box 14/31, MIT.

41. The group's difficulties are documented in "Programming Staff Notes," various dates, in AC 282, Box 2/CTSS Programming Staff Notes, MIT. The difficulties described here are discussed in notes 32, 34, and 63.

42. Norberg and O'Neill 1996, p. 111; University Committee on Computer Policy and Utilization, minutes (hereafter UCCPU minutes), 12 September 1966, UCCPU, Box 3/Minutes, BHL. There were other reasons for IBM's difficulties with TSS. In a recent interview with the author (transcript in author's possession), Bernard Galler suggests that IBM also tried to create a system that accommodated all the "important" customers the group at Michigan helped to assemble, leading to a very complicated set of specifications that was nearly impossible to implement.

43. UCCPU minutes, 19 January 1967; M. Alexander, "MTS" [1972], UM-CC, Box 1/Historical Files, BHL.

44. UCCPU minutes, 7 June and 20 November 1967, 26 March 1968; Arden, "The IBM 360/67 Development," UM-CC, Box 1/Historical Files, BHL.

45. Bernard Galler, transcript of an oral history interview by Enid Galler [1980]. Tape 1, pp. 2–26 (accessioned but unprocessed item, BHL); Arden to Bartels, 9 August 1965.

46. Arden to Bartels, 9 August 1965. A priority dispute did cause tension between the two groups. Fano to Bartels, 29 July 1965; and Bartels to Fano, 10 August 1965, UM-CC, Box 5/MAC Project Controversy, BHL.

47. UCCPU minutes, 20 November 1967 and 9 April 1968.

48. Committee on Computation, "MIT Computation—Present and Future," p. 30; D. C. Carroll, "The Educational Use of Computers. Resource Policy Study," 10 May 1969, AC 268, Box 14/19, MIT.

49. UCCPU minutes, 9 April 1968.

50. Mills to Brown, 8 May 1968, AC 268, Box 14/31, MIT.

51. Galler interview, Tape 1, pp. 6, 7.

52. Fano, a proposal to ARPA, 20 December 1966. As described by Norberg and O'Neill (1996, p. 110), the Multics system was made available for general use only in 1969, and efforts to improve the system remained an official part of Project MAC at least through 1973.

Notes to Conclusion

1. This is a term coined by Karin Knorr-Cetina (1999) to describe the distinct social and material cultures of knowledge production within the sciences. There is a close analogy between this phrase and the notion of "technological frame" introduced by Wiebe Bijker to describe specific engineering subcultures.

2. Again, the reference here is to the notion of "intercalation" presented by Peter Galison in chapter 9 of *Image and Logic* (1997).

3. Fujimura 1992.

4. Again, this is based on Rosenberg's (1997a/1979) arguments, especially on pp. 244 and 245.

5. Glaser 1967; Clarke 2005.

6. Hacking 1999, pp. 49, 50.

7. Sterne and Leach 2005, p. 190.

References

Published Primary Sources

Armer, Paul. 1980/1956. "SHARE—A Eulogy to Cooperative Effort." Reprinted in *Annals of the History of Computing* 2: 122–129.

Burks, Arthur, Hermann Goldstine, and John von Neumann. 1947. *Preliminary Discussion of the Logical Design of an Electronic Computing Instrument*, Part I, Volume 1, *Report on the Mathematical and Logical Aspects of an Electronic Computing Instrument*, second edition. Excerpt reprinted in *The Origins of Digital Computers, Selected Papers*, third edition, ed. B. Randell. Springer-Verlag, 1982.

Bush, Vannevar. 1945. "As We May Think." *Atlantic Monthly*, July: 101–108.

Compton, Karl. 1944. "Report of the President." *MIT Bulletin* 81, no. 1: 1–28.

Engineering Research Associates. 1983/1950. *High-Speed Computing Devices*, reprint edition. Tomash.

Forrester, Jay. 1961. *Industrial Dynamics*. MIT Press.

Forrester, Jay. 1969. *Urban Dynamics*. MIT Press.

Forrester, Jay. 1971. *World Dynamics*. Wright-Allen.

Hartree, Douglas. 1949. *Calculating Instruments and Machines*. University of Illinois Press.

Hurd, Cuthbert. 1985. "A Note on Early Monte Carlo Computations and Scientific Meetings." *Annals of the History of Computing* 7: 141–155.

Mauchly, John. 1942. "The Use of High Speed Vacuum Tube Devices for Computing." Reprinted in *The Origins of Digital Computers, Selected Papers*, third edition, ed. B. Randell. Springer-Verlag, 1982.

Mauchly, John. 1982. "Fireside Chat, 13 November 1973." *Annals of the History of Computing* 4: 248–254.

Proceedings of a Symposium on Large-Scale Digital Calculating Machinery. 1985/1948. MIT Press reprint edition.

Randell, Brian, ed. 1982. *The Origins of Digital Computers, Selected Papers*, third edition. Springer-Verlag.

Wolanski, H. S. 1957. "Applications of Computing in the Aircraft Industry." In *The Computing Laboratory in the University*, ed. P. Hammer. University of Wisconsin Press.

Secondary Sources

Abbott, Andrew. 1988. *The System of Professions: An Essay in the Division of Expert Labor.* University of Chicago Press.

Abbott, Andrew. 1989. "The New Occupational Structure: What Are the Questions?" *Work and Occupations* 16: 273–291.

Agar, Jon. 2003. *The Government Machine: A Revolutionary History of the Computer.* MIT Press.

Aitken, Hugh. 1985. *The Continuous Wave: Technology and American Radio, 1900–1932.* Princeton University Press.

Akera, Atsushi. 1998. Calculating a Natural World: Scientists, Engineers and Computers in the United States, 1937–1968. PhD dissertation, University of Pennsylvania.

Akera, Atsushi. 2000. "Engineers or Managers? The Systems Analysis of Electronic Data Processing in the Federal Bureaucracy." In *Systems, Experts and Computers*, ed. Agatha Hughes and Thomas Hughes. MIT Press.

Akera, Atsushi. 2005. "What Is Social about Social Construction? Understanding Internal Fissures in Constructivist Accounts of Technology." *Social Epistemology* 19, no. 3.

Akera, Atsushi. Forthcoming. "Constructing a Representation for an Ecology of Knowledge: Methodological Advances in the Integration of Knowledge and Its Social Context." *Social Studies of Science.*

Alder, Ken. 1997. *Engineering the Revolution: Arms and Enlightenment in France, 1763–1815.* Princeton University Press.

Anderson, Gregory. ed., 1988. *The White Blouse Revolution: Female Office Workers since 1870.* Manchester University Press.

Anderson, Warwick. 2002. "Postcolonial Technoscience." *Social Studies of Science* 32: 643–658.

Ashman, Keith, and Philip Baringer, eds. 2001. *After the Science Wars.* Routledge.

Aspray, William. 1985. "Introduction." In *Proceedings of a Symposium on Large-Scale Digital Calculating Machinery*. MIT Press.

Aspray, William. 1988. "The Emergence of Princeton as a World Center for Mathematical Research, 1896–1939." In *History and Philosophy of Modern Mathematics*, ed. W. Aspray and P. Kitcher. University of Minnesota Press.

Aspray, William. 1990. *John von Neumann and the Origins of Modern Computing*. MIT Press.

Austrian, Geoffrey. 1982. *Herman Hollerith, Forgotten Giant of Information Processing*. Columbia University Press.

Barley, Stephen, and Gideon Kunda. 2001. "Bringing Work Back In." *Organization Science* 12, no. 1: 76–95.

Barth, Fredrik, ed. 1969. *Ethnic Groups and Boundaries: The Social Organization of Culture Difference*. Little, Brown.

Becker, Howard. 1982. *Art Worlds*. University of California Press.

Becker, Howard, and James Carper. 1956. "The Development of an Identification with an Occupation." *American Journal of Sociology* 61: 289–298.

Benson, Susan Porter. 1986. *Counter Cultures: Saleswomen, Managers and Customers in American Department Stores, 1890–1940*. University of Illinois Press.

Berger, Peter, and Thomas Luckmann. 1966. *The Social Construction of Reality: A Treatise in the Sociology of Knowledge*. Doubleday.

Bijker, Wiebe. 1995. *Of Bicycles, Bakelites and Bulbs: Towards a Theory of Sociotechnical Change*. MIT Press.

Bijker, Wiebe, and John Law, eds. 1992. *Shaping Technology/Building Society: Studies in Sociotechnical Change*. MIT Press.

Bijker, Wiebe, Thomas Hughes, and Trevor Pinch, eds. 1987. *The Social Construction of Technological Systems: New Directions in the Sociology and History of Technology*. MIT Press.

Birkenstock, James. 1983. "Preliminary Planning for the 701." *Annals of the History of Computing* 5: 112–114.

Bloor, David. 1991/1976. *Knowledge and Social Imagery,* second edition. University of Chicago Press.

Bourdieu, Pierre. 1977. *Outline of a Theory of Practice*. Cambridge University Press.

Bourdieu, Pierre. 1984. *Distinction: A Social Critique of the Judgment of Taste*. Harvard University Press.

Brennan, Jean Ford. 1971. *The IBM Watson Laboratory at Columbia University*. IBM.

Bromley, Allan. 1990. "Analog Computing Devices." In *Computing before Computers*, ed. W Aspray. Iowa State University Press.

Bryant, Christopher, and David Jary. 1991. "Introduction: Coming to Terms with Anthony Giddens." In *Giddens' Theory of Structuration*, ed. C. Bryant and D. Jary. Routledge.

Bucher, Rue, and Anselm Strauss. 1961. "Professions in Process." *American Journal of Sociology* 66: 325–334.

Buchwald, Jed. 1985. *From Maxwell to Microphysics: Aspects of Electromagnetic Theory in the Last Quarter of the Nineteenth Century*. University of Chicago Press.

Burke, Colin. 1994. *Information and Secrecy: Vannevar Bush, Ultra, and the Other Memex*. Scarecrow.

Burks, Alice, and Arthur Burks. 1988. *The First Electronic Computer: The Atanasoff Story*. University of Michigan Press.

Burks, Arthur, and Alice Burks. 1981. "The ENIAC: First General Purpose Electronic Computer." *Annals of the History of Computing* 3: 310–389.

Byerly, Radford, and Roger Pielke. 1995. "The Changing Ecology of United States Science." *Science* 269: 1531–1532.

Campbell-Kelly, Martin, and William Aspray. 1996. *Computer: A History of the Information Machine*. Basic Books.

Carlson, W. Bernard. 1988. "Academic Entrepreneurship and Engineering Education: Dugald C. Jackson and the MIT-GE Cooperative Engineering Course, 1907–1932." *Technology and Culture* 29: 536–567.

Carlson, W. Bernard. 1990. "Understanding Invention as a Cognitive Process: The Case of Thomas Edison and Early Motion Pictures, 1888–1891." *Social Studies of Science* 20: 387–430.

Carlson, W. Bernard. 2000. "Invention and Evolution: The Case of Edison's Sketches of the Telephone." In *Technological Innovation as an Evolutionary Process*, ed. J. Ziman. Cambridge University Press.

Ceruzzi, Paul. 1997. "Crossing the Divide: Architectural Issues and the Emergence of the Stored Program Concept, 1935–1955." *IEEE Annals of the History of Computing* 19: 5–12.

Chandler, David. 2001. *Semiotics: The Basics*. Routledge.

Clarke, Adele. 1991. "Social Worlds/Arenas Theory as Organizational Theory." In *Social Organization and Social Process*, ed. D. Maines. Aldine de Gruyter.

Clarke, Adele. 2005. *Situational Analysis: Grounded Theory after the Postmodern Turn.* Sage.

Clarke, Adele, and Joan Fujimura, eds. 1992. *The Right Tools for the Job: At Work in Twentieth-Century Life Sciences.* Princeton University Press.

Cochrane, Rexmond. 1976. *Measures for Progress: A History of the National Bureau of Standards.* Arno.

Cohen, I. Bernard. 1999. *Howard Aiken: Portrait of a Computer Pioneer.* MIT Press.

Cortada, James. 1993. *IBM, NCR, Burroughs and Remington Rand and the Industry They Created, 1865–1956.* Princeton University Press.

Costello, John. 1996. "As the Twig Is Bent: The Early Life of John Mauchly." *IEEE Annals of the History of Computing* 18: 45–50.

Croarken, Mary. 1990. *Early Scientific Computing in Britain.* Clarendon.

Culick, Fred. 1983. *Guggenheim Aeronautical Laboratory at the California Institute of Technology: The First Fifty Years.* San Francisco Press.

Daston, Lorraine. 1994. "Enlightenment Calculations." *Critical Inquiry* 21: 182–202.

Dohrn, Susanne. 1988. "Pioneers in a Dead-End Profession." In *The White Blouse Revolution*, ed. G. Anderson. Manchester University Press.

Douglas, Susan. 1987. *Inventing American Broadcasting, 1899–1922.* Johns Hopkins University Press.

Dupree, A. Hunter. 1972/1970. "The Great Insaturation of 1940: The Organization of Scientific Research for War." In *The Twentieth-Century Sciences*, ed. G. Holton. Norton.

Dupree, A. Hunter. 1986/1957. *Science in the Federal Government: A History of Policies and Activities.* Johns Hopkins University Press.

Edwards, Paul. 1996. *The Closed World: Computers and the Politics of Discourse in Cold War America.* MIT Press.

Ensmenger, Nathan. n.d. "Building Castles in the Air": The Software Crisis and the Art of Programming, 1945–1968. Manuscript.

Etzkowitz, Henry. 2002. *MIT and the Rise of Entrepreneurial Science.* Routledge.

Etzkowitz, Henry. 2004. "The Evolution of the Entrepreneurial University." *International Journal of Technology and Globalization* 1: 64–77.

Etzkowitz, Henry, and Loet Leydesdorff. 1995. "The Triple Helix—University-Industry-Government Relations: A Laboratory for Knowledge Based Economic Development." *EASST Review* 14, no. 1: 14–19.

Etzkowitz, Henry, and Loet Leydesdorff, eds. 1997. *Universities and the Global Knowledge Economy: A Triple Helix of University-Industry-Government Relations*. Pinter.

Ferguson, Eugene. 1992. *Engineering and the Mind's Eye*. MIT Press.

Ferguson, Robert. 1996. Technology and Cooperation in American Aircraft Manufacture during World War II. PhD dissertation, University of Minnesota.

Fine, Lisa. 1990. *The Souls of the Skyscraper: Female Clerical Workers in Chicago, 1870–1930*. Temple University Press.

Fisher, Franklin, James McKie, and Richard Mancke. 1983. *IBM and the US Data Processing Industry: An Economic History*. Praeger.

Fowler, Bridget. 2000. "Introduction." In *Reading Bourdieu on Society and Culture*, ed. B. Fowler. Blackwell.

Friedman, Walter. 1998. "John H. Patterson and the Sales Strategy of the National Cash Register Company, 1884 to 1922." *Business History Review* 72: 552–584.

Fritz, Barkley. 1996. "The Women of the ENIAC." *IEEE Annals of the History of Computing* 18, no. 3: 13–28.

Fujimura, Joan. 1992. "Crafting Science: Standardized Packages, Boundary Objects, and 'Translation.'" In *Science as Practice and Culture*, ed. A. Pickering. University of Chicago Press.

Galison, Peter. 1987. *How Experiments End*. University of Chicago Press.

Galison, Peter. 1994. "The Ontology of the Enemy: Norbert Wiener and the Cybernetic Vision." *Critical Inquiry* 21: 228–266.

Galison, Peter. 1996. "Computer Simulations and the Trading Zone." In *The Disunity of Science*, ed. P. Galison and D. Stump. Stanford University Press.

Galison, Peter. 1997. *Image and Logic: A Material Culture of Microphysics*. University of Chicago Press.

Galison, Peter, and Bruce Hevly, eds. 1992. *Big Science: The Growth of Large-Scale Research*. Stanford University Press.

Geertz, Clifford. 1974. "Deep Play: Notes on the Balinese Cockfight." In *Myth, Symbol, and Culture*, ed. C. Geertz. Norton. Originally published in *Daedalus*, winter 1972.

Geiger, Roger. 1986. *To Advance Knowledge: The Growth of American Research Universities, 1900–1940*. Oxford University Press.

Geiger, Roger. 1993. *Research and Relevant Knowledge: American Research Universities since World War II*. Oxford University Press.

Genuth, Joel. 1988. "Microwave Radar, the Atomic Bomb, and the Background to US Research Priorities in World War II." *Science, Technology and Human Values* 13: 276–289.

Giddens, Anthony. 1976. *New Rules of Sociological Method: A Positive Critique of Interpretive Sociologies.* Hutchinson.

Giddens, Anthony. 1979. *Central Problems in Social Theory.* Macmillan.

Giddens, Anthony. 1984. *The Constitution of Society: Outline of the Theory of Structuration.* University of California Press.

Glaser, Barney, and Anselm Strauss. 1967. *The Discovery of Grounded Theory: Strategies for Qualitative Research.* Aldine.

Glaser, E. L. 1986. "The IBM 650 and the Woodenwheel." *Annals of the History of Computing* 8: 30–31.

Goldstein, Carolyn. 1997. "From Service to Sales: Home Economics in Light and Power, 1920–1940." *Technology and Culture* 38: 121–152.

Goldstine, Hermann. 1993/1972. *The Computer from Pascal to von Neumann.* Princeton University Press.

Golinski, Jan. 1998. *Making Natural Knowledge.* Cambridge University Press.

Graham, Margaret. 1986. *RCA and the VideoDisc: The Business of Research.* Cambridge University Press.

Green, Judy, and Jeanne LaDuke. 1988. "Women in American Mathematics: A Century of Contributions." In *A Century of Mathematics in America*, part II, ed. P. Duren. American Mathematical Society.

Grier, David Alan. 1996. "The ENIAC, the Verb 'to Program' and the Emergence of Digital Computers." *IEEE Annals of the History of Computing* 18: 51–55.

Grier, David Alan. 1997. "Gertrude Blanch of the Mathematical Tables Project." *IEEE Annals of the History of Computing* 19, no. 4: 18–27.

Grier, David Alan. 2005. *When Computers Were Human.* Princeton University Press.

Haber, Samuel. 1991. *Quest for Authority and Honor in the American Professions, 1750–1900.* University of Chicago Press.

Hacking, Ian. 1999. *The Social Construction of What?* Harvard University Press.

Hecht, Gabrielle. 1998. *The Radiance of France: Nuclear Power and National Identity After World War II.* MIT Press.

Heilbron, John, and Robert Seidel. 1989. *Lawrence and His Laboratory: A History of the Lawrence Berkeley Laboratory.* University of California Press.

Heims, Steve. 1980. *John von Neumann and Norbert Wiener: From Mathematics to the Technologies of Life and Death*. MIT Press.

Hewlett, Richard, and Oscar Anderson Jr. 1962. *The New World, 1939/1946*, volume I, *A History of the United States Atomic Energy Commission*. Pennsylvania State University Press.

Hounshell, David. 1992. "Du Pont and the Management of Large-Scale Research and Development." In *Big Science*, ed. P. Galison and B. Hevly. Stanford University Press.

Hounshell, David. 1996. "The Evolution of Industrial Research in the United States." In *Engines of Innovation*, ed. R. Rosenbloom and W. Spencer. Harvard Business School Press.

Hounshell, David, and John Kenly Smith Jr. 1988. *Science and Corporate Strategy: Du Pont R&D, 1902–1980*. Cambridge University Press.

Hughes, C. Everett. 1971a. "Ecological Aspects of Institutions." In *The Sociological Eye*. Aldine.

Hughes, C. Everett. 1971b. "Going Concerns: The Study of American Institutions." In Hughes, *The Sociological Eye*. Aldine.

Hughes, Thomas. 1983. *Networks of Power: Electrification in Western Society, 1870–1930*. Johns Hopkins University Press.

Hughes, Thomas. 1986. "The Seamless Web: Technology, Science, Etcetera, Etcetera." *Social Studies of Science* 16: 281–292.

Hughes, Thomas. 1987. "The Evolution of Large Technological Systems." In *The Social Construction of Technological Systems*, ed. W. Bijker, T. Hughes, and T. Pinch. MIT Press.

Hughes, Thomas. 1998. *Rescuing Prometheus*. Pantheon.

Hurd, Cuthbert. 1981. "Early IBM Computers: Edited Testimony." *Annals of the History of Computing* 3: 166–167.

Israel, Paul. 1998. *Edison: A Life of Invention*. Wiley.

Israel, Paul. 1992. *From Machine Shop to Industrial Research: Telegraphy and the Changing Context of American Invention, 1830–1920*. Johns Hopkins University Press.

Jakobson, Roman. 1990a. *On Language*, ed. L. Waugh and M. Monville-Burston. Harvard University Press.

Jakobson, Roman. 1990b. "Parts and Wholes in Language." In Jakobson, *On Language*, ed. L. Waugh and M. Monville-Burston. Harvard University Press.

Jakobson, Roman. 1990c. "Two Aspects of Language and Two Types of Aphasic Disturbances." In Jakobson, *On Language*, ed. L. Waugh and M. Monville-Burston. Harvard University Press.

Kaplan, Frank. 1983. *The Wizards of Armageddon*. Simon and Schuster.

Kevles, Daniel. 1975. "Scientists, the Military, and the Control of Postwar Defense Research: The Case of the Research Board for Natural Security, 1944–1946." *Technology and Culture* 16: 20–47.

Kevles, Daniel. 1979. *The Physicists: The History of a Scientific Community in Modern America*. Vintage.

Kidder, Tracy. 1981. *The Soul of a New Machine*. Little, Brown.

Kidwell, Margaret. 1990. "American Scientists and Calculating Machines—From Novelty to Commonplace." *Annals of the History of Computing* 12: 31–40.

Kleinmann, Daniel. 1995. *Politics on the Endless Frontier: Postwar Research Policy in the United States*. Duke University Press.

Knorr Cetina, Karin. 1999. *Epistemic Cultures: How the Sciences Make Knowledge*. Harvard University Press.

Kohler, Robert. 1991. *Partners in Science: Foundations and Natural Scientists, 1900–1945*. University of Chicago Press.

Latour, Bruno. 1992. "One More Turn after the Social Turn. . . ." In *The Social Dimensions of Science*, ed. E. McMullin. University of Notre Dame Press.

Latour, Bruno. 1993. *We Have Never Been Modern*. Harvard University Press.

Latour, Bruno. 1987. *Science in Action: How to Follow Scientists and Engineers Through Society*. Open University Press.

Law, John. 1987. "Technology and Heterogeneous Engineering: The Case of Portuguese Expansion." In *The Social Construction of Technological Systems*, ed. W. Bijker, T. Hughes, and T. Pinch. MIT Press.

Layton, Edwin. 1971. *Revolt of the Engineers: Social Responsibility and the American Engineering Profession*. Case Western Reserve University Press.

Leslie, Stuart. 1983. *Boss Kettering*. Columbia University Press.

Leslie, Stuart. 1993. *Cold War and American Science: The Military-Industrial Academic Complex at MIT and Stanford*. Columbia University Press.

Light, Jennifer. 1999. "When Computers Were Women." *Technology and Culture* 40: 455–483.

Lotchin, Roger. 1992. *Fortress California, 1910–1961: From Welfare to Warfare*. Oxford University Press.

Lukoff, Herman. 1979. *From Dits to Bits: A Personal History of the Electronic Computer*. Robotics Press.

Lynch, Michael. 1993. *Scientific Practice and Ordinary Action: Ethnomethodology and the Social Studies of Science*. Cambridge University Press.

MacKenzie, Donald. 1990. *Inventing Accuracy: A Historical Sociology of Nuclear Missile Guidance*. MIT Press.

MacLeod, Roy. 2000. "Introduction." *Osiris* 15: 1–13.

Macrae, Norman. 1992. *John von Neumann*. Pantheon.

Mahoney, Michael. 1997. "Computer Science: The Search for a Mathematical Theory." In *Science in the Twentieth Century*, ed. J. Krige and D. Pestre. Harwood.

Mapstone, Robina, and Morton Bernstein. 1980. "The Founding of SHARE." *Annals of the History of Computing* 2: 363–372.

Markusen, Ann, Peter Hall, Scott Campbell, and Sabina Dietrick. 1991. *The Rise of the Gunbelt: The Military Remapping of Industrial America*. Oxford University Press.

Mason, Daniel. 1983. "The 701 in the IBM Technical Computing Bureau." *Annals of the History of Computing* 5: 176–177.

Mauchly, Kathleen. 1984. "John Mauchly's Early Years." *Annals of the History of Computing* 6: 116–138.

McCartney, Scott. 1999. *ENIAC, the Triumphs and Tragedies of the World's First Computer*. Walker.

Mindell, David. 2002. *Between Human and Machine: Feedback, Control, and Computing before Cybernetics*. Johns Hopkins University Press.

Milkman, Ruth. 1987. *Gender at Work: The Dynamics of Job Segregation by Sex during World War II*. University of Illinois Press.

Morse, Philip. 1977. *In at the Beginnings: A Physicist's Life*. MIT Press.

Nagel, Ernest. 1963. "Wholes, Sums, and Organic Unities." In *Parts and Wholes*, ed. D. Lerner. Free Press of Glencoe.

Noble, David. 1984. *Forces of Production: A Social History of Industrial Automation*. Knopf.

Norberg, Arthur, and Judy O'Neill. 1996. *Transforming Computer Technology: Information Processing for the Pentagon*. Johns Hopkins University Press.

Owens, Larry. 1986. "Vannevar Bush and the Differential Analyzer: The Text and Context of an Early Computer." *Technology and Culture* 27: 63–95.

Owens, Larry. 1989. "Mathematicians at War: Warren Weaver and the Applied Mathematics Panel, 1942–45." In *Institutions and Applications*, Volume II of *The History of Modern Mathematics*. Academic Press.

Owens, Larry. 1994. "The Counterproductive Management of Science in the Second World War: Vannevar Bush and the Office of Scientific Research and Development." *Business History Review* 68: 515–576.

Parshall, Karen, and David Rowe. 1994. *The Emergence of the American Mathematical Research Community, 1876–1900: J. J. Sylvester, Felix Kline, and E. H. Moore*. American Mathematical Society.

Passer, Harold. 1953. *The Electrical Manufacturers, 1875–1900: A Study in Competition, Entrepreneurship, Technical Change and Electrical Growth*. Harvard University Press.

Peirce, Charles. 1991. *Peirce on Signs: Writings on Semiotics*, ed. J. Hoopes. University of North Carolina Press.

Perkin, Harold. 1996. *The Third Revolution: Professional Elites in the Modern World*. Routledge.

Pickering, Andrew. 1995. *The Mangle of Practice: Time, Agency and Science*. University of Chicago Press.

Pinch, Trevor, and Wiebe Bijker. 1987/1984. "The Social Construction of Technological Systems: Or How the Sociology of Science and the Sociology of Technology Might Benefit Each Other." In *The Social Construction of Technological Systems*, ed. W. Bijker, T. Hughes, and T. Pinch. MIT Press.

Porter, Theodore. 1995. *Trust in Numbers: The Pursuit of Objectivity in Science and Public Life*. Princeton University Press.

Powell, Walter, and Paul DiMaggio, eds. 1991. *New Institutionalism in Organizational Analysis*. University of Chicago Press.

Pugh, Emerson. 1995. *Building IBM*. MIT Press.

Redmond, Kent, and Thomas Smith. 1980. *Project Whirlwind: The History of a Pioneer Computer*. Digital Press.

Redmond, Kent, and Thomas Smith. 2000. *From Whirlwind to MITRE: The R&D Story of the SAGE Air Defense Computer*. MIT Press.

Regis, Ed. 1987. *Who Got Einstein's Office? Eccentricity and Genius at the Institute for Advanced Study*. Addison-Wesley.

Reich, Leonard. 1985. *The Making of American Industrial Research: Science and Business at GE and Bell, 1876–1926.* Cambridge University Press.

Reid, Constance. 1996/1972. *Courant.* Copernicus.

Reynolds, Terry. 1991. "The Engineer in 20th-Century America." In *The Engineer in America,* ed. T. Reynolds. University of Chicago Press.

Rose, Mark. 1995. *Cities of Light and Heat: Domesticating Gas and Electricity in Urban America.* Pennsylvania State University Press.

Rosenberg, Charles. 1997a/1979. "Toward an Ecology of Knowledge: On Discipline, Context, and History." In Rosenberg, *No Other Gods,* revised and expanded edition. Johns Hopkins University Press.

Rosenberg, Charles. 1997b/1988. "Woods or Trees? Ideas and Actors in the History of Science." In Rosenberg, *No Other Gods,* revised and expanded edition. Johns Hopkins University Press.

Ross, Andrew, ed. 1996. *Science Wars.* Duke University Press.

Sapolsky, Harvey. 1990. *Science and the Navy: The History of the Office of Naval Research.* Princeton University Press.

Schaffer, Simon. 1991. "The Eighteenth Brumaire of Bruno Latour." *Studies in the History and Philosophy of Science* 22: 174–192.

Schaffer, Simon. 1994. "Babbage's Intelligence: Calculating Engines and the Factory System." *Critical Inquiry* 21: 203–227.

Seely, Bruce. 1993. "Research, Engineering, and Science in American Engineering Colleges: 1900–1960." *Technology and Culture* 34: 344–386.

Spears, Timothy. 1993. " 'All Things to All Men': The Commercial Traveler and the Rise of Modern Salesmanship." *American Quarterly* 45: 524–557.

Star, Susan Leigh, ed. 1995. *Ecologies of Knowledge: Work and Politics in Science and Technology.* SUNY Press.

Star, Susan Leigh, and James R. Griesemer. 1989. "Institutional Ecology, 'Translations' and Boundary Objects: Amateurs and Professionals in Berkeley's Museum of Vertebrate Zoology, 1907–1939." *Social Studies of Science* 19: 387–420.

Strauss, Anselm. 1993. *Continual Permutations of Action.* Aldine de Gruyter.

Stern, Beatrice. n.d. A History of the Institute for Advanced Study, 1930–1950. Manuscript.

Stern, Nancy. 1981. *From ENIAC to Univac: An Appraisal of the Eckert-Mauchly Computers.* Digital Press.

Sterne, Jonathan and Joan Leach. 2005. "The Point of Social Construction and the Purpose of Social Critique." *Social Epistemology* 19: 189–198.

Stewart, Irvin. 1948. *Organizing Scientific Research for War: The Administrative History of the Office of Scientific Research and Development.* Little, Brown.

Strasser, Susan. 1993. "'The Smile That Pays': The Culture of Traveling Salesmen, 1880–1920." In *The Mythmaking Frame of Mind,* ed. J. Gilbert et al. Wadsworth.

Strom, Sharon. 1992. *Beyond the Typewriter: Gender, Class, and the Origins of Modern American Office Work, 1900–1930.* University of Illinois Press.

Taylor, Peter. 2005. *Unruly Complexity: Ecology, Interpretation, Engagement.* University of Chicago Press.

Thomas, Robert. 1994. *What Machines Can't Do: Politics and Technology in the Industrial Enterprise.* University of California Press.

Tomash, Erwin, and Arnold Cohen. 1979. "The Birth of an ERA: Engineering Research Associates, Inc. 1946–1955." *Annals of the History of Computing* 1, no. 2: 83–97.

Trice, Harrison. 1993. *Occupational Subcultures in the Workplace.* ILR.

Turner, Victor. 1967. *The Forest of Symbols: Aspects of Ndembu Ritual.* Cornell University Press.

Ulam, Stanislaw. 1976. *Adventures of a Mathematician.* Scribner.

Valley, George. 1985. "How the SAGE Development Began." *Annals of the History of Computing* 7: 196–226.

Vaughan, Diane. 1996. *The Challenger Launch Decision: Risky Technology, Culture, and Deviance at NASA.* University of Chicago Press.

Wang, Jessica. 1999. *American Science in an Age of Anxiety: Scientists, Anticommunism, and the Cold War.* University of North Carolina Press.

Warwick, Andrew. 1995. "The Laboratory of Theory, or What's Exact About the Exact Sciences?" In *The Values of Precision,* ed. M. Wise. Princeton University Press.

Weiss, Eric. 1997. "Eloge: Cuthbert Corwin Hurd (1911–1996)." *IEEE Annals of the History of Computing* 19, no. 1: 65–73.

Westrum, Ron. 1991. *Technologies & Society: The Shaping of People and Things.* Wadsworth.

Wildes, Karl, and Nilo Lindgren. 1985. *A Century of Electrical Engineering and Computer Science at MIT, 1882–1982.* MIT Press.

Williams, Michael. 1985. *A History of Computing Technology.* Prentice-Hall.

Winner, Langdon. 1986/1980. "Do Artifacts Have Politics?" In *The Whale and the Reactor*. University of Chicago Press.

Wise, George. 1985. *Willis R. Whitney, General Electric, and the Origins of US Industrial Research*. Columbia University Press.

Wise, M. Norton, ed. 1995. *The Values of Precision*. Princeton University Press.

Yates, JoAnne. 1993. "Co-evolution of Information-Processing Technology and Use: Interaction between the Life Insurance and Tabulating Industries." *Business History Review* 67: 1–51.

Zunz, Olivier. 1990. *Making America Corporate, 1870–1920*. University of Chicago Press.

Inside Technology
edited by Wiebe E. Bijker, W. Bernard Carlson, and Trevor Pinch

Index